Methods in Molecular Biology

Series Editor
John M. Walker
School of Life and Medical Sciences
University of Hertfordshire
Hatfield, Hertfordshire, UK

For further volumes:
http://www.springer.com/series/7651

For over 35 years, biological scientists have come to rely on the research protocols and methodologies in the critically acclaimed *Methods in Molecular Biology* series. The series was the first to introduce the step-by-step protocols approach that has become the standard in all biomedical protocol publishing. Each protocol is provided in readily-reproducible step-by-step fashion, opening with an introductory overview, a list of the materials and reagents needed to complete the experiment, and followed by a detailed procedure that is supported with a helpful notes section offering tips and tricks of the trade as well as troubleshooting advice. These hallmark features were introduced by series editor Dr. John Walker and constitute the key ingredient in each and every volume of the *Methods in Molecular Biology* series. Tested and trusted, comprehensive and reliable, all protocols from the series are indexed in PubMed.

Oncolytic Viruses

Edited by

Christine E. Engeland

Research Group Mechanisms of Oncolytic Immunotherapy, Clinical Cooperation Unit Virotherapy, National Center for Tumor Diseases (NCT), German Cancer Research Center (DKFZ), University Hospital Heidelberg, Heidelberg, Germany

Humana Press

Editor
Christine E. Engeland
Research Group Mechanisms of
Oncolytic Immunotherapy
Clinical Cooperation Unit Virotherapy
National Center for Tumor Diseases (NCT)
German Cancer Research Center (DKFZ)
University Hospital Heidelberg
Heidelberg, Germany

ISSN 1064-3745 ISSN 1940-6029 (electronic)
Methods in Molecular Biology
ISBN 978-1-4939-9793-0 ISBN 978-1-4939-9794-7 (eBook)
https://doi.org/10.1007/978-1-4939-9794-7

This Humana imprint is published by the registered company Springer Science+Business Media, LLC, part of Springer Nature.
The registered company address is: 233 Spring Street, New York, NY 10013, U.S.A.

Preface

Anecdotal clinical reports of tumor remissions after viral infections laid the foundation for the field of oncolytic virotherapy. Advances in molecular virology, tumor biology, and immunology have enabled more refined studies of tumor-selective viruses. Concomitant with the resurgence of cancer immunotherapy and after the approval of Talimogene laherparepvec by the FDA and EMA, oncolytic virotherapy has gained unprecedented momentum. The field has flourished in recent years, yielding many notable preclinical studies and clinical trials. This book aims to provide a guide for basic virologists, translational researchers, and clinician scientists in the field by providing reference protocols from vector development to clinical translation.

The initial chapter provides an introductory review of the field, followed by a series of chapters describing virus modifications to enhance tumor specificity and anti-tumor efficacy. Reflecting the increasing interest in immunotherapeutic effects of oncolysis, a number of chapters address different strategies for immunomodulation and immunomonitoring. The third section of the book covers methodologies for different model systems to study oncolytic viruses, including mouse tumor models, patient-derived samples, and also mathematical modeling.

A number of virus platforms and approaches are represented, providing a survey of state-of-the-art methods for study of this unique treatment approach. Therefore, I would like to take this opportunity to thank all authors who have made this possible with their contributions. Hopefully this book will serve the research community as a useful resource to further enhance progress in the field of oncolytic virotherapy.

Heidelberg, Germany *Christine E. Engeland*

Contents

Contributors

DANIEL ABATE-DAGA • *Department of Immunology, H. Lee Moffitt Cancer Center & Research Institute, Tampa, FL, USA*

CAROLE ACHARD • *CRCINA, INSERM, Université d'Angers, Université de Nantes, Nantes, France; Immunology Graft Oncology, Labex IGO, Nantes, France*

ASSIA L. ANGELOVA • *Division of Tumor Virology, German Cancer Research Center (DKFZ), Heidelberg, Germany*

NICOLA ANNELS • *Targeted Cancer Therapy, Department of Clinical and Experimental Medicine, Faculty of Health and Medical Science, University of Surrey, Guildford, Surrey, UK*

LUCA BAU • *Institute of Biomedical Engineering, University of Oxford, Oxford, UK*

JOHN C. BELL • *Centre for Innovative Cancer Therapeutics, Ottawa Hospital Research Institute, Ottawa, ON, Canada; Department of Biochemistry, Microbiology and Immunology, University of Ottawa, Ottawa, ON, Canada*

NICOLAS BOISGERAULT • *CRCINA, INSERM, Université d'Angers, Université de Nantes, Nantes, France; Immunology Graft Oncology, Labex IGO, Nantes, France*

CAROLINE J. BREITBACH • *Turnstone Biologics, Ottawa, ON, Canada*

BYRAM W. BRIDLE • *Department of Pathobiology, Ontario Veterinary College, University of Guelph, Guelph, ON, Canada*

ROBERT CARLISLE • *Institute of Biomedical Engineering, University of Oxford, Oxford, UK*

ROBERTO CATTANEO • *Department of Molecular Medicine, Mayo Clinic, Rochester, MN, USA*

KRISHNA DAS • *Division of Virology, Medical University of Innsbruck, Innsbruck, Austria; Christian Doppler Laboratory for Viral Immunotherapy of Cancer, Innsbruck, Austria*

TIPHAINE DELAUNAY • *CRCINA, INSERM, Université d'Angers, Université de Nantes, Nantes, France; Immunology Graft Oncology, Labex IGO, Nantes, France*

LAURA DIETZ • *Faculty of Biosciences, Institute of Pharmacy and Molecular Biotechnology (IPMB), Heidelberg University, Heidelberg, Germany; Research Group Mechanisms of Oncolytic Immunotherapy, Clinical Cooperation Unit Virotherapy, National Center for Tumor Diseases (NCT), German Cancer Research Center (DKFZ), University Hospital Heidelberg, Heidelberg, Germany*

HEIKO ENDERLING • *Department of Integrated Mathematical Oncology, H. Lee Moffitt Cancer Center & Research Institute, Tampa, FL, USA*

CHRISTINE E. ENGELAND • *Research Group Mechanisms of Oncolytic Immunotherapy, Clinical Cooperation Unit Virotherapy, National Center for Tumor Diseases (NCT), German Cancer Research Center (DKFZ), University Hospital Heidelberg, Heidelberg, Germany*

PATRIK ERLMANN • *ViraTherapeutics GmbH, Innsbruck, Austria*

JEAN-FRANÇOIS FONTENEAU • *CRCINA, INSERM, Université d'Angers, Université de Nantes, Nantes, France; Immunology Graft Oncology, Labex IGO, Nantes, France*

LINA S. FRANCO • *Center for Immunotherapy, Vaccines and Virotherapy (B-CIVV), Biodesign Institute, Arizona State University, Tempe, AZ, USA*

JOSHUA D. FREEDMAN • *Department of Oncology, University of Oxford, Oxford, UK*

SALLY FROST • *Department of Oncology, University of Oxford, Oxford, UK*

Marc Grégoire • *CRCINA, INSERM, Université d'Angers, Université de Nantes, Nantes, France; Immunology Graft Oncology, Labex IGO, Nantes, France*
Johannes P. W. Heidbuechel • *Research Group Mechanisms of Oncolytic Immunotherapy, Clinical Cooperation Unit Virotherapy, National Center for Tumor Diseases (NCT), German Cancer Research Center (DKFZ), University Hospital Heidelberg, Heidelberg, Germany; Faculty of Biosciences, Heidelberg University, Heidelberg, Germany*
Claudia A. P. Hill • *Institute of Biomedical Engineering, University of Oxford, Oxford, UK*
Elizabeth J. Ilett • *Leeds Institute of Medical Research at St James', St James' University Hospital, Leeds, UK*
Carolina S. Ilkow • *Centre for Innovative Cancer Therapeutics, Ottawa Hospital Research Institute, Ottawa, ON, Canada; Department of Biochemistry, Microbiology and Immunology, Faculty of Medicine, University of Ottawa, Ottawa, ON, Canada*
Alexandra Just • *Division of Tumor Virology, German Cancer Research Center (DKFZ), Heidelberg, Germany*
Brian A. Keller • *Department of Pathology and Laboratory Medicine, The Ottawa Hospital, Ottawa, ON, Canada; Centre for Innovative Cancer Therapeutics, Ottawa Hospital Research Institute, Ottawa, ON, Canada*
Hena Khalique • *Department of Oncology, University of Oxford, Oxford, UK*
Irina Kiprianova • *Division of Tumor Virology, German Cancer Research Center (DKFZ), Heidelberg, Germany; BIORON GmbH, Ludwigshafen, Germany*
Marcos Kostalas • *Targeted Cancer Therapy, Department of Clinical and Experimental Medicine, Faculty of Health and Medical Science, University of Surrey, Guildford, Surrey, UK*
Florian Kühnel • *Department of Gastroenterology, Hepatology, and Endocrinology, Hannover Medical School, Hannover, Germany*
Janet Lei-Rossmann • *Department of Oncology, University of Oxford, Oxford, UK*
Barbara Leuchs • *Division of Tumor Virology, German Cancer Research Center (DKFZ), Heidelberg, Germany*
Brian D. Lichty • *Department of Pathology and Molecular Medicine, McMaster Immunology Research Centre, McMaster University, Hamilton, ON, Canada; Turnstone Biologics, Ottawa, ON, Canada*
Grant McFadden • *Center for Immunotherapy, Vaccines and Virotherapy (B-CIVV), Biodesign Institute, Arizona State University, Tempe, AZ, USA*
Philipp Mueller • *Boehringer Ingelheim Pharma GmbH & Co. KG, Biberach a.d. Riss, Germany*
Michael D. Mühlebach • *Section Product Testing of Immunological Veterinary Medicinal Products, Division of Veterinary Medicine, Paul-Ehrlich-Institut, Langen, Germany*
Julia Niemann • *Department of Gastroenterology, Hepatology, and Endocrinology, Hannover Medical School, Hannover, Germany*
Hardev Pandha • *Targeted Cancer Therapy, Department of Clinical and Experimental Medicine, Faculty of Health and Medical Science, University of Surrey, Guildford, Surrey, UK*
Jonathan G. Pol • *Gustave Roussy Comprehensive Cancer Institute, Villejuif, France; INSERM, U1138, Paris, France; Equipe 11 Labellisée par la Ligue Nationale Contre le Cancer, Centre de Recherche des Cordeliers, Paris, France; Université de Paris, Paris, France; Sorbonne Université, Paris, France*
Samuel D. Rabkin • *Department of Neurosurgery, Massachusetts General Hospital, Harvard Medical School, Boston, MA, USA*

Kate Relph • *Targeted Cancer Therapy, Department of Clinical and Experimental Medicine, Faculty of Health and Medical Science, University of Surrey, Guildford, Surrey, UK*

Jean Rommelaere • *Division of Tumor Virology, German Cancer Research Center (DKFZ), Heidelberg, Germany*

Dipongkor Saha • *Department of Immunotherapeutics and Biotechnology, School of Pharmacy, Texas Tech University Health Sciences Center, Abilene, TX, USA; Department of Neurosurgery, Massachusetts General Hospital, Harvard Medical School, Boston, MA, USA*

Liesa-Marie Schreiber • *Division of Virology, Medical University of Innsbruck, Innsbruck, Austria; Christian Doppler Laboratory for Viral Immunotherapy of Cancer, Innsbruck, Austria*

Autumn J. Schulze • *Department of Molecular Medicine, Mayo Clinic College of Medicine, Rochester, MN, USA*

Eleanor M. Scott • *Department of Oncology, University of Oxford, Oxford, UK*

Len W. Seymour • *Department of Oncology, University of Oxford, Oxford, UK*

Chris Smith • *Targeted Cancer Therapy, Department of Clinical and Experimental Medicine, Faculty of Health and Medical Science, University of Surrey, Guildford, Surrey, UK*

Bart Spiesschaert • *Division of Virology, Medical University of Innsbruck, Innsbruck, Austria; Christian Doppler Laboratory for Viral Immunotherapy of Cancer, Innsbruck, Austria; ViraTherapeutics GmbH, Innsbruck, Austria*

Abera Surendran • *Centre for Innovative Cancer Therapeutics, Ottawa Hospital Research Institute, Ottawa, ON, Canada; Department of Biochemistry, Microbiology and Immunology, Faculty of Medicine, University of Ottawa, Ottawa, ON, Canada*

Frédéric Tangy • *CNRS 3569, Institut Pasteur, Paris, France*

Carles Urbiola • *Division of Virology, Medical University of Innsbruck, Innsbruck, Austria; Christian Doppler Laboratory for Viral Immunotherapy of Cancer, Innsbruck, Austria*

Rūta Veinalde • *Latvian Biomedical Research and Study Centre, Riga, Latvia*

Gayathri Vijayakumar • *Department of Microbiology, Icahn School of Medicine at Mount Sinai, New York, NY, USA; Memorial Sloan Kettering Cancer Center, New York, NY, USA*

Nancy Y. Villa • *Center for Immunotherapy, Vaccines and Virotherapy (B-CIVV), Biodesign Institute, Arizona State University, Tempe, AZ, USA*

Marie-Ève Wedge • *Centre for Innovative Cancer Therapeutics, Ottawa Hospital Research Institute, Ottawa, ON, Canada; Department of Cellular and Molecular Medicine, Faculty of Medicine, University of Ottawa, Ottawa, ON, Canada*

Guido Wollmann • *Division of Virology, Medical University of Innsbruck, Innsbruck, Austria; Christian Doppler Laboratory for Viral Immunotherapy of Cancer, Innsbruck, Austria*

Dmitriy Zamarin • *Department of Medicine, Memorial Sloan Kettering Cancer Center, New York, NY, USA; Department of Medicine, Weil Cornell Medical College, New York, NY, USA; Parker Institute for Cancer Immunotherapy, Memorial Sloan Kettering Cancer Center, New York, NY, USA*

Chapter 1

Introduction to Oncolytic Virotherapy

Christine E. Engeland and John C. Bell

Abstract

Oncolytic viruses exploit key hallmarks of cancer for replication in malignant cells, leading to tumor cell lysis, modulation of the tumor microenvironment and in situ vaccination effects. Diverse virus platforms have been developed as oncolytic vectors and designed for improved tumor specificity, intratumoral spread, therapeutic gene delivery and especially as targeted cancer immunotherapeutics. This chapter provides a concise overview of the basic principles as well as current progress in preclinical and clinical studies of oncolytic virotherapy.

Key words Oncolytic viruses, Viral vectors, Cancer immunotherapy, Tumor targeting, Cancer gene therapy

1 Principles of Oncolytic Virotherapy: Exploiting Hallmarks of Cancer and Turning Cold Tumors Hot

Treating cancer patients with replicating viruses may seem an outrageous idea—which was actually inspired by clinical observations of tumor remissions after natural virus infections [1]. Indeed, these experiments of nature were followed up by clinicians and researchers, who deduced the following principles of oncolytic virotherapy (Fig. 1):

On a cellular level, viruses with oncolytic properties show tumor-selective infection, replication, and spread—supported by inherent characteristics of cancer cells, the "hallmarks of cancer." As such, cancer cells show many properties conducive to viral replication including sustained proliferation, resistance to apoptosis, and immune evasion [2, 3]. Malignant transformation can include upregulation of viral entry receptors (e.g., CD46, a complement regulator) and proliferative signaling pathways usurped by viruses (e.g., Wnt/ß-Catenin and EGFR) as well as downregulation of antiproliferative and antiviral signaling (especially interferon) [4].

Christine E. Engeland (ed.), *Oncolytic Viruses*, Methods in Molecular Biology, vol. 2058,
https://doi.org/10.1007/978-1-4939-9794-7_1,

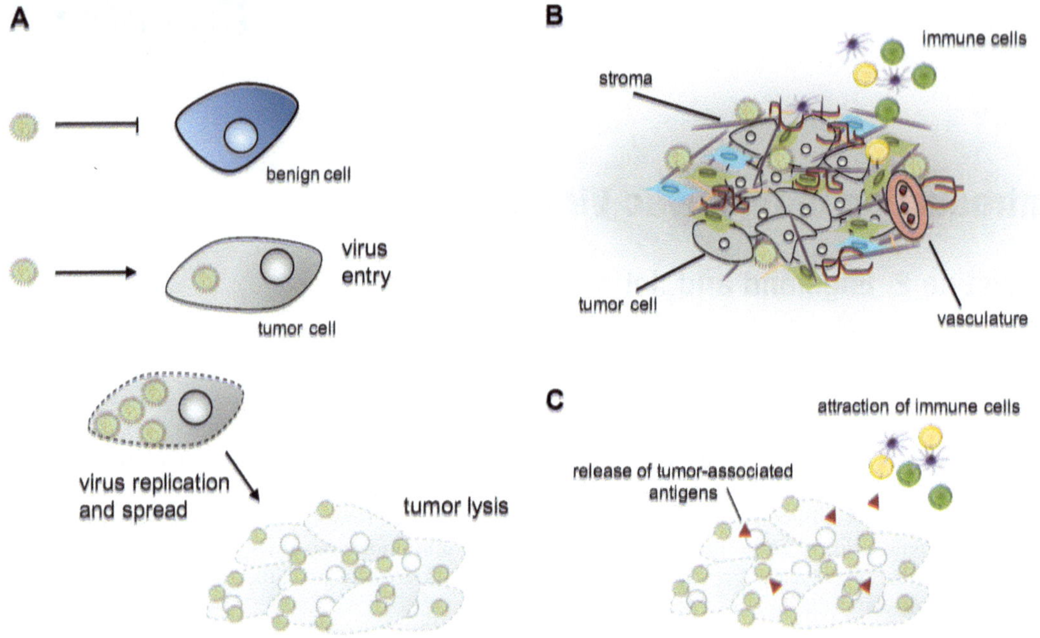

Fig. 1 Principles of oncolytic virotherapy. (**a**) Oncolytic viruses replicate selectively in malignant cells. (**b**) Oncolysis reshapes the tumor microenvironment. (**c**) Exposure of tumor antigens in the context of oncolysis can elicit tumor vaccination effects

A tumor comprises not only individual malignant cells but a complex microenvironment composed of stroma, vasculature, and leukocytes, typically characterized by immunosuppression. Oncolytic virotherapy can act to reshape the local milieu. An acute viral infection serves as a potent stimulus for the immune system. Local inflammation, innate immune activation, and danger signals (DAMPs and PAMPs) arise during viral replication which can change the immune contexture, thereby "turning cold tumors hot" [5].

During oncolysis, tumor-associated antigens are released in this context, which provides adjuvants for induction of adaptive antitumor immune responses. Thus, on a systemic level, oncolytic virotherapy can act as an in situ tumor vaccine, inducing therapeutic and protective antitumor immunity [6].

Preclinical and clinical data have provided proof of these principles. However, the role and contribution of these mechanisms of action to efficacy of oncolytic virotherapy has been a subject of debate. Moreover, this may depend on the specific oncolytic vector and the therapeutic setting.

2 Oncolytic Vector Platforms: From Adeno to Zika

These principle mechanisms of action outlined above are common to a diverse set of viruses which have been developed as oncolytic vector platforms (Fig. 2). These include the following:

- Small (e.g., parvovirus, approximately 25 nm and 5 kb), large (Vaccinia virus, 300 nm and 200 kb).
- Enveloped (herpes) and nonenveloped (PVSRIPO, derived from polio).
- DNA (adeno), RNA positive (Coxsackie) and negative (Maraba), and double-stranded (reovirus) RNA viruses as well as retroviruses (Toca 511, derived from amphotropic murine leukemia virus).
- Human (mumps), animal (Newcastle disease, vesicular stomatitis, myxoma).
- Pathogenic (influenza, Zika) and live-attenuated (measles) viruses.

These diverse viruses have been tested in preclinical studies and many have advanced to clinical trials. Overall, the clinical data have

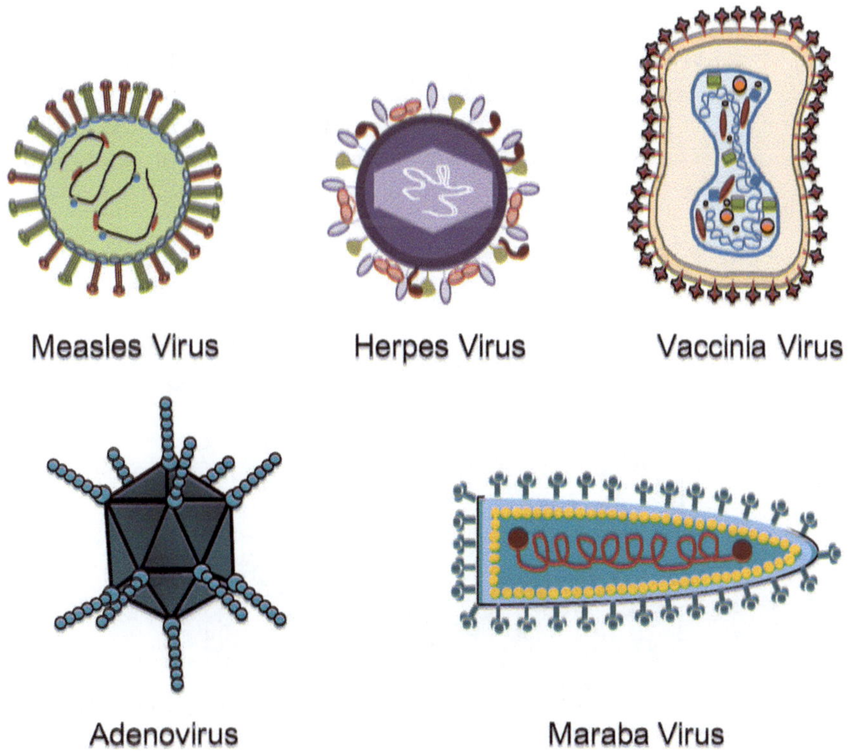

Fig. 2 Schematic depictions of five representative oncolytic viruses

demonstrated safety and typically mild, often flu-like symptoms as adverse events as well as some promising results in terms of anti-tumor efficacy [7]. While the adenovirus Oncorine H101 has been licensed for treatment of nasopharyngeal cancer in China since 2005, 2015 marked the approval of the herpes virus talimogene laherparepvec for treatment of advanced melanoma in the USA and Europe. Thus, the paradigm of using replicating viral vectors for cancer treatment has entered clinical practice.

To date, systematic head-to-head comparisons of these diverse viruses have not been performed. Viruses which have evolved a specific tissue tropism, conceivably, may be especially adapted to replicate in tumors originating from these tissues. In addition to the range of naturally occurring oncolytic viruses, the possibilities opened by genetic engineering offer a plethora of treatment options with vectors designed for specific therapeutic purposes.

3 Vector Design: Tumor Targeting and Spread, Tracers, Therapeutic Genes

Progress in molecular biology including the development of reverse genetics systems has enabled the design of oncolytic therapeutics with improved properties (Fig. 3) [8]. Main arenas of vector design include *tumor targeting* to increase specificity, which can be achieved on the entry level by modifying receptor tropism or incorporating matrix metalloproteinase cleavage sites into viral surface proteins. Targeting on the post-entry level can be achieved by placing viral genes under transcriptional control of a tumor-specific promoter, inserting target sites for microRNAs with differential expression in healthy and malignant cells or deletion of virulence

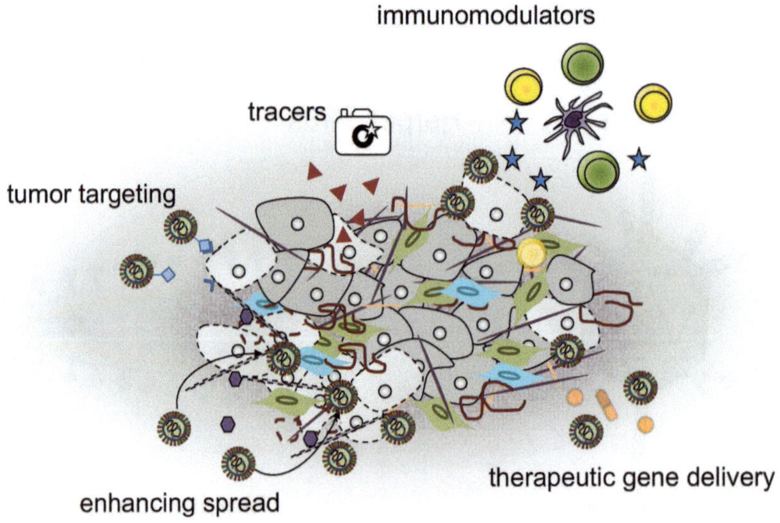

Fig. 3 Strategies to improve efficacy of oncolytic viruses by rational vector design

genes such as thymidine kinase. Measures to enhance viral *spread* within the tumor include encoding factors to degrade the extracellular matrix or addition of fusogenic proteins into the virus genome. Encoding *tracers* such as secreted marker proteins or imaging reporters allows for monitoring of viral spread and pharmacokinetics. As oncolytic viruses amplify specifically in tumors they are appealing vehicles for targeted delivery of *therapeutic genes* for enhanced therapeutic efficacy, for instance prodrug convertases, toxins, the sodium iodide symporter for radiotherapy, and especially *immunomodulators*. Immunomodulators targeting different phases of the „cancer immunity cycle“ have been incorporated into oncolytic vectors, including cytokines, chemokines, tumor antigens, immune checkpoint antibodies, and bispecific T cell engagers [9].

4 Current Challenges and Future Perspectives

With the many opportunities offered by the range of vector platforms and design options arise a multitude of research questions for basic and clinician scientists. As optimized virotherapies become more and more efficacious, how can *safety* be ensured? What is the optimal *route* of delivery—intratumoral, intravenous, or intracavital—for a given clinical situation? This requires precise analyses of viral *pharmacokinetics and pharmacodynamics*. Due to their pleiotropic modes of action, both unmodified and modified oncolytic viruses are attractive combination partners for different cancer treatment modalities, in particular immunotherapy. How can *combination schedules and doses* be optimized to achieve maximum antitumor efficacy? What is the balance of antiviral vs. antitumor immunity? In the era of *personalized medicine*, oncolytic virotherapies may be tailored to tumor expression profiles or immune signatures. A thorough understanding of *mechanisms of action* as well as identifying biomarkers of response and also resistance mechanisms are key toward achieving this goal. These efforts require *relevant models* that yield clinically translatable results, both in laboratory animals and with primary samples. The challenges associated with *clinical implementation* of virotherapy studies including vector production, clinical protocol design, regulatory questions, pharmacology, and toxicology studies should not be underestimated [10] but can be met with interdisciplinary team efforts.

These are exciting times to be working in the field of oncolytic virotherapy at the intersection of virology, cancer biology, immunology, and clinical oncology, as we await the results of ongoing studies and anticipate widespread application of oncolytic virotherapy for the benefit of cancer patients.

Acknowledgments

The authors are grateful to Luisa Henkel for creating the excellent artwork accompanying this chapter. C.E.E. is supported by the German Research Foundation (Grants EN-1119/2-1 and EN-1119/2-2), the Else Kröner-Fresenius Stiftung (Grant 2015_A78), and the Wilhelm Sander-Stiftung (Grant 2018.058.1).

References

1. Kelly E, Russell SJ (2007) History of oncolytic viruses: genesis to genetic engineering. Mol Ther 15(4):651–659. https://doi.org/10.1038/sj.mt.6300108
2. Hanahan D, Weinberg RA (2011) Hallmarks of cancer: the next generation. Cell 144 (5):646–674. https://doi.org/10.1016/j.cell.2011.02.013
3. Seymour LW, Fisher KD (2016) Oncolytic viruses: finally delivering. Br J Cancer 114 (4):357–361. https://doi.org/10.1038/bjc.2015.481
4. Pikor LA, Bell JC, Diallo JS (2015) Oncolytic viruses: exploiting cancer's deal with the devil. Trends Cancer 1(4):266–277. https://doi.org/10.1016/j.trecan.2015.10.004
5. Achard C, Surendran A, Wedge ME, Ungerechts G, Bell J, Ilkow CS (2018) Lighting a fire in the tumor microenvironment using oncolytic immunotherapy. EBioMedicine 31:17–24. https://doi.org/10.1016/j.ebiom.2018.04.020
6. Russell SJ, Barber GN (2018) Oncolytic viruses as antigen-agnostic cancer vaccines. Cancer Cell 33(4):599–605. https://doi.org/10.1016/j.ccell.2018.03.011
7. Pol JP, Levesque S, Workenhe ST, Gujar S, Le Boeuf F, Clements DR, Fahrner JE, Fend L, Bell JC, Mossman KL, Fucikova J, Spisek R, Zitvogel L, Kroemer G, Galluzzi L (2018) Trial Watch: Oncolytic viro-immunotherapy of hematologic and solid tumors. Oncoimmunology 7(12):e1503032. https://doi.org/10.1080/2162402x.2018.1503032
8. Maroun J, Munoz-Alia M, Ammayappan A, Schulze A, Peng KW, Russell S (2017) Designing and building oncolytic viruses. Fut Virol 12 (4):193–213. https://doi.org/10.2217/fvl-2016-0129
9. Twumasi-Boateng K, Pettigrew JL, Kwok YYE, Bell JC (2018) Oncolytic viruses as engineering platforms for combination immunotherapy. Nat Rev Cancer 18(7):419–432. https://doi.org/10.1038/s41568-018-0009-4
10. Ungerechts G, Bossow S, Leuchs B, Holm PS, Rommelaere J, Coffey M, Coffin R, Bell J, Nettelbeck DM (2016) Moving oncolytic viruses into the clinic: clinical-grade production, purification, and characterization of diverse oncolytic viruses. Mol Ther Meth Clin Dev 3:16018. https://doi.org/10.1038/mtm.2016.18

Chapter 2

Methods for Modification of Therapeutic Viruses

Claudia A. P. Hill, Luca Bau, and Robert Carlisle

Abstract

The optimal clinical exploitation of viruses as gene therapy or oncolytic vectors will require them to be administered intravenously. Strategies must therefore be deployed to enable viruses to survive the harsh neutralizing environment of the bloodstream and achieve deposition within and throughout target tissues or tumor deposits. This chapter describes the genetic and chemical engineering approaches that are being developed to overcome these challenges.

Key words Oncolytic virus, Capsid engineering, Adenovirus, Capsid chemical modification, Capsid surface modification, PEGylation, Polymer coating, Polymer stealthing, Polymer shielding

1 Introduction

The application of viruses in the treatment of human disease has attracted much attention, in addition to their long-recognized potential as gene therapy vectors there has been a recent surge of interest in harnessing the specific cancer-lysis (oncolytic) properties of a range of viruses. The four key benefits of such "oncolytic virotherapy" can be summarized as selectivity, self-amplification, potential for arming with therapeutic transgenes, and cancer cell kill through mechanisms impervious to resistance [1]. These benefits provide a powerful rationale for the development of oncolytic viruses (OV) for cancer treatment, and preclinical and clinical testing continues to provide compelling supportive data. Indeed, recent success using herpes simplex virus (HSV) [2–6], vaccinia virus (VV) [7–14], and adenovirus (Ad) [15–24] in clinical trials for cancer treatment has led to intensified clinical and commercial interest [25] in virotherapy.

Currently, there are two oncolytic viruses that have been approved for clinical use, H101 and Talimogene laherparepvec (Imlygic®). H101 is an E1B-deleted serotype 5 adenovirus, which was approved by the China Food and Drug Administration (CFDA) in 2005 for intratumoral (IT) administration in the

Christine E. Engeland (ed.), *Oncolytic Viruses*, Methods in Molecular Biology, vol. 2058,
https://doi.org/10.1007/978-1-4939-9794-7_2,

treatment of head and neck cancer in combination with chemotherapy [26]. Talimogene laherparepvec (Imlygic®) is an HSV type 1 (HSV-1) derived by functional deletion of ICP34.5 and ICP47 and insertion of the coding sequence for human granulocyte macrophage colony-stimulating factor (GM-CSF), and in 2015 was the first oncolytic virus to gain FDA approval, when it was approved for IT administration in patients with inoperable malignant melanoma [27, 28]. Many other OVs are being investigated in clinical trials, including oncolytic VV, which has also shown promise. VV has been used to treat hundreds of cancer patients in late stage clinical trials to date. However, it is notable that out of the 24 reported studies investigating oncolytic VV clinically, only eight used intravenous (IV) administration. Robust antitumour response has only been shown with IT administration of oncolytic VV, and this may be the reason that prominent ongoing or recruiting trials are using IT administration (NCT02562755, NCT02977156, NCT03071094). Hence, while IT administration of oncolytic VV has shown a safe profile with promising antitumor response, these trials have also strongly indicated that effective biodistribution is still an area which needs improvement. Despite the clinical advances seen over the past 10 years, most of the successes with OVs have therefore been limited to situations where direct delivery presents no obstacle. Hence, there are still important challenges that need to be overcome before the use of viruses as therapeutic agents becomes a standard clinical approach for treatment of systemic diseases. First, there is the challenge of delivering the virus via the bloodstream to achieve dissemination throughout the body and enable distant disease sites to be reached. This is because after administration, viruses are recognized as xenogens and are attacked and rapidly cleared from the body, often before they are able to reach the desired locations, which dramatically impedes the efficiency of gene therapies and oncolytic virotherapies [14, 29, 30]. Second, any virus which does manage to reach its target site may have suffered losses in infectivity during its journey.

Chemical modification, often using hydrophilic polymers, has been utilized to prolong the half-life of many biological therapeutics [31, 32]. This chapter will explore how hydrophilic polymer coatings have been used to decrease interactions with blood components, as a means of improving bloodstream stability and target organ accumulation of therapeutic viruses. The text explores the aims and findings of the approaches taken and the tables provide a guide to the methodology used in the cited papers.

2 Polymer-Coated Viruses

Due to its infection efficiency and well defined and stable genome, adenovirus type 5 (Ad5) is commonly used as a viral vector for gene therapy and oncolytic virotherapy. Clinical translation has, however, been limited by the strong innate and adaptive immune responses raised against Ad5 in many patients. Approaches to decreasing recognition of Ad5 by the immune system have included genetic modification of viral capsid proteins, or their replacement with capsid proteins from serotypes that are less prevalent in the human population [33–35]. These techniques have only had moderate success due to the complexity of Ad5's cellular tropism and cell-entry mechanisms, which are determined by several capsid proteins (namely, the fiber [36], penton base [37], and hexon [38] proteins). Furthermore, preexisting immunity directed against Ad5 is widespread, and is targeted against a range of capsid epitopes [39].

The concept of polymer coating of Ad5 was first reported by O'Riordan et al. [40] and Fisher et al. [30]. O'Riordan et al. used three polymers with different amine-reactive end groups (discussed in further detail in Subheading 3). In theory such coating provided a relatively simple and comparatively inexpensive alternative to genetic modification. Fisher et al. demonstrated successful coating and redirection of Ad5 tropism ("retargeting") of Ad5 with an amine reactive polymer capable of attachment to the virus at multiple points ("multivalent polymer"). In these initial studies, although the coating strategy employed was designed to attach polymers to any available lysine on the capsid surface, the infectivity of the virus was still either retained or redeemable. Since these reports, a range of different optimized coating, retargeting, and bioresponsive uncoating strategies have been reported. The use of both covalent and noncovalent chemical modifications of viruses (Fig. 1a, b respectively), will be discussed in the next sections, with a particular focus on adenovirus.

3 Covalent Chemical Modification

3.1 Reactive Polymers Used in Virus Coating

Viral polymer coating has taken advantage of many existing polymers, and at the same time stimulated the development of new ones. Details of common synthetic methods for the synthesis of some of those polymers have been reviewed in detail [41]. The most common polymer backbones used for the covalent coating of viruses are based on poly[*N*-(2-hydroxypropyl)methacrylamide] (PHPMA) and poly(ethyleneglycol) (PEG). PHPMA and PEG-based polymers are known to extend the circulation of an array of proteins and small-molecule drugs, thereby enhancing their therapeutic benefit, while having little negative impact on their toxicity [31, 32].

3.1.1 Monofunctional and Bifunctional Polymers

The majority of viral modification work performed to achieve polymer stealthing has used monofunctional polymer chains with amine reactive groups. The three common functional groups which have

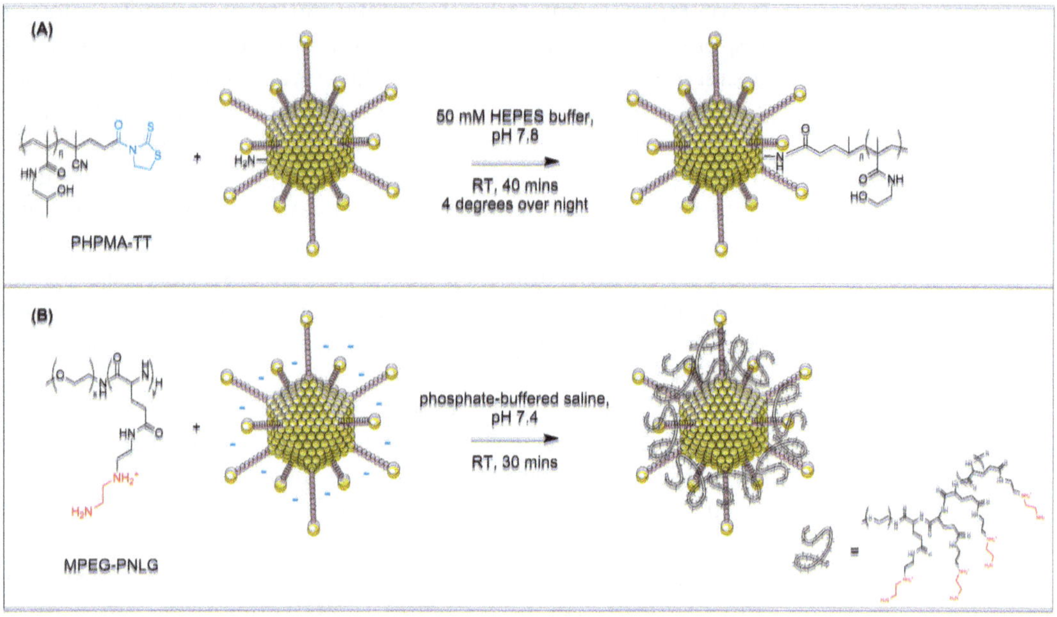

Fig. 1 Schematic displaying examples of chemical conjugation strategies for viral modification. (**a**) Chemical polymer coating of Ad using a PHPMA polymer with an amine-reactive end group carbonylthiazolidine-2-thione (TT) (PHPMA-TT). Coating conditions taken from Morrison et al. [105]. (**b**) Noncovalent polymer coating of Ad using a cationic PEG polymer (PEG-PNLG). Coating conditions taken from Kim et al. [79]. Adenovirus image adapted from Smart Servier Medical Art [106]

been used to acylate lysines on the Ad5 viral capsid are thiazolidine-2-thiones (TT), *p*-nitrophenyl esters (ONp), and *N*-hydroxysuccinimidyl esters (NHS) (Fig. 2a(i), (ii), and (iii) respectively). Subr et al. [42] synthesized an HPMA-based polymer with a TT end group (Fig. 2a(i)). Furthermore, additional modification of the p(HPMA)-TT polymer enabled the inclusion of a reduction-cleavable disulfide (S-S) bond between the polymer backbone and the TT group [43].

PEGs are the most widely used polymers in drug delivery [44–49] and have been utilized in several viral coating strategies. Some of the early studies describing the synthesis of PEGs with amine reactive groups include the work of Herman et al. [49], who activated hydroxyl end groups of PEGs as dichlorotriazines and carboxy end groups as succinimidyl esters (both of which form stable amide bonds with most proteins), and Zalipsky et al. [50], who converted a hydroxyl end group into a succinimidyl carbonate. Today, a multitude of monofunctional and bifunctional PEGs are readily available commercially. It is worth noting that the amine reactive groups on these polymers are prone to hydrolysis. The relative rate of hydrolysis to aminolysis is thus key to how effective the reactive ester is for viral capsid coating and is an important consideration when reaction timings and buffers are chosen.

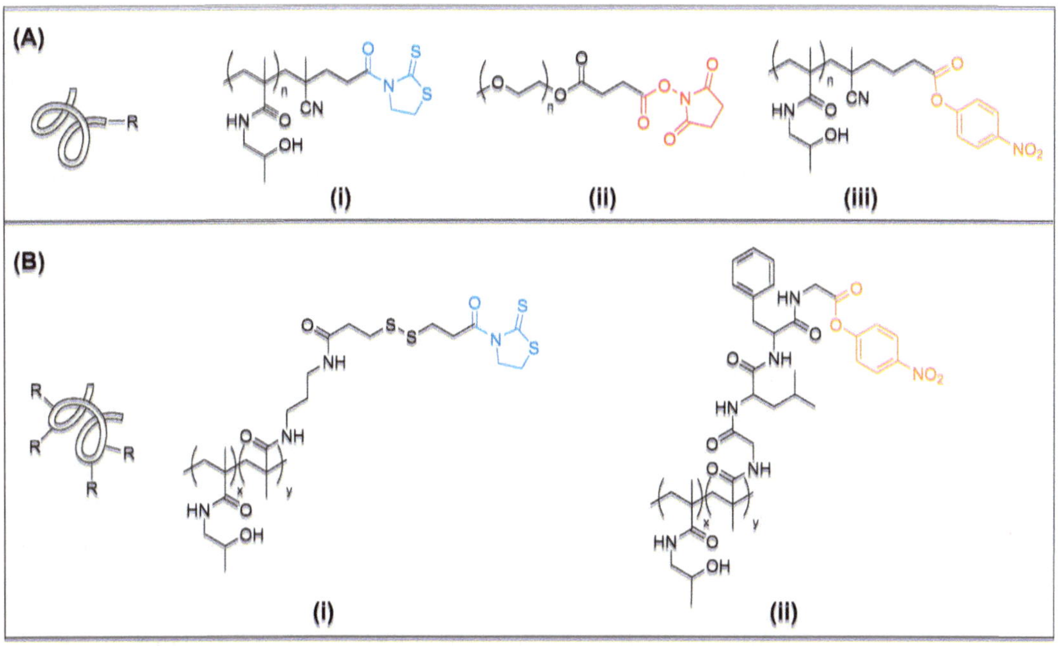

Fig. 2 General structures of polymers used in chemical viral coating methods. (**a**) Monofunctional polymers: (i) PHPMA-TT, (ii) PEG-SS, (iii) PHPMA-ONp. (**b**) Linear multivalent polymers: (i) PHPMA-S-S-TT where S-S is a disulfide bond (reductively cleavable linker), (ii) PHPMA-GLFG-ONp (GLFG is an enzymatically cleavable peptide linker)

3.1.2 Linear Multivalent Reactive Polymers

Linear multivalent reactive polymers are polymer chains that contain multiple reactive sites either randomly or regularly distributed along the polymer backbone. These have a relatively high content of functional groups, as in contrast to monofunctional or bifunctional polymers, the functional groups are located within the repeating unit, not just at the polymer end(s). The presence of a high number of functional groups leads to two major benefits; first, the polymer layer provides a more complete coverage of the surface of the virus particle, and second, there is scope for further modification of the unreacted functional groups [51, 52].

In 1989, Rihova et al. [53] reported the synthesis of an HPMA-based multivalent copolymer (PHPMA) with side chains ending in amino reactive ONp groups (Fig. 2b(ii)). This copolymer was directly conjugated with polylysine/DNA complexes [54]. Later, it was also shown that further modification of the unreacted ONp groups of the PHPMA coated complexes could be used for the attachment of targeting ligands for cell-specific delivery [51].

As aforementioned, the relative rate of hydrolysis to aminolysis can be very different for different amine-reactive groups, and can significantly affect the coating process. Subr et al. [42] determined that the relative rate of hydrolysis to aminolysis was significantly lower during coating for an HPMA-based polymer with TT functional end group than ONp. TT-functionalized polymers with disulfide linkages to an HPMA polymer backbone (Fig. 2b(i)) were synthesized by direct copolymerization of HPMA with TT-functionalized monomers by Kostka et al. [55].

3.2 PEGylation

Modifications of Ad5 with PEG were initially focused on enhancing the efficacy of Ad5 as a vector for gene therapy. O'Riordan et al.'s first studies of Ad5 PEGylation compared and optimized the coating method for PEG with a molecular weight of 5 kDa (PEG_{5000}), and with three different amine-reactive groups [40]. Of the three, methoxypolyethylene glycol tresylate (TM-PEG) was determined to be the most efficient at coating, and it was shown, via ELISA detection of a biotinylated analogue, that capsid modification increased as the ratio of TM-PEG to Ad5 increased. As mentioned above, the Ad5 hexon, penton, and fiber capsid proteins have been shown to be targets of preexisting immunity against Ad5 [39], thus it is notable that at high concentrations of polymer (ratios of 20,000 polymer to virus), modification by polymer attachment occurred on all three of these proteins [40]. Polymer coating of Ad5, comparing the same three PEGs as in O'Riordan et al.'s study, was further explored by Croyle et al. [56]. They showed that after PEGylation, the Ad5 negative zeta potential was reduced (from −48 mV to around −20 mV). Furthermore, when virus proteins were separated by gel electrophoresis (SDS-PAGE) and compared to their unmodified counterparts, they showed that there was modification of the hexon, penton, and fiber capsid proteins [44]. In order to determine which of the polymers produced the most efficient coating, a fluorescamine assay was developed which quantified the extent of polymer modification, as the fluorescence detected was proportional to the number of non-polymer-modified amino acids on the viral capsid [57]. It was concluded that modification with a TM amine reactive functional group was optimal, as it produced highly modified viruses with a high titer. It was subsequently shown in in vitro studies that these PEGylated Ad5 were able to effectively evade binding of neutralizing antibodies (NAbs) in the serum of mice which had been immunized against Ad5 [44].

These studies showed that not only can polymer modification of the capsid proteins of Ad5 be achieved and protect the virus from subsequent neutralization, but that infectivity can be maintained. This unexpected result suggested either one of two things: first, the epitopes required for cell infection were only modified to the extent that they were able to evade detection from NAbs but were still able to infect cells, or second, that modification only occurred on epitopes that are not required for cell infection and that the antibodies present in the mouse serum were only raised against these epitopes. The latter could be due to the studies being performed in mice that were given preexisting immunity by preadministration of nonreplicating Ad5, which is known to preferentially produce a high number of anti-hexon antibodies [34]. It was therefore hypothesized that incomplete PEGylation of the fiber protein could allow for CAR or HSPG interaction in serum which contains low concentrations of anti-fiber antibodies. The impact that the modification of different capsid proteins may have on infection in murine models was subsequently supported by further studies which showed that infection of HeLa cells by a PEGylated retargeted Ad5 could still be inhibited by the presence of excess fiber protein [58].

To determine the influence of PEGylation on reducing the induction of an anti-Ad5 adaptive immune response, studies were carried out comparing modified and unmodified Ad5 after intratracheal administration in immune competent mice [59]. The study reported that the number of cytotoxic T lymphocytes (CTLs) and anti-Ad5 NAbs detected were reduced for Ad5 conjugated with PEG. Significantly, it was shown that the decrease in NAbs detected after administration was different depending on the reactive group used for PEGylation. Ad5 conjugated with both succinimidyl succinate PEG (SSPEG) and cyanuric chloride-activated PEG (CC-PEG) induced a higher production of NAbs when compared to the very low titer of NAbs produced against TM-PEG conjugated Ad5. Both this work and the work of O'Riordan et al. [40] demonstrates the ability of the virus to express transgenes after intratracheal administration in murine models and hence, the ability of PEGylated Ad5 to escape binding by NAbs. Of further importance, this work was the first to show that repeat dosing of Ad5 PEGylated with the same amine reactive linker leads to a knockdown in transgene expression. This knockdown was, however, not seen for repeat doses of PEGylated Ad5 when each dose employed a virus that was coated with a different amine reactive linker. It was hypothesized that the mechanism by which the Ad5 is PEGylated produces antibodies specific for new antigenic epitopes dependent on the linker used. It follows that, in this case, the PEGylation strategy directly changes the way the adaptive immune system responds to the viral vector [59]. This production of new antigenic epitopes from PEGylation must be noted as a potential limitation and has been previously noted as a limitation in its use for therapeutic proteins. Given the number of amine reactive groups that could be used as linkers, Croyle et al.'s work suggests one possible approach to overcoming this limitation—although the use of a range of different reactive groups would add manufacturing and regulatory complexities.

Mok et al. provided a detailed analysis of the impact of PEGylation on the innate immune response and liver toxicity [60]. The PEGylation strategy that these authors applied to Ad5 used *N*-succinimidyl propionate (SPA) as the reactive linker, leading to the complete eradication of infectivity of the PEGylated Ad5 virus in vitro but not in vivo. This same study first showed that Kupffer cell capture of Ad5 was reduced for the PEGylated Ad5, but that this did not translate to reduced hepatic damage [60].

Further studies investigating the degree of Ad5 modification and its impact on circulation and liver capture shed light on the potential impact that polymer coating could have in murine models. Specifically, when heavy polymer modification of Ad5 was achieved (as determined by a 1000-fold decrease in infection of A549 cells), the blood clearance rate in vivo was reduced fourfold when compared to the unmodified virus [61].

A study comparing the infectivity of Ad5 coated with 5, 20 or 35 kDa PEG showed in vitro that the ablation of infectivity was independent of polymer length used [62]. The infectivity could be restored by the addition of FX and hence it was hypothesized that its binding sites on hexon were still available. Interestingly, in vivo high levels of hepatic infection were seen for Ad5 coated with the lowest molecular weight polymer ($AdPEG_{5000}$), but not with either the $AdPEG_{20000}$ or $AdPEG_{35000}$. The reported sizes for the three constructs were 132.7, 145.9, and 158.5 nm respectively, with increasing PEG molecular weight (MW) [62]. Given that murine liver endothelial fenestrae are approximately 140 nm [63] in diameter, and access to hepatocytes is only possible for constructs below this cut-off size, the size of the construct, and hence the MW of PEG used for coating, is an important factor for enhanced circulation. However, the size of human liver endothelial fenestrae is significantly smaller than that of mice; 107 nm compared with 140 nm. This could explain why, in human trials, infection of hepatocytes after IV administration of Ad5 was minimal, when compared to the levels seen in murine models [64, 65]. Hence, the reduction in hepatocyte binding seen with PEGylated Ad5 in mouse models may not be a good predictor of the extent to which this benefit will be seen in humans.

Recently, Nguyen et al. [66] PEGylated Ad6 using a $mPEG_{5000}$-SPA at increasing concentrations. Again, it was shown that infection was reduced in vitro but maintained in vivo after IV injection in murine models. Liver damage was also reduced. Krutzke et al. [67] used a combined genetic and chemical approach to replace the coagulation factor X (FX) shield that occurs naturally on Ad5 with the aim of further reducing the clearance while maintaining its infectivity. They used a maleimide-activated linear polymer PEG (MeO-PEG-mal) of molecular weight 2K, 5K, and 20K. When MeO-PEG-mal was bound to an engineered cysteine residue in hexon hypervariable region 1 of Ad5, the vectors were able to evade detection from NAbs and complement. It was also shown that this modification was able to extend circulation and reduce hepatocyte infection in vivo [67].

Adeno-associated virus (AAV) was also successfully modified using a Biotin-PEG-SPA [68]. Lee et al. investigated a range of polymer chain lengths (2K–20K) and polymer to lysine ratios

($1{:}1$–$10^4{:}1$) and found that for each polymer chain length there was a unique polymer to virus critical ratio at which AAV's infectivity was ablated. For PEG 5K and 20K the key ratio was above 2500:1; however, for PEG 2K the critical ratio was 5000:1. At these critical ratios the polymer coated AAV's showed a two- to threefold protection from NAbs compared with naked AAV [68]. For an overview of conditions used for PEGylating methods of virus coating, *see* Table 1.

3.3 PHPMA

As discussed in Subheading 3.1.1, HPMA is another synthetic polymer which has been used to successfully coat viruses. HPMA is multivalent and allows for attachment of the polymer to the virus at multiple points along the polymer backbone. Unlike PEGylation, which forms a brush-like coating of the virion, HPMA attaches to the virion surface in a much more complete fashion [32]. Despite the number of binding sites available on each polymer, there is no virus-to-virus cross-linking due to cooperative binding effects and partial shielding.

Fisher et al. [30] were the first to show successful modification of Ad5 with PHPMA (16.5 kDa), reporting that 70% of capsid amino groups on Ad5 were modified and that the coating increased the hydrodynamic diameter by 20 nm. Removal of natural tropism was confirmed by the vector's inability to infect human lung carcinoma A549 cells (which are known to be readily infected by unmodified Ad5). Some of the reactive sites on HPMA were left unconjugated [30] and subsequently reacted with FGF2 and $VEGF_{165}$, which were then able to successfully retarget the virus to cell lines with the appropriate receptors. Reestablishment of Ad infectivity selectively via the vascular endothelial growth factor receptor (VEGFR) was shown through infection of a coculture of VEGFR-rich HUVEC cells and VEGFR-poor SUIT2 cells.

Furthermore, an in vivo study by Green et al. [69] in Balb/c mice showed PHPMAylated vectors to have increased blood stability compared to an unmodified vector. When transgene expression in the liver was analyzed following an IV dose of 1×10^9 virus particles there was 1000-fold less transgene expression from the coated virus than the naked virus. This showed that detargeting the virus from the liver was possible using a PHPMA coating and that the consequent enhancement of circulation duration may permit more opportunity for enhanced permeability and retention (EPR) mediated accumulation in tumors [69]. In response, Fisher et al. followed this study with an investigation into whether the increased circulation achieved with PHPMAylated Ad5 would lead to increased passive tumor accumulation [70]. Quantitative PCR was used to show a 40-fold increase in the amount of tumor-associated Ad5 genomes when Ad5-PHPMA was used vs. nonmodified Ad5.

Despite these impressive findings there were still limitations to this approach. Notably there was imprecise control of the viral

Table 1
Summary of conditions used for PEGylating methods of virus coating

Paper	Polymer	Amount virus, VP/ml	Amount polymer	Buffer	pH	Extent of modification	Modification assay	Time (min)	Purification method
O'Riordan et al. [40]	TM-PEG CC-MPEG SS-MPEG (5K)	1e12	5–20% w/v	TM: 130 mM NaH_2PO_4, 5% sucrose CC: 0.1 M $Na_2B_4O_7$, 0.15 M NaCl SS: phosphate-buffered saline (PBS) ± 5% sucrose	TM: 7 CC: 9.2 SS: 7.4	~18,000 PEG/virus	Biotin peg and ELISA	30	CsCl centrifugation
Alemany et al. [61]	TM-PEG	1e12	41 mg/ml	130 mM NaH_2PO_4, 5% sucrose	7	NA	NA	30	NA
Croyle et al. [57, 59]	TM-PEG CC-MPEG SS-MPEG (5K)	NA	PEG: virus, 10:1	TM: 10 mM K_2HPO_4 buffer CC: 0.1 M $Na_2B_4O_7$ SS: 0.2 M NaH_2PO_4	TM: 7.4 CC: 9.2 SS: 7.2	NA	Fluorescamine assay	90	Sephadex G-50 column
Lanciotti et al. [98]	Tresyl-PEG-mal	1e10	1, 5, 10% w/v	130 mM NaH_2PO_4, 5% sucrose	7	75–100 FGF2/virus	NA	1%: 45 5%: 30 10%: 60	Size exclusion chromatography

Romanczuk et al. [58]	Tresyl-PEG-thiol	1.5e12	5% w/v	130 mM NaH_2PO_4, 5% sucrose	7	NA	NA	60	CsCl centrifugation
Kreppel et al. [101]	mPEG-SPA (2K)	5e11	PEG: virus cystine 100:1	50 mM $C_8H_{18}N_2O_4S$ (HEPES) buffer	7.3	NA	NA	180	NA
Mok et al. [60]	mPEG-SPA (5K)	1e12	10–0.01 mg/ml	0.1 mM K_2HPO_4 buffer	8.2	~85% capsid	fluorescamine assay	60	300 kDa dialysis
Hofherr et al. [62]	mPEG-SPA (5K) m-SCM-(20K) m-SCM-(35K)	NA	NA	0.5 mol/l sucrose potassium PBS	8	65–74% capsid	CBQCA Protein Quantitation Kit	120	Sephadex G100
Nguyen et al. [66]	mPEG-SPA (5K)	1e12	10–0.01 mg/ml	0.1 mM K_2HPO_4 buffer	8.2	NA	NHS-Oregon Green staining	60	300 kDa dialysis
Krutzke et al. [67]	MeO-PEG-mal (2K, 5K, 20K)	NA	PEG: virus cystine 20:1	50 mM HEPES buffer	7.2	~700 PEG/virus	SDS-PAGE and silver staining	60	CsCl centrifugation

Room temperature was used in all methods described here

epitopes being modified and there was a loss of infectivity associated with an incomplete uncoating and removal of the polymer at its target site. To address this Prill et al. performed a study investigating the infectivity and stability of Ad after coating with a "bioresponsive" shield, using PHPMA polymers activated with either maleimide (mal) groups or pyridyl-dithio (PySS) groups [71]. These groups were designed to react with cysteine groups genetically engineered into Ad viruses to create either an irreversible (PHPMA with mal group) or a bioresponsive cleavable (PHPMA with PySS group) polymer coating. It was determined that the bioresponsive coating was able to protect the virus as effectively as the irreversible coating and that viruses coated with either polymer maintained their ability to enter cells after 5 min. Modification with an irreversible coating did impact the intracellular trafficking of the virus to the nucleus, but the bioresponsive coating, after a delay, did allow for particle trafficking. Next, the effect of the charge of the bioresponsive polymer coating of the Ad was investigated in vivo. It was reported that a positively charged bioresponsive polymer coat brought about high levels of liver infection.

Carlisle et al. [72] used a PHMPA-based polymer with bioresponsive side chains containing an acid labile hydrazone and an amine-reactive TT group to coat an oncolytic adenovirus, with the intention of providing an uncoating process which was triggered by the slightly acidic environment within the tumor. Ad5 coated with this low pH cleavable polymer gave similar levels of infection to Ad5 coated with noncleavable polymer following subcutaneous injection but showed raised levels when injected intratumorally. In vivo experiments showed a 50-fold increase in circulation half-life and an 8000-fold lower hepatic sequestration compared to naked Ad. The favorable circulation profile of the coated virus was exploited to increase its tumor accumulation by ultrasound-assisted delivery. When ultrasound, microbubbles and the coated virus were combined, infectivity was enhanced 30-fold, with a corresponding 125-fold increase of the tumor-to-liver ratio. Furthermore, a 40-fold increase in infectivity was measured at >100 mm from the vasculature compared to a control without ultrasound [72].

Beyond the application to Ad5, Carlisle et al. also tested the impact of PHPMA coating on adeno-associated virus (AAV) [73]. In these studies AAV5 and AAV8 were modified. Modification of AAV5 using PMPMA-TT (2–20 mg/ml) was only possible following viral modification with 1-ethyl-3-[3-dimethylaminopropyl]carbodiimide hydrochloride (EDC) mediated coupling of hexanediamine. In comparison AAV8 was amenable to coating directly with PHPMA-TT. The modification mediated successful retargeting via EGFR and a level of enhanced protection against neutralizing antiserum. For an overview of conditions used for PHPMAylating methods of virus coating, *see* Table 2.

Table 2
Summary of conditions used for PHPMAylating methods of virus coating

Paper	Polymer	Amount virus (VP/ml)	Amount polymer	Buffer	pH	Extent of modification	Determining modification	Temp	Time	Purification
Fisher et al. [30, 70]	PHPMA-ONp	1e12	10 mg/ml	10% glycerol/PBS	7.8	>900 polymers/virus	Fluorescamine, 14C-trace	4 °C	12 h	SizeSep 400 Spun Columns
Green et al. [69]	PHPMA-ONp	1e11–1e12	NA	HEPES/PBS	7.8	90% amino groups on surface modified	Fluorescamine, 14C-trace	RT	Over night	NA
Stevenson et al. [103]	PHPMA-ONp	1e11–1e12	60, 90, 120 mg/ml	PBS containing 50 mM HEPES	8	NA	NA	RT, then 4 °C	1 h, then over night	Sepharose S-400 spin column
Green et al. [104]	PHPMA-ONp	4e11	2, 5, 10, 20 mg/ml	HEPES/PBS	7.8	NA	NA	RT, then on ice	40 min, then over night	Sepharose S-400 spin column
Morrison et al. [105]	PHPMA-TT	NA	20 mg/ml or 5 mg/ml	50 mM HEPES buffer	7.8	NA	NA	RT, then on ice	40 min, then over night	NA
Prill et al. [71]	PHPMA-mal/PHPMA-PySS	5e11	PEG: virus cystine, 20:1	50 mM HEPES buffer	7.2	NA	SDS-PAGE and western blot	RT	Over night	NA

Adenovirus was used in all coating methods described here

4 Noncovalent Modification

An argument can be made that the covalent modification of capsid epitopes means full infectivity can never be recovered despite attempts to incorporate bio-responsive cleavage mechanisms. An alternative approach to capsid modification is the addition of cationic polymers which can interact electrostatically with the capsid surface; such an approach does not therefore create an irreversible modification of protein residues that may be required for infection. Traditionally used as nonviral gene delivery vectors, cationic polymers are known to enhance cellular uptake and can even promote lysosomal escape via the proton sponge effect [74]. Two potential drawbacks identified in early work using cationic polymers were a generally unfavorable toxicological profile and the in vivo instability of such complexes as a result of nonspecific binding and displacement by serum proteins. Although these problems seem to have been overcome in the newest generations of cationic polymers used for virus coating, care should still be taken when developing new formulations.

Fasbender et al. [75] were among the first to explore the concept of coating adenovirus with cationic species. They reported an increased transgene expression in vitro with poly(L-lysine)-coated compared to naked adenovirus, which was however abolished in the presence of serum. Similar results were obtained by incubating adenovirus with the cationic lipid GL67, a carbamate of cholesterol and spermine. Since this early work, many different cationic polymer classes have been used for viral coating, including polyacrylates [76], branched polyamine copolymers [77], polypeptides [78–81], polysaccharides [82] cross-linked [83–86] and non-cross-linked [87, 88] poly(ethylene imine) derivatives, and poly(*N*,*N*′-bis(acryloyl)cystamine) derivatives [89–91]. Nonpolymeric coatings like dendrimers [92, 93], cationic lipid envelopes [94], and cationic micelles [91] have also been used to modify viruses.

The serum instability of some of the early cationic polymer-coated viruses was overcome by modulation of the complex surface charge through inclusion of PEG or oligoethylene glycol blocks into the polymers. For example, Jiang et al. [78] showed that adenovirus coated with a copolymer of poly(L-lysine) (PLL) and 2-[2-(methoxyethoxy)ethoxy]acetylated poly(L-lysine) was less prone to aggregation compared to PLL-coated adenovirus, and could inhibit NAbs and FX binding in vitro. However, the inhibition of FX binding did not translate into reduced murine liver transduction in vivo compared to naked Ad.

More promising in vivo results were obtained with a PEGylated copolymer, PEG-b-poly(*N*-[*N*-(2-aminoethyl)-2-aminoethyl]-L-glutamate) (PNLG), which was shown by Kim et al. [79] to increase the tumor-to-liver ratio 1229-fold compared to naked

Ad, and suppress tumor growth with minimal liver toxicity. PNLG was further optimized in vitro by Choi et al. [81] by comparing different PEG molecular weights and numbers of side-chain amines. A PEG molecular weight of 5 kDa and a side chain with five amines was found to result in the lowest serum inhibition, lowest innate immune response (as measured by IL-6 secretion), and highest cell killing effect on both CAR negative MCF7 and CAR positive A549 cells.

Reduction of liver toxicity was also obtained with PEGylated bioresponsive cationic polymers. For example, Kim et al. [95] coated an oncolytic adenovirus with a graft copolymer of a hetero-bifunctional PEG and the reductively cleavable poly(cystaminebisacrylamide-diaminohexane), which was reacted with a cRGD targeting peptide through the end group of the PEG side chains. They reported both innate and adaptive immune response evasion, no signs of hepatotoxicity and increased tumor growth suppression compared to naked Ad. In another example of the effect of PEGylation, a remarkable 10^5-fold increase in tumor-to-liver ratio compared to naked Ad was reported by Kwon et al. [82] with a PEGylated chitosan coating that was targeted to the folate receptor.

Adenovirus was not the only virus to be successfully coated with physical methods. Interestingly, the coating of an oncolytic measles virus (MV) by Nosaki et al. [88] using layer-by-layer deposition of a linear polyethyleneimine hydrochloride (PEI) and negatively charged chondroitin sulfate was also able to protect it from NAbs in vitro in human HEp2, A549 and WiDr cells and murine LL/2 cells modified to express human CD46 (LL/2-CD46). Protection from NAbs in vivo was demonstrated after intratumoral administration of the modified virus in immunocompetent preimmunized mice bearing subcutaneous LL/2-CD46 tumors, which resulted in enhanced oncolysis compared to naked MV. These impressive results with MV have yet to be followed up. Indeed, it is intriguing that while a huge amount of research effort has been dedicated to the modification of nonenveloped viruses such as Ad, little attention has been given to enveloped viruses such as HSV, measles, and vaccinia virus which are similarly challenged by the neutralizing environment within the blood stream due to the action of complement and the high prevalence of preexisting immunity [29] as a result of natural infection or vaccination programs. Where covalent modification strategies are concerned this may relate to the difficulty in achieving controlled, predictable modification of a lipid envelope, but noncovalent charge mediated modification seems like an area where important developments could be made. Notably, the challenge of systemic delivery of enveloped viruses has not been totally ignored with a range of studies reporting the use of cell carrier systems, as reviewed by [96], while Rojas et al. [97] demonstrated enzymatic treatment of a thymidine kinase gene deleted Western Reserve vaccinia virus (VV) surface had important ramifications for virus delivery and immunogenicity. In particular it was shown that when VV has its N-linked and simple O-linked glycans removed using a mixture of *N*- and *O*-glycanases and sialidase A, there was a reduction of TLR2 activation and anti-VV neutralizing antibody production. In vivo, systemic delivery of deglycosylated VV saw an increase in viral gene expression in Renca xenografts after 24 h when compared with unmodified VV. After 5 days this increase in infection was over tenfold.

5 Retargeting

In addition to the benefits derived from increased circulation time, PEGylation of adenovirus using bifunctional PEG molecules can also provide a platform for the attachment of targeting ligands to redirect the tropism of the virus. This method of retargeting has benefits over retargeting by the genetic modification of viral capsid proteins: it enables the use of full length proteins and the inclusion of ligands which would otherwise be too large to fit without disrupting virus structure and/or to be encoded genetically. The link between the extent of PEG modification and extent of detargeting by disrupting CAR binding was further confirmed by Mok et al. [60] in vitro where the conjugation of 15,000 PEG molecules per particle reduced the infectivity of the Ad5 particle by three orders of magnitude. Retargeting of these detargeted vectors will be discussed here.

An early investigation by Romanczuk et al. [58] of polymer coating and retargeting of Ad2 utilized PEG chains with amine reactive groups on one end and thiol reactive groups on the other. The amine-reactive tresyl moiety bound to the Ad2 capsid (as described above), leaving the thiol-reactive vinyl sulfone free to react with a cysteine-containing targeting peptide. The viral vector was still able to evade Nabs in vitro and despite only a small decrease in baseline infectivity, the inclusion of targeting peptides gave rise to a four- to fivefold rise in cancer cell infection. Following from this, Lanciotti et al. [98] employed a tresyl-PEG-maleimide polymer to bind fibroblast growth factor (FGF2)-targeting ligands to Ad2. This enabled a tenfold increase in infectivity in both in vitro and in vivo studies using SKOV3.ipl cells. The contrast in binding to CAR between Lanciotti et al. and O'Riordan et al.'s pioneering study is noteworthy: Lanciotti reported that CAR binding was severely reduced whereas O'Riordan reported that infectivity was retained. This could be due to small differences in conjugation protocol, O'Riordan et al.'s polymer being of higher quality or an overestimated amount of PEGylation described in O'Riordan et al. [40]. Furthermore, it is noted that FGF2 is a large protein, and its binding to the adenovirus surface is likely to also protect capsid proteins and thereby prevent CAR binding. This was supported by the stronger decrease reported in both types of immune responses for the retargeted Ad compared to Ad modified only with PEG.

Kreppel et al. [99] used a combined genetic and chemical approach to retarget Ad modified with SPA-PEG. It involved genetically modifying the capsid so as to introduce cysteine residues into the HI-loop of the fiber protein while using PEGylation to remove its ability to infect. The result was the production of an Ad vector which had its natural tropism removed but with the cysteines on the tip of the fiber protein unmodified. Full length transferrin was conjugated onto these fiber proteins, using a heterobifunctional PEG. It was reported that this produced vectors with unique selectivity for transferrin receptors that were still able to replicate in in vitro experiments. Another of the benefits of Kreppel et al.'s method is the specificity of the target ligand binding to the fiber protein; the site of the virus which is usually responsible for binding target sites and cell entry. A study performed by Campos and Barry [100] determined that targeting may only be effective when it is achieved through the fiber protein. It follows that modification of other capsid proteins, such as pIX and hexon which travel to the nucleus upon cell entry, may actually hinder intracellular trafficking [101, 102]. Despite potential limitations born out of heavy PEGylation of some capsid proteins, it has been shown that a combination of polymer coating and retargeting is effective and can overcome some of the challenges of systemic viral delivery.

After Fisher et al.'s initial study using FGF and VEGF targeting ligands (Subheading 3.3), further studies exploring the retargeting of Ad5 coated with PHPMA were performed. When Ad5-PHPMA was retargeted using SIKVAV peptide, strong selective infection of PC-3 cells was shown in vitro. This however did not lead to increased transgene expression between Ad-PHPMA and Ad-PHPMA-SIKVAV in PC-3 xenografts in vivo [103]. When FGF ligands were used in a study investigation Ad-PHPMA-FGF2 for intraperitoneal (i.p.) delivery by Green et al. [104] reduced toxicity and efficacy was reported with no peritoneal adhesion formation seen. However, when tested intravenously, Ad5-PHPMA-FGF2 did have a strong association with murine erythrocytes and, consequentially, a decrease in tumor infection was seen. Of note, the conjugation strategy used in Morrison et al.'s [105] study looking at i.p. delivery differed from those used previously: the EGF targeting ligand was conjugated directly onto PHPMA before attachment to the viral capsid. This one-step coating and retargeting method was able to produce a vector which retained its ability to infect cells and reduced peritoneal adhesion following i.p. delivery. This one-step methodology has potential scale-up and GMP preparation benefits.

6 Summary

Chemical modification to enhance therapeutic index has been successfully applied to a range of drug classes. Viruses hold huge potential as gene therapy and oncolytic vectors but can be severely limited by bloodstream neutralization. It is no surprise therefore that the idea of chemical modification of viruses initially attracted a huge amount of research focus. However, although the scientific rationale for this approach remained strong, clinical and commercial imperatives perhaps dictated that demonstration of safety and therapeutic efficacy to achieve rapid regulatory approval of non-modified, nonoptimal viruses, delivered by nonoptimal routes, took precedence over dedicating time and expense to developing more sophisticated fit-for-purpose modified vectors. In essence, for many viruses the issue of poor bloodstream compatibility has been sidestepped for now by a focus on using local delivery. It may be that as a raft of viruses approach and obtain approval, a desire to broaden the clinical indications and patient populations that can be treated with these now approved vectors will drive a resurgence of interest in chemical modification. We hope the information and technical detail we provide here will help guide such a resurgence.

References

1. Cattaneo R, Miest T, Shashkova EV, Barry MA (2008) Reprogrammed viruses as cancer therapeutics: targeted, armed and shielded. Nat Rev Microbiol 6:529–540. https://doi.org/10.1038/nrmicro1927
2. Markert JM, Medlock MD, Rabkin SD et al (2000) Conditionally replicating herpes simplex virus mutant, G207 for the treatment of malignant glioma: results of a phase I trial. Gene Ther 7:867–874. https://doi.org/10.1038/sj.gt.3301205
3. Markert JM, Liechty PG, Wang W et al (2009) Phase Ib trial of mutant herpes simplex virus G207 inoculated pre-and post-tumor resection for recurrent GBM. Mol Ther 17:199–207. https://doi.org/10.1038/mt.2008.228
4. Harrington KJ, Hingorani M, Tanay MA et al (2010) Phase I/II study of oncolytic HSVGM-CSF in combination with radiotherapy and cisplatin in untreated stage III/IV squamous cell cancer of the head and neck. Clin Cancer Res 16:4005–4015. https://doi.org/10.1158/1078-0432.CCR-10-0196
5. Senzer NN, Kaufman HL, Amatruda T et al (2009) Phase II clinical trial of a granulocyte-macrophage colony-stimulating factor-encoding, second-generation oncolytic herpesvirus in patients with unresectable metastatic melanoma. J Clin Oncol 27:5763–5771. https://doi.org/10.1200/JCO.2009.24.3675
6. Kaufman HL, Kim DW, DeRaffele G et al (2010) Local and distant immunity induced by intralesional vaccination with an oncolytic herpes virus encoding GM-CSF in patients with stage IIIc and IV melanoma. Ann Surg Oncol 17:718–730. https://doi.org/10.1245/s10434-009-0809-6
7. Conry RM, Khazaeli MB, Saleh MN et al (1999) Phase I trial of a recombinant vaccinia virus encoding carcinoembryonic antigen in metastatic adenocarcinoma: comparison of intradermal versus subcutaneous administration. Clin Cancer Res 5:2330–2337
8. Mastrangelo MJ, Maguire HC, Eisenlohr LC et al (1999) Intratumoral recombinant GM-CSF-encoding virus as gene therapy in patients with cutaneous melanoma. Cancer Gene Ther 6:409–422. https://doi.org/10.1038/sj.cgt.7700066
9. Park B-H, Hwang T, Liu T-C et al (2008) Use of a targeted oncolytic poxvirus, JX-594, in patients with refractory primary or metastatic liver cancer: a phase I trial. Lancet Oncol

9:533–542. https://doi.org/10.1016/S1470-2045(08)70107-4
10. Hwang T-H, Moon A, Burke J et al (2011) A mechanistic proof-of-concept clinical trial with JX-594, a targeted multi-mechanistic oncolytic poxvirus, in patients with metastatic melanoma. Mol Ther 19:1913–1922. https://doi.org/10.1038/mt.2011.132
11. Zeh HJ, Downs-Canner S, McCart JA et al (2015) First-in-man study of western reserve strain oncolytic vaccinia virus: safety, systemic spread, and antitumor activity. Mol Ther 23:202–214. https://doi.org/10.1038/mt.2014.194
12. Cripe TP, Ngo MC, Geller JI et al (2015) Phase 1 study of intratumoral Pexa-Vec (JX-594), an oncolytic and immunotherapeutic vaccinia virus, in pediatric cancer patients. Mol Ther 23:602–608. https://doi.org/10.1038/mt.2014.243
13. Mansfield D, Kyula J, Rosenfelder N et al (2016) Oncolytic vaccinia virus as a vector for therapeutic sodium iodide symporter gene therapy in prostate cancer. Gene Ther 23:357–368. https://doi.org/10.1038/gt.2016.5
14. Breitbach CJ, Burke J, Jonker D et al (2011) Intravenous delivery of a multi-mechanistic cancer-targeted oncolytic poxvirus in humans. Nature 477:99–102. https://doi.org/10.1038/nature10358
15. Liikanen I, Ahtiainen L, Hirvinen ML et al (2013) Oncolytic adenovirus with temozolomide induces autophagy and antitumor immune responses in cancer patients. Mol Ther 21:1212–1223. https://doi.org/10.1038/mt.2013.51
16. Morse MA, Chaudhry A, Gabitzsch ES et al (2013) Novel adenoviral vector induces T-cell responses despite anti-adenoviral neutralizing antibodies in colorectal cancer patients. Cancer Immunol Immunother 62:1293–1301. https://doi.org/10.1007/s00262-013-1400-3
17. Nemunaitis J, Nemunaitis J (2011) Head and neck cancer: response to p53-based therapeutics. Head Neck 33:131–134. https://doi.org/10.1002/hed.21364
18. Chang J, Zhao X, Wu X et al (2009) A Phase I study of KH901, a conditionally replicating granulocyte-macrophage colony-stimulating factor: armed oncolytic adenovirus for the treatment of head and neck cancers. Cancer Biol Ther 8:676–682
19. Li J-L, Liu H-L, Zhang X-R et al (2009) A phase I trial of intratumoral administration of recombinant oncolytic adenovirus overexpressing HSP70 in advanced solid tumor patients. Gene Ther 16:376–382. https://doi.org/10.1038/gt.2008.179
20. Barton KN, Paielli D, Zhang Y et al (2006) Second-generation replication-competent oncolytic adenovirus armed with improved suicide genes and ADP gene demonstrates greater efficacy without increased toxicity. Mol Ther 13:347–356. https://doi.org/10.1016/j.ymthe.2005.10.005
21. Reid TR, Freeman S, Post L et al (2005) Effects of Onyx-015 among metastatic colorectal cancer patients that have failed prior treatment with 5-FU/leucovorin. Cancer Gene Ther 12:673–681. https://doi.org/10.1038/sj.cgt.7700819
22. Morley S, MacDonald G, Kirn D et al (2004) The dl1520 virus is found preferentially in tumor tissue after direct intratumoral injection in oral carcinoma. Clin Cancer Res 10:4357–4362. https://doi.org/10.1158/1078-0432.CCR-03-0443
23. Hecht JR, Bedford R, Abbruzzese JL et al (2003) A phase I/II trial of intratumoral endoscopic ultrasound injection of ONYX-015 with intravenous gemcitabine in unresectable pancreatic carcinoma. Clin Cancer Res 9:555–561
24. Nemunaitis J, Cunningham C, Buchanan A et al (2001) Intravenous infusion of a replication-selective adenovirus (ONYX-015) in cancer patients: safety, feasibility and biological activity. Gene Ther 8:746–759. https://doi.org/10.1038/sj.gt.3301424
25. Sheridan C (2015) First oncolytic virus edges towards approval in surprise vote. Nat Biotechnol 33:569–570. https://doi.org/10.1038/nbt0615-569
26. Garber K (2006) China approves world's first oncolytic virus therapy for cancer treatment. J Natl Cancer Inst 98:298–300. https://doi.org/10.1093/jnci/djj111
27. Bommareddy PK, Patel A, Hossain S, Kaufman HL (2017) Talimogene Laherparepvec (T-VEC) and other oncolytic viruses for the treatment of melanoma. Am J Clin Dermatol 18:1–15. https://doi.org/10.1007/s40257-016-0238-9
28. Rehman H, Silk AW, Kane MP, Kaufman HL (2016) Into the clinic: talimogene laherparepvec (T-VEC), a first-in-class intratumoral oncolytic viral therapy. J Immunother Cancer 4:53. https://doi.org/10.1186/s40425-016-0158-5
29. Evgin L, Acuna SA, Tanese de Souza C et al (2015) Complement inhibition prevents oncolytic vaccinia virus neutralization in

immune humans and cynomolgus macaques. Mol Ther 23:1066–1076. https://doi.org/10.1038/mt.2015.49
30. Fisher KD, Stallwood Y, Green NK et al (2001) Polymer-coated adenovirus permits efficient retargeting and evades neutralising antibodies. Gene Ther 8:341–348. https://doi.org/10.1038/sj.gt.3301389
31. Delgado C, Francis GE, Fisher D (1992) The uses and properties of PEG-linked proteins. Crit Rev Ther Drug Carr Syst 9 (3–4):249–304
32. Rihova B, Kubackova K (2003) Clinical implications of N-(2-hydroxypropyl)methacrylamide copolymers. Curr Pharm Biotechnol 4 (5):311–322
33. Wickham TJ, Tzeng E, Shears LL et al (1997) Increased in vitro and in vivo gene transfer by adenovirus vectors containing chimeric fiber proteins. J Virol. https://doi.org/10.1684/mst.2016.0568
34. Roberts DM, Nanda A, Havenga MJE et al (2006) Hexon-chimaeric adenovirus serotype 5 vectors circumvent pre-existing anti-vector immunity. Nature. https://doi.org/10.1038/nature04721
35. Raki M, Hakkarainen T, Bauerschmitz GJ et al (2007) Utility of TK/GCV in the context of highly effective oncolysis mediated by a serotype 3 receptor targeted oncolytic adenovirus. Gene Ther. https://doi.org/10.1038/sj.gt.3302992
36. Dechecchi MC, Melotti P, Bonizzato A et al (2001) Heparan sulfate glycosaminoglycans are receptors sufficient to mediate the initial binding of adenovirus types 2 and 5. J Virol. https://doi.org/10.1128/JVI.75.18.8772-8780.2001
37. Wickham TJ, Carrion ME, Kovesdi I (1995) Targeting of adenovirus penton base to new receptors through replacement of its RGD motif with other receptor-specific peptide motifs. Gene Ther
38. Buckley SMK, Waddington SN, Jezzard S et al (2008) Intra-amniotic delivery of CFTR-expressing adenovirus does not reverse cystic fibrosis phenotype in inbred CFTR-knockout mice. Mol Ther. https://doi.org/10.1038/mt.2008.26
39. Sumida SM, Truitt DM, Lemckert AAC et al (2005) Neutralizing antibodies to adenovirus serotype 5 vaccine vectors are directed primarily against the adenovirus hexon protein. J Immunol. https://doi.org/10.4049/jimmunol.174.11.7179
40. O'Riordan CR, Lachapelle A, Delgado C et al (1999) PEGylation of adenovirus with retention of infectivity and protection from neutralizing antibody in vitro and in vivo. Hum Gene Ther. https://doi.org/10.1089/10430349950018021
41. Laga R, Carlisle R, Tangney M et al (2012) Polymer coatings for delivery of nucleic acid therapeutics. J Control Release
42. Šubr V, Koňák Č, Laga R, Ulbrich K (2005) Coating of DNA/poly(l-lysine) complexes by covalent attachment of poly[N-(2-hydroxypropyl)methacrylamide]. https://doi.org/10.1021/BM050524X
43. Laga R, Koňák Č, Šubr V, Ulbrich K (2009) New, hydrophilic, HPMA-based polymers for bioresponsive shielding of gene-delivery vectors. Macromol Chem Phys. https://doi.org/10.1002/macp.200900078
44. Wonganan P, Croyle MA (2010) PEGylated adenoviruses: from mice to monkeys. Viruses 2:468–502. https://doi.org/10.3390/v2020468
45. Amoozgar Z, Yeo Y (2012) Recent advances in stealth coating of nanoparticle drug delivery systems. Wiley Interdiscip Rev Nanomed Nanobiotechnol 4:219–233. https://doi.org/10.1002/wnan.1157
46. Dozier JK, Distefano MD (2015) Site-specific PEGylation of therapeutic proteins. Int J Mol Sci 16:16. https://doi.org/10.3390/ijms161025831
47. Veronese FM, Pasut G (2005) PEGylation, successful approach to drug delivery. Drug Discov Today 10:1451–1458. https://doi.org/10.1016/S1359-6446(05)03575-0
48. Francini N (2016) Polymer coating for the systemic delivery of oncolytic adenoviruses. The British Library Board, London
49. Herman S, Hooftman G, Schacht E (1995) Poly(ethylene glycol) with reactive end-groups: I. Modification of proteins. J Bioact Compat Polym 10:145–187. https://doi.org/10.1177/088391159501000205
50. Zalipsky S (1995) Chemistry of polyethylene glycol conjugates with biologically active molecules. Adv Drug Deliv Rev
51. Fisher KD, Ulbrich K, Subr V et al (2000) A versatile system for receptor-mediated gene delivery permits increased entry of DNA into target cells, enhanced delivery to the nucleus and elevated rates of transgene expression. Gene Ther. https://doi.org/10.1038/sj.gt.3301230
52. Oupický D, Parker AL, Seymour LW (2002) Laterally stabilized complexes of DNA with linear reducible polycations: Strategy for triggered intracellular activation of DNA delivery

vectors. J Am Chem Soc. https://doi.org/10.1021/ja016440n

53. Rihova B, Bilej M, Vetvicka V et al (1989) Biocompatibility of N-(2-hydroxypropyl) methacrylamide copolymers containing adriamycin. Immunogenicity, and effect on haematopoietic stem cells in bone marrow in vivo and mouse splenocytes and human peripheral blood lymphocytes in vitro. Biomaterials. https://doi.org/10.1016/0142-9612(89)90075-6
54. Dash PR, Read ML, Fisher KD et al (2000) Decreased binding to proteins and cells of polymeric gene delivery vectors surface modified with a multivalent hydrophilic polymer and retargeting through attachment of transferrin. J Biol Chem. https://doi.org/10.1074/jbc.275.6.3793
55. Kostka L, Koňák Č, Šubr V et al (2011) Removable nanocoatings for siRNA polyplexes. Bioconjug Chem. https://doi.org/10.1021/bc100197e
56. M a C, Yu QC, Wilson JM (2000) Development of a rapid method for the PEGylation of adenoviruses with enhanced transduction and improved stability under harsh storage conditions. Hum Gene Ther. https://doi.org/10.1089/10430340050111368
57. Croyle MA, Callahan SM, Auricchio A et al (2004) PEGylation of a vesicular stomatitis virus G pseudotyped lentivirus vector prevents inactivation in serum. J Virol. https://doi.org/10.1128/JVI.78.2.912-921.2004
58. Romanczuk H, Galer CE, Zabner J et al (1999) Modification of an adenoviral vector with biologically selected peptides: a novel strategy for gene delivery to cells of choice. Hum Gene Ther. https://doi.org/10.1089/10430349950016654
59. Croyle MA, Chirmule N, Zhang Y, Wilson JM (2001) 'Stealth' adenoviruses blunt cell-mediated and humoral immune responses against the virus and allow for significant gene expression upon readministration in the lung. J Virol. https://doi.org/10.1128/JVI.75.10.4792-4801.2001
60. Mok H, Palmer DJ, Ng P, Barry MA (2005) Evaluation of polyethylene glycol modification of first-generation and helper-dependent adenoviral vectors to reduce innate immune responses. Mol Ther 11:66–79. https://doi.org/10.1016/j.ymthe.2004.09.015
61. Alemany R, Suzuki K, Curiel DT (2000) Blood clearance rates of adenovirus type 5 in mice. J Gen Virol. https://doi.org/10.1099/0022-1317-81-11-2605
62. Hofherr SE, Shashkova EV, Weaver EA et al (2008) Modification of adenoviral vectors with polyethylene glycol modulates in vivo tissue tropism and gene expression. Mol Ther 16:1276–1282. https://doi.org/10.1038/mt.2008.86
63. Lievens J, Snoeys J, Vekemans K et al (2004) The size of sinusoidal fenestrae is a critical determinant of hepatocyte transduction after adenoviral gene transfer. Gene Ther. https://doi.org/10.1038/sj.gt.3302326
64. Reid T, Warren R, Kirn D (2002) Intravascular adenoviral agents in cancer patients: lessons from clinical trials. Cancer Gene Ther
65. Au T, Thorne S, Korn WM et al (2007) Minimal hepatic toxicity of Onyx-015: spatial restriction of coxsackie-adenoviral receptor in normal liver. Cancer Gene Ther. https://doi.org/10.1038/sj.cgt.7700988
66. Nguyen TV, Heller GJ, Barry MAME et al (2016) Evaluation of polymer shielding for adenovirus serotype 6 (Ad6) for systemic virotherapy against human prostate cancers. Mol Ther Oncolytics 3:15021. https://doi.org/10.1038/MTO.2015.21
67. Krutzke L, Prill JM, Engler T et al (2016) Substitution of blood coagulation factor X-binding to Ad5 by position-specific PEGylation: preventing vector clearance and preserving infectivity. J Control Release 235:379–392. https://doi.org/10.1016/J.JCONREL.2016.06.022
68. Lee GK, Maheshri N, Kaspar B, Schaffer DV (2005) PEG conjugation moderately protects adeno-associated viral vectors against antibody neutralization. Biotechnol Bioeng. https://doi.org/10.1002/bit.20562
69. Green NK, Herbert CW, Hale SJ et al (2004) Extended plasma circulation time and decreased toxicity of polymer-coated adenovirus. Gene Ther 11:1256–1263. https://doi.org/10.1038/sj.gt.3302295
70. Fisher KD, Green NK, Hale A et al (2007) Passive tumour targeting of polymer-coated adenovirus for cancer gene therapy. J Drug Target 15:546–551. https://doi.org/10.1080/10611860701501014
71. Prill JM, Šubr V, Pasquarelli N et al (2014) Traceless bioresponsive shielding of adenovirus hexon with HPMA copolymers maintains transduction capacity in vitro and in vivo. PLoS One 9:e82716. https://doi.org/10.1371/journal.pone.0082716
72. Carlisle R, Choi J, Bazan-Peregrino M et al (2013) Enhanced tumor uptake and penetration of virotherapy using polymer stealthing

and focused ultrasound. J Natl Cancer Inst 105:1701–1710. https://doi.org/10.1093/jnci/djt305

73. Carlisle RC, Benjamin R, Briggs SS et al (2008) Coating of adeno-associated virus with reactive polymers can ablate virus tropism, enable retargeting and provide resistance to neutralising antisera. J Gene Med. https://doi.org/10.1002/jgm.1161
74. Behr JP, Lehn JM (1981) Design of synthetic molecular receptors and catalysts. In: Eggerer H, Huber R (eds) Structural and functional aspects of enzyme catalysis. Colloquium der Gesellschaft für Biologische Chemie, vol 32. Springer, Berlin, pp 23–25
75. Fasbender A, Zabner J, Chillón M et al (1997) Complexes of adenovirus with polycationic polymers and cationic lipids increase the efficiency of gene transfer in vitro and in vivo. J Biol Chem 272:6479–6489. https://doi.org/10.1074/jbc.272.10.6479
76. Bonsted A, Engesaeter BO, Hogset A et al (2006) Photochemically enhanced transduction of polymer-complexed adenovirus targeted to the epidermal growth factor receptor. J Gene Med 8:286–297. https://doi.org/10.1002/jgm.853
77. Kasman LM, Barua S, Lu P et al (2009) Polymer-enhanced adenoviral transduction of CAR-negative bladder cancer cells. In: Molecular pharmaceutics. American Chemical Society, Washington, DC, pp 1612–1619
78. Jiang ZK, Koh SBS, Sato M et al (2013) Engineering polypeptide coatings to augment gene transduction and in vivo stability of adenoviruses. J Control Release 166:75–85. https://doi.org/10.1016/j.jconrel.2012.10.023
79. Kim J, Li Y, Kim SW et al (2013) Therapeutic efficacy of a systemically delivered oncolytic adenovirus – biodegradable polymer complex. Biomaterials 34:4622–4631. https://doi.org/10.1016/j.biomaterials.2013.03.004
80. Choi J-W, Jung S-J, Kasala D et al (2015) pH-sensitive oncolytic adenovirus hybrid targeting acidic tumor microenvironment and angiogenesis. J Control Release 205:134–143. https://doi.org/10.1016/J.JCONREL.2015.01.005
81. Choi JW, Kim J, Bui QN et al (2015) Tuning surface charge and PEGylation of biocompatible polymers for efficient delivery of nucleic acid or adenoviral vector. Bioconjug Chem 26:1818–1829. https://doi.org/10.1021/acs.bioconjchem.5b00357
82. Kwon O-J, Kang E, Choi J-W et al (2013) Therapeutic targeting of chitosan–PEG–folate-complexed oncolytic adenovirus for active and systemic cancer gene therapy. J Control Release 169:257–265. https://doi.org/10.1016/J.JCONREL.2013.03.030
83. Han J, Zhao D, Zhong Z et al (2010) Combination of adenovirus and cross-linked low molecular weight PEI improves efficiency of gene transduction. Nanotechnology 21:105106. https://doi.org/10.1088/0957-4484/21/10/105106
84. Jung S-J, Kasala D, Choi J-W et al (2015) Safety profiles and antitumor efficacy of oncolytic adenovirus coated with bioreducible polymer in the treatment of a CAR negative tumor model. Biomacromolecules 16:87–96. https://doi.org/10.1021/bm501116x
85. Choi JW, Nam JP, Nam K et al (2015) Oncolytic adenovirus coated with multidegradable bioreducible core-cross-linked polyethylenimine for cancer gene therapy. Biomacromolecules 16:2132–2143. https://doi.org/10.1021/acs.biomac.5b00538
86. Fan G, Fan M, Wang Q et al (2016) Bio-inspired polymer envelopes around adenoviral vectors to reduce immunogenicity and improve in vivo kinetics. Acta Biomater 30:94–105. https://doi.org/10.1016/J.ACTBIO.2015.11.005
87. Lee C-H, Kasala D, Na Y et al (2014) Enhanced therapeutic efficacy of an adenovirus-PEI-bile-acid complex in tumors with low coxsackie and adenovirus receptor expression. Biomaterials 35:5505–5516. https://doi.org/10.1016/J.BIOMATERIALS.2014.03.060
88. Nosaki K, Hamada K, Takashima Y et al (2016) A novel, polymer-coated oncolytic measles virus overcomes immune suppression and induces robust antitumor activity. Mol Ther Oncolytics 3:16022. https://doi.org/10.1038/mto.2016.22
89. Kim P-H, Kim T, Yockman JW et al (2010) The effect of surface modification of adenovirus with an arginine-grafted bioreducible polymer on transduction efficiency and immunogenicity in cancer gene therapy. Biomaterials 31:1865–1874. https://doi.org/10.1016/j.biomaterials.2009.11.043
90. Kim J, Nam HY, il KT et al (2011) Active targeting of RGD-conjugated bioreducible polymer for delivery of oncolytic adenovirus expressing shRNA against IL-8 mRNA.

Biomaterials 32:5158–5166. https://doi.org/10.1016/j.biomaterials.2011.03.084

91. Kasala D, Lee S-H, Hong JW et al (2017) Synergistic antitumor effect mediated by a paclitaxel-conjugated polymeric micelle-coated oncolytic adenovirus. Biomaterials 145:207–222. https://doi.org/10.1016/J.BIOMATERIALS.2017.08.035
92. Vetter A, Virdi KS, Espenlaub S et al (2013) Adenoviral vectors coated with PAMAM dendrimer conjugates allow CAR independent virus uptake and targeting to the EGF receptor. Mol Pharm 10:606–618. https://doi.org/10.1021/mp300366f
93. Yoon AR, Kasala D, Li Y et al (2016) Antitumor effect and safety profile of systemically delivered oncolytic adenovirus complexed with EGFR-targeted PAMAM-based dendrimer in orthotopic lung tumor model. J Control Release 231:2–16. https://doi.org/10.1016/j.jconrel.2016.02.046
94. Singh R, Tian B, Kostarelos K (2008) Artificial envelopment of nonenveloped viruses: enhancing adenovirus tumor targeting in vivo. FASEB J 22:3389–3402. https://doi.org/10.1096/fj.08-103275
95. Kim J, Nam HY, Choi JW et al (2014) Efficient lung orthotopic tumor-growth suppression of oncolytic adenovirus complexed with RGD-targeted bioreducible polymer. Gene Ther 21:476–483. https://doi.org/10.1038/gt.2014.18
96. Willmon C, Harrington K, Kottke T et al (2009) Cell carriers for oncolytic viruses: fed ex for cancer therapy. Mol Ther 17:1667–1676. https://doi.org/10.1038/MT.2009.194
97. Rojas JJ, Sampath P, Bonilla B et al (2016) Manipulating TLR signaling increases the anti-tumor T cell response induced by viral cancer therapies. Cell Rep 15:264–273. https://doi.org/10.1016/j.celrep.2016.03.017
98. Lanciotti J, Song A, Doukas J et al (2003) Targeting adenoviral vectors using heterofunctional polyethylene glycol FGF2 conjugates. Mol Ther 8:99–107. https://doi.org/10.1016/S1525-0016(03)00139-4
99. Kreppel F, Kochanek S (2008) Modification of adenovirus gene transfer vectors with synthetic polymers: a scientific review and technical guide. Mol Ther
100. Campos SK, Barry MA (2006) Comparison of adenovirus fiber, protein IX, and hexon capsomeres as scaffolds for vector purification and cell targeting. Virology 349:453–462. https://doi.org/10.1016/j.virol.2006.01.032
101. Kreppel F, Gackowski J, Schmidt E, Kochanek S (2005) Combined genetic and chemical capsid modifications enable flexible and efficient de- and retargeting of adenovirus vectors. Mol Ther 12:107–117. https://doi.org/10.1016/j.ymthe.2005.03.006
102. Greber UF, Willetts M, Webster P, Helenius A (1993) Stepwise dismantling of adenovirus 2 during entry into cells. Cell. https://doi.org/10.1016/0092-8674(93)90382-Z
103. Stevenson M, Hale ABH, Hale SJ et al (2007) Incorporation of a laminin-derived peptide (SIKVAV) on polymer-modified adenovirus permits tumor-specific targeting via α6-integrins. Cancer Gene Ther 14:335–345. https://doi.org/10.1038/sj.cgt.7701022
104. Green NK, Morrison J, Hale S et al (2008) Retargeting polymer-coated adenovirus to the FGF receptor allows productive infection and mediates efficacy in a peritoneal model of human ovarian cancer. J Gene Med 10:280–289. https://doi.org/10.1002/jgm.1121
105. Morrison J, Briggs SS, Green N et al (2008) Virotherapy of ovarian cancer with polymer-cloaked adenovirus retargeted to the epidermal growth factor receptor. Mol Ther 16:244–251. https://doi.org/10.1038/sj.mt.6300363
106. Adenovirus – Servier Medical Art – 3000 free medical images. https://smart.servier.com/smart_image/adenovirus/. Accessed 21 Jan 2019

Chapter 3

Tumor Targeting of Oncolytic Adenoviruses Using Bispecific Adapter Proteins

Julia Niemann and Florian Kühnel

Abstract

Tumor-selectively replicating "oncolytic" adenoviruses based on serotype 5 are promising tools for the treatment of solid tumors. However, their effective delivery to the tumor by systemic administration remains challenging. Several strategies of molecular retargeting have been pursued to equip adenoviruses with molecular features that facilitate their efficient uptake by tumors and to protect healthy tissue from damage. Transductional retargeting can be conveniently achieved using bispecific molecular adapter proteins based on the ectodomain of the coxsackievirus and adenovirus receptor linked to tumor ligands of choice. In this chapter, we describe methods for their design, purification, and application.

Key words Bispecific adapter proteins, Soluble CAR, Oncolytic adenovirus, Tumor targeting, Hepatic detargeting

1 Introduction

Tumor-selectively replicating "oncolytic" adenoviruses based on serotype 5 have been established during the last two decades and first promising results with advanced oncolytic adenovirus variants have been recently observed in clinical trials in glioblastoma patients [1, 2]. The strong infectious and lytic capabilities of adenovirus serotype 5 (Ad5) require tight regulation of infection in a tumor-cell specific manner to minimize damage to healthy tissue. Key characteristics of oncolytic adenoviruses are genetic modifications that have been introduced to facilitate selective virus replication in response to carcinogenesis-related molecular alterations in tumor cells. Such genetic modifications have been defects in the early genes E1A or E1B thus restricting replication to tumor cells with dysfunctional p53 or activated E2F [3, 4]. Furthermore, adenovirus can be transcriptionally retargeted by introduction of tumor-specific promoters to restrict E1A expression and the onset of adenoviral replication in a tumor-selective manner [5]. The first prototypes of Ad5-based oncolytic viruses did not meet the high

Christine E. Engeland (ed.), *Oncolytic Viruses*, Methods in Molecular Biology, vol. 2058,
https://doi.org/10.1007/978-1-4939-9794-7_3,

expectations in clinical trials, particularly when applied systemically [6]. This is due to significant obstacles that adenoviruses are facing at many levels: delivery to the tumor, intratumor spread of infection, and preexisting and induced immune responses. Consequently, the virus capsid has been modified in many ways, as part of a strategy referred to as "transductional retargeting" to overcome these hurdles. Ad5 infects target cells in a two-step process, involving a primary receptor needed for attachment of the viral particle, and a second receptor responsible for intracellular signaling and endocytotic uptake of the particle. It has been assumed for longer, that the coxsackievirus and adenovirus receptor (CAR) is the primary receptor which recognizes the fiber knob of the virus capsid [7]. Subsequently, cellular integrins bind to RGD motifs in penton proteins located at the vertices of the viral capsid and virus uptake is initiated. In vivo, several serum factors such as coagulation factors IX, X and complement binding protein C4 recognize the capsid proteins fiber and hexon resulting in predominant delivery to the liver and low bioavailability for tumor infection [8, 9]. Additionally, the lack of CAR on tumor cells is responsible for insufficient delivery of Ad5 to the target tumor and low intratumor spread [10]. To achieve effective retargeting in vivo, particularly after systemic injection, it might be therefore necessary to delete binding sites of serum factors to reduce the hepatotropism of Ad5 [11]. Nevertheless, the two-step nature of the infection process provides interesting options to redirect the virus to target cells of choice. Retargeting can be achieved by reducing or ablating the tropism-related motifs in capsid proteins, and by providing the viral capsid with alternative molecular "address labels" for specific binding of the virus on tumor cells. Three broad strategies exist to realize transductional retargeting of Ad5 and oncolytic variants thereof. A first strategy uses the genetically stable integration of tumor cell-specific peptides or ligands into capsid proteins. A second approach is the replacement of complete domains or proteins by equivalents from different adenovirus serotypes, or even from other virus species. A third approach, which will be the focus of the methods described in this chapter, is the use of bispecific adapter molecules binding to both the adenoviral capsid and the target cell. This method allows a rapid and convenient switching of the primary receptor for virus attachment resulting in a modified target cell tropism for the initial application round. The three retargeting strategies (as illustrated in Fig. 1 with the adenoviral fiber as example) and their particular pros and cons will be briefly introduced in the following.

1.1 Transductional Retargeting by Integration of Peptides into Capsid Proteins

The first strategy is the integration of peptides into surface-exposed sites in capsid proteins to mediate binding to alternative cell surface structures. Mainly, the HI-loop of fiber, the C-termini of fiber and protein IX, and the hypervariable regions (HVR) of hexon are suitable sites for incorporation of ligands, such as the integrin-

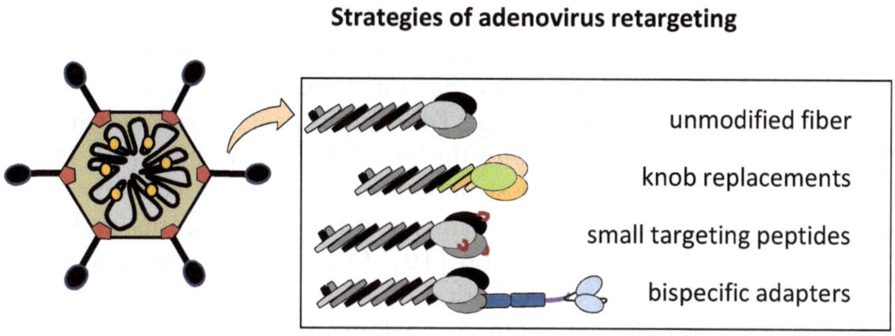

Fig. 1 Fundamental strategies of adenovirus retargeting via fiber modifications

binding RGD motif. The insertion of the RGD motif into the knob domain of fiber improves the ability to directly bind to integrins on CAR-deficient tumor cells thereby improving tumor delivery and intratumor virus spread [12, 13]. Whereas the insertion of small peptides is mostly well tolerated by the fiber knob protein, insertion of larger, more complex ligands may severely interfere with virus stability and generation of infectious progeny. Furthermore, many promising candidates for tumor ligands, such as antibody-based scFvs, are not compatible with the cytoplasmic synthesis of the capsid proteins and nuclear assembly of progeny virions [14]. Therefore, narrow limits are set for the selection of suitable ligands and possible targets.

1.2 Transductional Retargeting by Using Fiber Knob Replacements

A further retargeting approach is the molecular replacement of the entire knob domain, eventually including variable proportions of the fiber shaft, by analogous structures that facilitate the recognition of alternative primary receptors. The knob domains of alternative adenovirus serotypes recognize their target cells via CD46, desmoglein, or sialic acids which are more commonly expressed on cancer cells compared with CAR. Preserving the favorable features of already existing oncolytic Ad5 these can be pseudotyped using knob domains of alternative serotypes with more desired targeting properties. The structural similarity of the fiber knob proteins facilitates their genetic switch. Chimeric fibers harboring the knob domain of group B serotypes (such as Ad3) have been used to reduce uptake by the liver and to improve tumor infection [15]. At least the N-terminal tail domain of Ad5 fiberknob is a minimum requirement to stably connect the chimeric fiber with penton base at the vertices of the Ad5 capsid. However, the spectrum of primary receptors available by using knob domains of alternative serotypes limits this approach and does not facilitate true tumor-selective targeting. Genetic replacement of the knob domain by tumor cell-selective ligands is challenging since the knob domain is required for proper folding of the fiber shaft and plays a role in virus stability and generation of infectious progeny

[16]. Knob replacements for tumor-selective retargeting must therefore support trimerization of the fiber and deliver a tumor-targeting ligand at the same time. Trimerization motifs like fibritin from bacteriophage T4 or sigma protein from reovirus can be used to complement for the loss of the knob domain [17, 18]. Alternative ligands that are structurally more compatible with adenovirus generation, such as camelid antibodies, Darpins and Aptamers are under investigation [19]. Occasionally, xenogeneic proteins that combine trimerization, folding, and retargeting capacities can be used to replace the knob domain such as the phage-derived trimeric enzyme endosialidase NF for targeting polysialic acid [20]. However, those developments depend on the identification of suitable proteins meeting all requirements of a knob replacement with retargeting properties.

1.3 Transductional Retargeting Using Bispecific Adapters

A rapid and convenient approach to modify the cellular tropism of adenovirus is the use of bispecific adapters. These adapters contain a capsid-binding domain at one end, and a tumor-selective ligand at the other end. Like a molecular bridge, bispecific adapters redirect the virus to tumor cells thus bypassing the need of CAR as primary receptor. This principle has been shown with knob-specific FAB antibodies for virus binding linked to a variety of retargeting ligands recognizing fibroblast growth factor receptor 2 (FGF2) or epithelial cell adhesion molecule (EpCAM) [21]. Diabodies, consisting of two scFv against the knob domain and a tumor target are promising alternatives that can be expressed as single agents [22].

A highly versatile approach for the generation of bispecific adapters that do not require a virus-specific scFv, is the use of a truncated, soluble form of the CAR receptor corresponding to the fiber knob binding ectodomain (referred to as sCAR or CARex). This approach has been initially introduced by the Curiel group who established a sCAR-EGF adapter using the natural ligand of the epidermal growth factor receptor (EGFR) for improved adenovirus infection of tumor cells [23]. The efficacy of soluble CAR-based adapters in retargeting of adenoviral vectors and oncolytic adenoviruses has been broadly confirmed using various tumor targets such as Her2/neu, CEA, and polysialic acid [24–26]. The principle of infection mediated by sCAR-based adapters is illustrated in Fig. 2. Importantly, studies on adapter retargeting report a profound reduction of hepatic infection and prevention of hepatotoxicity after intravenous delivery of adapter-treated oncolytic adenoviruses [26, 27]. The technique offers more striking advantages. Soluble CAR adapters can be conveniently generated by PCR amplification of the CAR ectodomain from available templates and subsequent ligation to the coding sequence of the selected tumor ligand. Successful retargeting can be easily assessed by combining the adapter with an adenovirus expressing a reporter gene for infection trials. For detailed characterization and subsequent

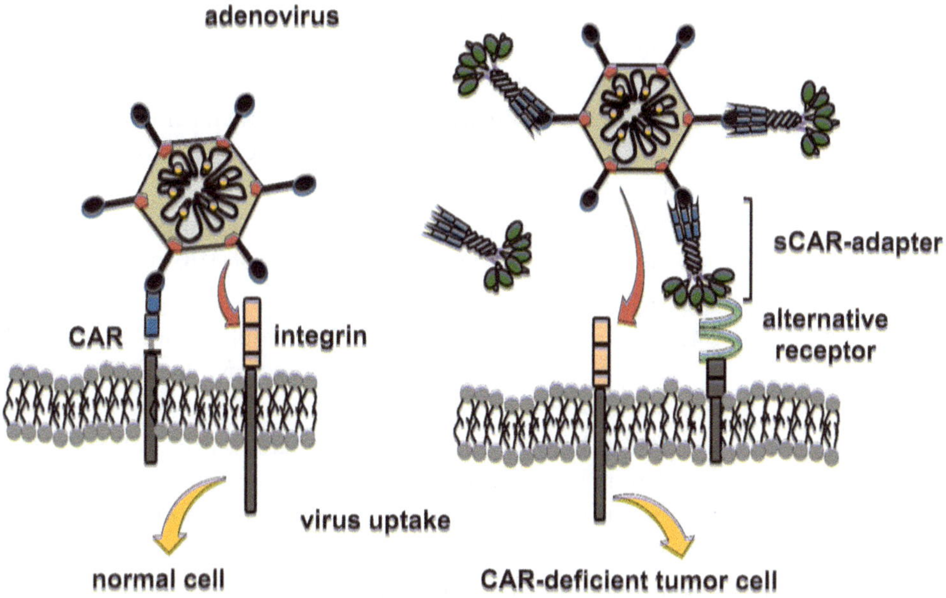

Fig. 2 Principle of adenovirus retargeting using sCAR-based adapter proteins. Left side: canonical mechanism of cell infection by adenovirus using the coxsackievirus and adenovirus receptor (CAR) as primary receptor for attachment. Attachment of the particle to the cell surface facilitates binding of the capsid to integrins subsequently leading to internalization of the viral particle. Right side: Tumor retargeting of adenovirus using bispecific adapters based on sCAR and a tumor ligand (here a trimeric variant is shown). Bispecific adapters facilitate virus attachment to alternative receptors without the need of CAR and simultaneously inhibit CAR-dependent virus uptake

studies in vivo, the protein can be expressed and purified at large scale using conventional techniques. We will provide according protocols and recommendations in Subheading 3.

2 Materials

General recommendations: Use deionized water to prepare solutions. Dissolve solid chemicals in max. 80% final volume and subsequently add water/buffer to the final volume.

2.1 Protein Purification

1. 1 M phosphate buffer, pH 8.0: Dissolve 11.998 g sodium dihydrogen phosphate (NaH_2PO_4) in water, adjust to a total volume of 100 ml to obtain a 1 M sodium dihydrogen phosphate (NaH_2PO_4) solution. For a 1 M solution of disodium hydrogen phosphate (Na_2HPO_4) dissolve 14.196 g Na_2HPO_4 in water and adjust to a total volume of 100 ml. To obtain 1 M phosphate buffer pH 8.0 add 1 M NaH_2PO_4 solution dropwise to 1 M Na_2HPO_4 solution until the solution reaches pH 8.0. Sterilize buffer by filtration through a 0.2 μm filter and store at

room temperature. Before use, check buffer for precipitations/crystallizations and if necessary heat buffer to 50 °C to dissolve any precipitate.

2. 5 M NaCl: Dissolve 29.22 g sodium chloride (NaCl) in water and adjust solution to a final volume of 100 ml. Store at room temperature.
3. 150 mM imidazole: Dissolve 1.02 g imidazole in water and adjust the final volume to a total of 100 ml. Store at room temperature.
4. Protein binding buffer: 0.5 M phosphate buffer, 1.5 M NaCl. Mix phosphate buffer (50% v/v), 5 M NaCl (30% v/v), and water (20% v/v) (*see* **Note 1**).
5. Wash buffer: 50 mM phosphate buffer, 300 mM NaCl, 1.5 mM imidazole. Mix phosphate buffer (5% v/v), 5 M NaCl (6% v/v), 150 mM imidazole (1% v/v), and water (88% v/v) (*see* **Note 1**).
6. Elution buffer: 50 mM phosphate buffer, 300 mM NaCl, 150 mM L-histidine. Weigh 0,233 g L-histidine in a 15 ml falcon tube, add 0.5 ml phosphate buffer, 0.6 ml 5 M NaCl and water to a final volume of 10 ml. Mix by vortexing until L-histidine is completely resolved. Store at 4 °C.
7. Ni-NTA (nickel-nitrilotriacetic acid) agarose beads (*see* **Note 2**).
8. Cell scraper.

2.2 Adenovirus Preparation

1. DMEM GlutaMAX™-I (Dulbecco's Modified Eagle Medium containing 4.5 g/l glucose and pyruvate, *see* **Note 3**).
2. FCS (fetal calf serum). Prepare 50 and 10 ml aliquots and store at −20 °C.
3. Penicillin/streptomycin (10,000 U/ml). Prepare 5 ml aliquots and store at −20 °C.
4. T25 and T75 cell culture flasks, TC-treated.
5. 1 M Tris pH 8.0.
6. 1 M Tris pH 7.5.
7. Mouse serum albumin.
8. Cesium chloride (CsCl).
9. Glycerol.
10. 1 M magnesium chloride ($MgCl_2$) solution: Weigh 4.76 g $MgCl_2$ into a 50 ml falcon tube and add water to a final volume of 50 ml. Store at room temperature.
11. Ten percentage FCS (v/v) cell culture growth medium: sterilize 50 ml FCS and 5 ml penicillin/streptomycin by filtration (0.2 μm) and add to 500 ml DMEM medium. Store medium at 4 °C.

12. Two percentage FCS medium for adenovirus infection: sterilize 10 ml FCS and 5 ml penicillin/streptomycin by filtration (0.2 μm) and add to 500 ml DMEM medium. Store medium at 4 °C.
13. Dialysis buffer: 20 mM Tris pH 8.0, 25 mM NaCl, 1.25 mM $MgCl_2$. Mix 20 ml 1 M Tris pH 8.0, 5 ml 5 M NaCl solution and 1.25 ml 1 M $MgCl_2$ solution and add water to a final volume of 1 l.
14. Cesium chloride (CsCl) solutions for CsCl-gradient ultracentrifugation: To prepare 50 ml 1.2-CsCl solution (with a specific weight of 1.2 g/ml) weigh 1.325 g CsCl in a 50 ml falcon tube, add 0.5 ml 1 M Tris pH 7.5 and add water to a final volume of 50 ml. To prepare 50 ml 1.4-CsCl solution (specific weight of 1.4 g/ml) weigh 2.675 g CsCl in a 50 ml falcon tube, add 0.5 ml 1 M Tris pH 7.5 and add water to a final volume of 50 ml. Store both solutions at 4 °C.
15. 2× glycerol storage buffer: 10 mM Tris pH 7.5, 100 mM NaCl, 50% glycerol, 1% mouse serum albumin. Weigh 0.05 g mouse serum albumin (*see* **Note 4**) in a 50 ml falcon tube, add 25 ml glycerol, 0.5 ml 1 M Tris pH 7.5, 1 ml 5 M NaCl solution, and add water to a final volume of 50 ml. Mix well and store buffer at −20 °C.
16. Syringes and 24-G (ø 0.55) needles.

3 Methods

In the simplest form of CAR-based bispecific adapters, the soluble CAR domain is either directly connected to a tumor-selective targeting ligand or by a functionally neutral spacer region for improved spatial separation of the two domains. The affinity of soluble CAR-based adapters to fiber knob and the efficacy of retargeting can be significantly improved when trimerization motifs, derived from T4 fibritin or GCN4, are integrated between the two functional domains [24, 28]. Generally, the adenovirus/adapter complexes are remarkably stable. Adenovirus/adapter complexes can be purified by gel filtration or using CsCl density gradient ultracentrifugation [23, 29].

Interestingly, the sCAR adapter method leads to effective hepatic detargeting even in case adenovirus vectors with unmodified hexon are used. When considering the use of adapters for oncolytic adenoviruses, it is certainly a limitation that the retargeting function is only transiently conferred to the viral capsid. The targeting properties of adapter-modified adenoviruses therefore only apply for the first administration whereas the generated virus progeny will again infect target cells in a CAR-dependent fashion. It can be considered to generate oncolytic adenoviruses secreting their own adapter after target cell infection. It has been reported that this approach indeed resulted in improved and prolonged retargeting properties of the corresponding viruses [30]. However, it has also been observed that adapter expression by the virus may reduce the oncolytic efficacy of viral replication indicating that further improvement in technical details is required [31].

3.1 The Genetic Setup of Adapter Constructs

The genetic setup of bispecific adapters based on soluble CAR is illustrated in Fig. 3. sCAR is used as the adenovirus binding domain. Furthermore, a suitable ligand for tumor targeting has to be selected. For instance, this can be the natural ligand of a protein or molecule that is preferentially present on the surface of tumor cells, such as ligands of growth factor receptors, or single chain fragments (scFv) from antibodies binding to these molecular targets. As illustrated in Fig. 3a, we recommend fusing both domains using a functionally neutral spacer peptide for clear spatial separation of the two functional domains. For this purpose, a glycine-serine linker can be used (following amino acid sequence: GGGGSGGGGSGGGGS, corresponding DNA sequence: GGA GGA GGA GGG TCC GGA GGG GGC GGA TCT GGA GGG GGA GGC TCT).

To enhance the binding capacity of the adapter, we recommend to introduce a trimerization motif (*see* Fig. 3b) such as the C-terminus of the bacteriophage T4 fibritin protein coding

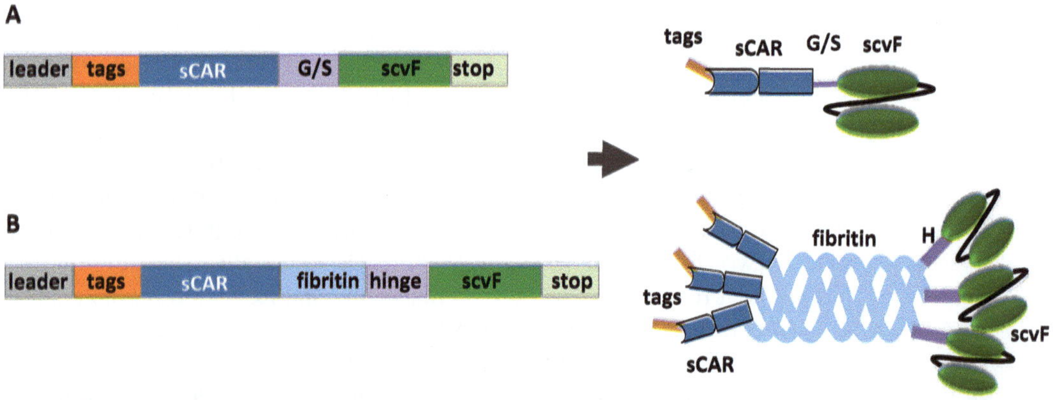

Fig. 3 Adapter construction—genetic setup. (**a**) The genetic setup for a simple, monomeric adapter construct based on soluble CAR is shown on the left side. From 5′ to 3′ terminus the construct contains a leader sequence, tags for purification and detection (e.g., a c-myc/His_6 tag), the soluble ectodomain of the coxsackie and adenovirus receptor (sCAR), a linker (e.g., glycin-serin linker), and a retargeting domain (e.g., a single chain variable fragment (scvF) followed by a stop codon. The right panel shows a schematic illustration of the translated adapter. (**b**) Left: The genetic setup of a trimeric adapter based on sCAR additionally including a trimerization motif and a hinge region. The correspondingly translated and trimerized adapter protein is illustrated in the right panel

for VEERGLTNSIKANETNIASVTQEVNTAKGNISSLQGDV-QALQEA**GYIPEAPRDGQAYVRKDGEWVFLSTFLSPA**. The amino acids printed in bold refer to the trimerization foldon domain which reflects a minimum that should be introduced into an adapter construct. The corresponding DNA sequence of the foldon domain is:

GGT TAT ATT CCT GAA GCT CCA AGA GAT GGG CAA GCT TAC GTT CGT AAA GAT GGC GAA TGG GTA TTC CTT TCT ACC TTT TTA TCA CCA GCA.

Place the motif between the functional domains of the bispecific protein (*see* Fig. 2). Since a trimerization of the adapter requires more space and flexibility for the scFv domains to function properly, a proline-rich hinge region (amino acid sequence NRSQNLNPSHNLSPNLSLNRFK; corresponding DNA sequence: AAC CGC AGC CAA AAC CTC AAC CCC AGC CAC AAC CTC AGC CCA AAC CTC AGC CTA AAC CGG TTT AAA) should be integrated between the trimerization motif and the scFv domain. To enable detection and purification of the adapter protein, include a c-myc/His_6-tag (amino acid sequence: EQKLISEEDLNMHTGHHHHHH; corresponding DNA sequence: GAA CAG AAA CTC ATC TCA GAA GAG GAT CTG AAT ATG CAT ACC GGA CAT CAT CAC CAT CAC CAT) into the construct (*see* **Note 5**). We recommend including these tags downstream of the cleavage site of the CAR leader peptide. For instance, a CAR fragment lacking the leader sequence and the transmembrane and intracellular domain can be amplified and subsequently ligated with oligonucleotides encoding the sequence for the CAR leader peptide and c-myc/His_6-tag (tags) to achieve secretion of the adapter protein into the supernatant for convenient purification (the amino acid sequence of the leader peptide of the Coxsackie and adenovirus receptor (CAR-L) is: MALLLCFVLLCGVVDFARSLS; corresponding DNA sequence respectively: ATG GCC CTC CTG CTG TGC TTC GTG CTC CTG TGC GGA GTG GTG GAT TTC GCC AGA AGT TTG AGT).

3.2 Protein/Adapter Expression

For first experimentation, protein expression can be achieved by transient transfection. For this purpose, we use PEI transfection reagent and HEK293 producer cells. We recommend to transfect at least ten 10 cm cell culture plates or an equivalent amount of cell culture flasks and start protein purification after 48–72 h post transfection. For the generation of large-scale stocks with defined properties, we recommend to generate transgenic cells which stably express the adapter protein. These cells can be expanded and subsequently used for large-scale protein isolation whenever needed. For the generation of adapter-transgenic cells we use either retroviral transduction or transposon-based strategies (*see* **Note 6**).

When using adapter-transgenic cell lines harboring a resistance gene for the purpose of selection (e.g., neomycin or puromycin) we recommend initial expansion of cells in selection medium until splitting cells to the final number of flasks, to ensure high transgene expression levels. Wait at least 48 h after final splitting before starting the protein isolation.

3.3 Protein/Adapter Purification

1. Collect medium from transfected or transgenic, adapter-expressing cells and pass through a 0.45 μm filter. Keep filtrate on ice.
2. Remove cells from cell culture plates using a cell scraper and rinse plates with filtered medium (use approximately 10 ml medium to rinse five plates). Discard cell culture plates and transfer suspended cells to a 50 ml falcon tube. For cell lysis and release of protein freeze suspension in liquid nitrogen and carefully thaw in a 37 °C water bath while shaking the tube until the lysate has been completely thawed (*see* **Note 7**). Vigorously vortex the lysate for 30 s and repeat freezing and thawing once. Centrifuge the tube at 1700 × *g* and 4 °C for 10 min to pellet cell debris. Pass supernatant through a 0.45 μm filter to remove remaining cell debris and add to the adapter-containing medium (**step 1**). Add 10% binding buffer (v/v) to the combined medium/cell extract (e.g., at 5 ml binding buffer to 45 ml suspension) and transfer the buffered suspension to 50 ml falcon tubes (*see* **Note 8**). Add 250 μl Ni-NTA agarose beads to each tube (*see* **Note 9**). Close falcons tightly and place tubes in an overhead shaker. Invert tubes slowly (max. 1 rotation/s) over night at 4 °C.
3. Remove tubes from the overhead shaker and spin down beads for 10 min at 300 × *g* and 4 °C (*see* **Note 10**). Prepare wash buffer during centrifugation step. Remove the supernatant carefully leaving approx. 1 cm of liquid above the pellet to avoid loss of beads. If more than one tube has been used for the previous step, pool all beads in one 50 ml tube using a 2 ml transfer pipette (*see* **Note 11**). Wash empty tubes with 2 ml wash buffer to collect remaining beads and add to the pooled beads. Spin down beads for 10 min at 300 × *g* and 4 °C (*see* **Note 10**). Remove the supernatant carefully as described before. Wash the beads by adding 20 ml wash buffer and resuspending by careful shaking. Let the tube sit at room temperature for 5 min and spin down beads again (300 × *g*, 10 min, 4 °C). Remove supernatant as described above, wash beads again with 20 ml wash buffer, and incubate for 5 min at room temperature. Spin down beads (300 × *g*, 10 min, 4 °C), remove supernatant carefully but as completely as possible without removing any beads. Dissolve beads in the remaining liquid and transfer to 2 ml tubes (using approx. 1 ml/ tube)

using a 2 ml transfer pipette (*see* **Note 11**). Use a volume of wash buffer equivalent to the volume of the previously transferred beads to wash the empty 50 ml falcon and combine with the transferred beads. Centrifuge the 2 ml tubes at 2000 × *g* for 5 min and 4 °C with breaks switched off. Remove supernatant carefully but completely using a 200 μl pipette. Add an equivalent amount of elution buffer to the pelleted beads (e.g., if the tube is filled with approx. 0.5 ml beads, add 0.5 ml elution buffer). Resuspend beads by vortexing and place tubes in an overhead shaker. Invert tubes slowly (maximum 1 rotation/s) for 4–6 h at 4 °C.

4. Remove tubes from the shaker and spin down tubes at ≥10,000 × *g* for 10 min at 4 °C. Transfer the supernatant (eluate) in a fresh tube without transferring any beads, aliquot as desired and keep aliquots at −80 °C.

3.4 Adenovirus Preparation

General recommendations: During adenovirus preparation, avoid handling other viruses in clean bench/incubator to prevent contaminations. Use cell culture medium (DMEM) containing 10% FCS as growth medium for producer cells. Before infecting producer cells, change medium to medium containing 2% FCS. If needed, you may pause with virus production after each infection cycle by freezing virus containing cells in dialysis buffer at −20 °C. This freezing step also serves as the first freezing step from following repeated freeze–thaw cycles to release the virus (*see* below **step 2**).

1. Linearize and transfect adenoviral plasmid (if working with AdEasy or comparable adenovirus cloning systems [32], linearize the plasmid containing the virus genome via enzymatic digestion with PacI). After linearization use the digestion mix directly for transfection without any purification or inactivation step. Mix the digestion mix very carefully with the transfection reagent to avoid precipitation of the viral DNA (*see* **Notes 12** and **13**). After incubation of transfection reagent and viral DNA, transfer the transfection mix to producer cells (70–80% confluent) in a T25 cell culture flask (*see* **Note 14**) and place cells in an incubator (37 °C, 5% CO_2) for 5 days.
2. After 5 days of incubation, check for viral cell lysis (round shaped cells detached from the surface of the flask). Resuspend cells in the cell culture medium by carefully pipetting up and down with a 5 or 10 ml pipette. Transfer the cell suspension to a 15 ml falcon tube. Spin down cells by centrifugation (1200 × *g*, 10 min, 4 °C). Remove supernatant and add 0.5 ml dialysis buffer. Resuspend cells by vortexing and release viral particles from the cells by three cycles of freezing and thawing (*see* **Note 15**). After the final freeze–thaw cycle, remove cell debris by centrifugation (2000 × *g*, 10 min,

4 °C). Use the supernatant to infect new cells. Once complete cell lysis in a T25 flask can be observed (almost all cells rounded up and completely detached), virus production can be upscaled and continued with **step 3**. If no or only partial lysis was detectable, transfer the lysate again to a new T25 flask with fresh producer cells (90–95% confluent). Incubate flask in a cell culture incubator for 48 h (37 °C, 5% CO_2). Repeat **step 2**.

3. Transfer the virus containing supernatant/lysate to a T75 cell culture flask with fresh producer cells (90–95% confluent). Incubate flask in a cell culture incubator for 48 h (37 °C, 5% CO_2).
4. Remove flask from the incubator and confirm that cell lysis is complete. Remove cells with the medium from the flask and transfer to a 50 ml falcon tube. Spin down cells (1200 × *g*, 10 min, 4 °C), resuspend in dialysis buffer using a total volume of 1–2 ml buffer and transfer to a 15 ml falcon. Perform 3× freeze–thaw cycles to release viral particles, and pellet the cell debris (for details *see* **step 2**). Infect 4T75 flasks with producer cells using the virus-containing supernatant and incubate cells (37 °C, 5% CO_2) for 48 h. Repeat this step with the following scale-up scheme: 4 × T75 → 16 × T75 → 60 × T75.
5. Collect cells from all flasks (50 ml falcons), spin down cells and discard the medium. Use approx. 4 ml (max. 5 ml) dialysis buffer to resuspend and to pool all cells in a 15 ml falcon. Perform two freeze–thaw cycles. Spin down cell debris at 2000 × *g* 10 min 4 °C. Transfer supernatant to a fresh 15 ml falcon tube and wash the cell debris with 2 ml dialysis buffer (add and vortex). Centrifuge both tubes again (2000 × *g*, 10 min, 4 °C). Pool supernatant of both tubes in a fresh 15 ml falcon (discard falcons with remaining pellets). Centrifuge again (2000 × *g*, 10 min, 4 °C). Transfer supernatant to a fresh 15 ml falcon or directly to the CsCl gradient (*see* **step 6**). Assure that no cell debris gets transferred.
6. Prepare cesium chloride (CsCl) gradient as follows: fill 3.5 ml 1.4-CsCl solution into a centrifugation tube. Very slowly overlay with 3.5 ml of the 1.2-CsCl solution to form a discontinuous gradient with a sharp interphase. Slowly overlay with the supernatant obtained in **step 5**.
7. Centrifuge CsCl-gradient in a swing-out rotor at 28,000 × *g* for 4 h at 4 °C in an ultracentrifuge.
8. Post centrifugation, the gradient should look similar to the one shown in Fig. 4. The upper band contains partially assembled and defective particles, while the bottom band contains complete, infectious viral particles. To remove this band, prepare a 2 ml syringe with a 24-gauge (ø 0.55) needle and puncture the tube wall approx. 3 mm below the band (*see* scheme in Fig. 4).

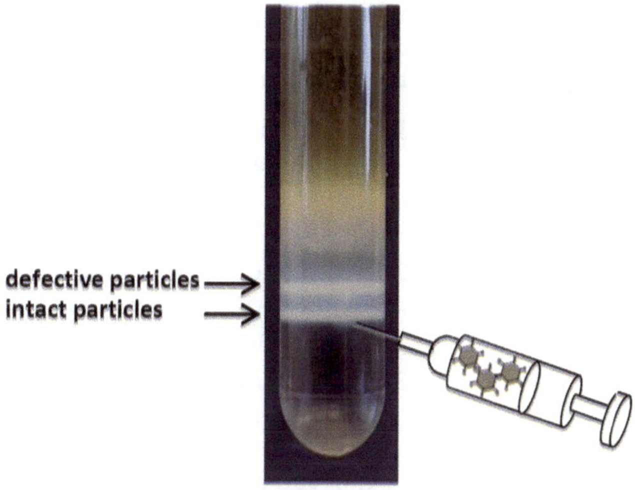

Fig. 4 Adenovirus purification by CsCl-gradient ultracentrifugation. The figure shows the characteristic banding pattern after purification of an adenovirus preparation following a CsCl-gradient ultracentrifugation. Two distinct bands represent defective (upper band) and intact (lower band) viral particles

Remove the correct band very slowly. To generate virus glycerol stocks for long-term storage proceed with **step 9**. To form virus-adapter complexes proceed with the instructions under "Virus-Adapter complexation."

9. Transfer viral particles to a fresh 15 ml falcon and add 1:1 (v/v) 2× glycerol stock solution which will protect virus stocks from crystallizing at −20 °C. Mix by pipetting up and down or careful vortexing. Store virus stocks at −20 °C until dialysis.
10. For determination of virus titers, dilute a sample of the virus stock 1:20 (v/v) in 0.1% SDS and measure the OD260 in an optical spectrophotometer using appropriate controls (obtained from "empty" gradients). OD = 1 is equivalent to 1×10^{12} total viral particles/ml. For the determination of infectious viral titers we recommend methods based on immunocytochemical staining of hexon in infected cells (*see* **Note 16**).
11. Virus dialysis: Prepare 2 l of dialysis buffer and precool to 4 °C. Prepare a 2 ml syringe with a 24-G (ø 0.55) needle. Fill the syringe with approximately 1 ml virus stock solution and transfer solution into a dialysis cassette following the instructions provided by the manufacturer (*see* **Note 17**). Dialyze in a volume of 1 l buffer at 4 °C for 2 h in total. Renew the buffer after 1 h. Remove the dialyzed virus solution from the cassette (using a syringe and a 24-G (ø 0.55) needle), aliquot and store at −80 °C until usage. Do not use dialyzed virus stocks older than 6 months.

3.5 Virus–Adapter Complexation for Large-Scale Experimentation

Follow these instructions when larger amounts of virus-adapter complexes are needed (e.g., for in vivo experiments). For first functional testing *see* alternative method in Subheading 3.6.

1. Obtain adenovirus directly (*see* **Note 18**) from a CsCl-gradient after ultracentrifugation of a fresh adenovirus preparation (Subheading 3.4, **step 8**) and dialyze against dialysis buffer for 2 h (*see* Subheading 3.4, **step 10**).
2. Determine the total particle titer by measuring the OD260. Proceed with the next step but also collect a sample to determine the infectious titer (*see* **Note 19**).
3. Take an amount of virus needed for subsequent experimentation and transfer to a new tube. Make sure that the amount of virus is sufficiently dimensioned to facilitate convenient detection and recovery after later CsCl gradient ultracentrifugation as described below.
4. Add adapter-containing eluate (*see* Subheading 3.3) to the tube and place it in an overhead shaker at 4 °C. Invert the tube for 1 h. For calculation of the required adapter amount use the total viral particle number to be treated and consider 12 fiber molecules/virus particle. Additionally, we recommend using a 10- to 50-fold excess of the adapter to allow for complete complexation and to compensate for adapter molecules that are bound by free fiber protein present in Ad5 preparations.
5. Subject virus-adapter complexes to a second CsCl-gradient centrifugation (*see* Subheading 3.4, **steps 6** and 7). Usually, this centrifugation should yield a single band containing the virus-adapter complexes. Remove the band according to Subheading 3.4, **step 8**.
6. Dialyze the complexes again (*see* Subheading 3.4, **step 10**), determine the OD260, and subject the obtained solution to the planned in-vivo application (*see* **Note 20**).

3.6 Virus–Adapter Complexation for Small-Scale Experimentation

1. Thaw a frozen aliquot of dialyzed virus stock (*see* Subheading 3.4, **step 11**), take a volume/amount of virus sufficient for subsequent experimentation and transfer to a new tube (e.g., 1×10^9 infectious viral particles for infection of 1×10^8 target cells at low multiplicity of infection (MOI) of 10) (*see* **Note 21**).
2. Calculate the amount of required adapter as described in Subheading 3.5, **step 4**.
3. Dilute with dialysis buffer and add adapter eluate to reach a recommended volume of at least 250 μl. Adapter binding can also be performed in cell growth medium (2% FCS).
4. Place tube in an overhead shaker and invert the tube slowly at 4 °C for 1 h.
5. Dilute with the required volume of cell growth medium (2% FCS) and proceed with infection experiments.

3.7 Validation In Vitro

1. For infection experiments select cells which express the molecular target, but are deficient for the natural entry receptor CAR (nonpermissive for infection with naked Ad5). Ideally, these cells express the target receptor endogenously. Alternatively, prior to infection experiments, target cells can be transduced to express the target molecule to obtain an isogenic experimental setting.
2. For infection, withdraw growth medium from target cells and add the prepared dilutions of adapter-coated viruses or naked viral particles as controls, respectively. Incubate for a short period to allow for infection. Infection time can be varied between seconds and 1 h maximum.
3. Withdraw the infection solution and wash several times with medium. Finally aspirate the wash medium.
4. Renew the growth medium and incubate for a time period depending on the desired readout needed to compare the results of infection (*see* **Note 22**).

3.8 Application In Vivo and Validation

1. To confirm adenovirus retargeting (in general: detargeting of the liver and infection of tumor tissue) we recommend to administer the adapter–virus complexes or the untreated ("naked") virus, diluted in dialysis buffer by intravenous injection using a maximum volume per mouse according to the local guidance of animal care.
2. Confirm liver detargeting and specific infection of the target tumor. We recommend to analyze virus distribution by taking tissue samples between 48 and 72 h after infection (at least from liver and target tissue) and by (a) performing histological analyses (e.g., immunohistochemical staining for viral proteins) and (b) extracting DNA and performing real-time PCR to measure the amount of viral DNA. Alternatively, liver toxicity can be measured by taking blood samples and determining transaminase activity in serum using commercially available alanine or aspartate aminotransferase assay systems (*see* **Note 23**).

4 Notes

1. Prepare freshly before use and keep at room temperature. Any cooling may lead to precipitates.
2. We used Ni-NTA beads from Qiagen, but comparable material by other suppliers should work as well.
3. We use DMEM GlutaMAX™-I (ThermoFisher), but equivalent media might work as well.
4. Since we are predominantly working with immunocompetent murine models, we use mouse serum albumin as supplement in

our virus storage buffer to avoid toxicity. In different settings alternative serum albumin can be used (e.g., bovine serum albumin).

5. This is difficult to generalize. We found that N-terminal location of the c-myc/His_6-tag at the CAR ectodomain worked well. This is therefore our preferred solution for CAR-adapters since we observed at least in one case that C-terminal location proximal to the scFv resulted in nonfunctional adapters after purification. We also had mixed results with integrating the tags in between the two functional domains. This position is probably less accessible to the affinity matrix and purification of the construct might be affected. However, this approach has been successfully pursued by others as reported in the literature [23, 24].
6. For stable integration and strong, reliable transgene expression we frequently use retroviral transduction with PQCXIN/P-based retroviral vectors and select transduced cells via puromycin/neomycin resistance. Alternatively, transposon-based vectors can be used for transgene integration (e.g., pT3-based expression plasmids mixed with a plasmid for expression of sleeping beauty transposase). In this case, we recommend the integration of a resistance gene for selection of transgenic cells and control of adapter protein expression (e.g., by western blot analysis) on a regular basis.
7. Thaw cell suspension just until all solid parts have been dissolved but do not allow the suspension to warm up to 37 °C.
8. Avoid tubes with less than 50 ml solution volume. If the remaining solution is less than 50 ml, add serum free medium (DMEM) to a total volume of 50 ml.
9. Vortex beads vigorously and make sure that beads are well suspended before pipetting.
10. Make sure there are no beads remaining in the supernatant before proceeding. If not, centrifuge again for 5 min at 750 × *g* at 4 °C. Alternatively, decrease deceleration speed of the centrifuge to minimize that beads get resuspended by forced deceleration.
11. Beads strongly adhere to the walls of the pipet tip. Therefore fill the pipet tip max. Until 0.5 ml and transfer beads stepwise to reduce the loss.
12. For transfection of viral DNA we use Xtreme gene 9 transfection reagent (Sigma-Aldrich). Due to the low-toxic properties no subsequent change of medium is required (neither before nor after transfection). Furthermore, we found that Lipofectamine 2000 transfection reagent, though resulting in very good transfection efficiency (as controlled using an adenovirus

plasmid equipped with a GFP reporter gene), did not lead to the production of functional viral particles by unknown reasons.

13. Add the transfection reagent dropwise (20–50 μl drops) to the viral DNA (digestion mix) and mix very gently by slowly pipetting up and down with a 1 ml pipet tip.
14. To produce adenoviral particles and achieve high virus titers we use HEK293 cells (e.g., from ATCC: CRL-1573). Alternatively, A549 cells (e.g., from ATCC: CCL-185) can be used when using E1A-containing oncolytic adenoviruses. For initial transfection of viral DNA we use 70–80% confluent T25 cell culture flasks. For subsequent infection/passaging of viral particles we use 90–95% confluent flasks (flasks should be almost confluent, with small spaces between the cells, but still in the logarithmic growth phase).
15. Freeze cells in liquid nitrogen and thaw in a 37 °C water bath, make sure that samples do not warm up to 37 °C after thawing, immediately repeat freezing or continue with the next step. Incubation at warm temperatures will reduce the virus titer.
16. For determination of infectious adenovirus titers by hexon-immunocytochemistry we recommend to use the Adeno-X™ Rapid Titer Kit (Takara). This method and comparable stainings require a full replication cycle and more than 2 days. Therefore, total viral particle and infectious titers of new virus preparations must be determined once to calculate the ratio of infectious/total particles. Assuming this ratio remains constant during the following procedures, the infectious titer can be recalculated at any step by a measurement of the OD260.
17. We recommend the use of Slide-A-Lyzer dialysis cassettes (ThermoFisher).
18. Do not use virus from frozen glycerol stocks since these particles do not form a sharp band in a CsCl gradient after ultracentrifugation.
19. At this stage, we recommend working with the total particle count since the determination of the infectious titer needs more than two days. The infectious titer should be determined in parallel to adapter complexation and retargeting experiments.
20. For optional freezing and long-term-storage, we recommend testing the stability of retargeting complexes first.
21. For a very quick functional testing of adapter-mediated retargeting, supernatants from cells expressing the adapter can be simply mixed with an adenovirus and transferred to suitable target cells for infection.

22. To compare infectivity and therefore the efficacy of adenovirus retargeting in vitro we use adenoviruses encoding a reporter gene that enables convenient detection of infected cells (e.g., adenoviruses containing a LacZ or GFP reporter gene). Alternatively, infected cells can be detected by immunocytochemical detection of hexon-positive cells or by using the commercially available kits using this method (*see* **Note 16**).
23. We usually used the alanine aminotransferase assay purchased from Catachem.

Acknowledgments

This work was supported by Else-Kröner-Fresenius-Stiftung and Deutsche Forschungsgemeinschaft (DFG).

References

1. Larson C, Oronsky B, Scicinski J, Fanger GR, Stirn M, Oronsky A, Reid TR (2015) Going viral: a review of replication-selective oncolytic adenoviruses. Oncotarget 6:19976–19989
2. Lang FF, Conrad C, Gomez-Manzano C, Yung WKA, Sawaya R, Weinberg JS, Prabhu SS, Rao G, Fuller GN, Aldape KD et al (2018) Phase I study of DNX-2401 (delta-24-RGD) oncolytic adenovirus: replication and immunotherapeutic effects in recurrent malignant glioma. J Clin Oncol 36:1419–1427
3. Bischoff JR, Kirn DH, Williams A, Heise C, Horn S, Muna M, Ng L, Nye JA, Sampson-Johannes A, Fattaey A et al (1996) An adenovirus mutant that replicates selectively in p53-deficient human tumor cells. Science 274:373–376
4. Fueyo J, Gomez-Manzano C, Alemany R, Lee PS, McDonnell TJ, Mitlianga P, Shi YX, Levin VA, Yung WK, Kyritsis AP (2000) A mutant oncolytic adenovirus targeting the Rb pathway produces anti-glioma effect in vivo. Oncogene 19:2–12
5. Ko D, Hawkins L, Yu DC (2005) Development of transcriptionally regulated oncolytic adenoviruses. Oncogene 24:7763–7774
6. Aghi M, Martuza RL (2005) Oncolytic viral therapies – the clinical experience. Oncogene 24:7802–7816
7. Bergelson JM, Cunningham JA, Droguett G, Kurt-Jones EA, Krithivas A, Hong JS, Horwitz MS, Crowell RL, Finberg RW (1997) Isolation of a common receptor for Coxsackie B viruses and adenoviruses 2 and 5. Science 275:1320–1323
8. Shayakhmetov DM, Gaggar A, Ni S, Li ZY, Lieber A (2005) Adenovirus binding to blood factors results in liver cell infection and hepatotoxicity. J Virol 79:7478–7491
9. Kalyuzhniy O, Di Paolo NC, Silvestry M, Hofherr SE, Barry MA, Stewart PL, Shayakhmetov DM (2008) Adenovirus serotype 5 hexon is critical for virus infection of hepatocytes in vivo. Proc Natl Acad Sci U S A 105:5483–5488
10. Douglas JT, Kim M, Sumerel LA, Carey DE, Curiel DT (2001) Efficient oncolysis by a replicating adenovirus (ad) in vivo is critically dependent on tumor expression of primary ad receptors. Cancer Res 61:813–817
11. Uusi-Kerttula H, Davies JA, Thompson JM, Wongthida P, Evgin L, Shim KG, Bradshaw A, Baker AT, Rizkallah PJ, Jones R et al (2018) Ad5NULL-A20: a tropism-modified, alphavbeta6 integrin-selective oncolytic adenovirus for epithelial ovarian cancer therapies. Clin Cancer Res 24:4215–4224
12. Wickham TJ, Tzeng E, Shears LL, Roelvink PW, Li Y, Lee GM, Brough DE, Lizonova A, Kovesdi I (1997) Increased in vitro and in vivo gene transfer by adenovirus vectors containing chimeric fiber proteins. J Virol 71:8221–8229
13. Bauerschmitz GJ, Lam JT, Kanerva A, Suzuki K, Nettelbeck DM, Dmitriev I, Krasnykh V, Mikheeva GV, Barnes MN, Alvarez RD et al (2002) Treatment of ovarian cancer with a tropism modified oncolytic adenovirus. Cancer Res 62:1266–1270
14. Magnusson MK, Hong SS, Henning P, Boulanger P, Lindholm L (2002) Genetic

retargeting of adenovirus vectors: functionality of targeting ligands and their influence on virus viability. J Gene Med 4:356–370
15. Kawakami Y, Li H, Lam JT, Krasnykh V, Curiel DT, Blackwell JL (2003) Substitution of the adenovirus serotype 5 knob with a serotype 3 knob enhances multiple steps in virus replication. Cancer Res 63:1262–1269
16. Henning P, Lundgren E, Carlsson M, Frykholm K, Johannisson J, Magnusson MK, Tang E, Franqueville L, Hong SS, Lindholm L et al (2006) Adenovirus type 5 fiber knob domain has a critical role in fiber protein synthesis and encapsidation. J Gen Virol 87:3151–3160
17. Krasnykh V, Belousova N, Korokhov N, Mikheeva G, Curiel DT (2001) Genetic targeting of an adenovirus vector via replacement of the fiber protein with the phage T4 fibritin. J Virol 75:4176–4183
18. Schagen FH, Wensveen FM, Carette JE, Dermody TS, Gerritsen WR, van Beusechem VW (2006) Genetic targeting of adenovirus vectors using a reovirus sigma1-based attachment protein. Mol Ther 13:997–1005
19. Baker AT, Aguirre-Hernandez C, Hallden G, Parker AL (2018) Designer oncolytic adenovirus: coming of age. Cancers (Basel) 10:pii: E201
20. Martin NT, Wrede C, Niemann J, Brooks J, Schwarzer D, Kuhnel F, Gerardy-Schahn R (2018) Targeting polysialic acid-abundant cancers using oncolytic adenoviruses with fibers fused to active bacteriophage borne endosialidase. Biomaterials 158:86–94
21. Curiel DT (1999) Strategies to adapt adenoviral vectors for targeted delivery. Ann N Y Acad Sci 886:158–171
22. Haisma HJ, Grill J, Curiel DT, Hoogeland S, van Beusechem VW, Pinedo HM, Gerritsen WR (2000) Targeting of adenoviral vectors through a bispecific single-chain antibody. Cancer Gene Ther 7:901–904
23. Dmitriev I, Kashentseva E, Rogers BE, Krasnykh V, Curiel DT (2000) Ectodomain of coxsackievirus and adenovirus receptor genetically fused to epidermal growth factor mediates adenovirus targeting to epidermal growth factor receptor-positive cells. J Virol 74:6875–6884
24. Kashentseva EA, Seki T, Curiel DT, Dmitriev IP (2002) Adenovirus targeting to c-erbB-2 oncoprotein by single-chain antibody fused to trimeric form of adenovirus receptor ectodomain. Cancer Res 62:609–616
25. Li HJ, Everts M, Pereboeva L, Komarova S, Idan A, Curiel DT, Herschman HR (2007) Adenovirus tumor targeting and hepatic untargeting by a coxsackie/adenovirus receptor ectodomain anti-carcinoembryonic antigen bispecific adapter. Cancer Res 67:5354–5361
26. Kloos A, Woller N, Gurlevik E, Ureche CI, Niemann J, Armbrecht N, Martin NT, Geffers R, Manns MP, Gerardy-Schahn R et al (2015) PolySia-specific retargeting of oncolytic viruses triggers tumor-specific immune responses and facilitates therapy of disseminated lung cancer. Cancer Immunol Res 3:751–763
27. Li HJ, Everts M, Yamamoto M, Curiel DT, Herschman HR (2009) Combined transductional untargeting/retargeting and transcriptional restriction enhances adenovirus gene targeting and therapy for hepatic colorectal cancer tumors. Cancer Res 69:554–564
28. Kim J, Smith T, Idamakanti N, Mulgrew K, Kaloss M, Kylefjord H, Ryan PC, Kaleko M, Stevenson SC (2002) Targeting adenoviral vectors by using the extracellular domain of the coxsackie-adenovirus receptor: improved potency via trimerization. J Virol 76:1892–1903
29. Kuhnel F, Schulte B, Wirth T, Woller N, Schafers S, Zender L, Manns M, Kubicka S (2004) Protein transduction domains fused to virus receptors improve cellular virus uptake and enhance oncolysis by tumor-specific replicating vectors. J Virol 78:13743–13754
30. van Beusechem VW, Mastenbroek DC, van den Doel PB, Lamfers ML, Grill J, Wurdinger T, Haisma HJ, Pinedo HM, Gerritsen WR (2003) Conditionally replicative adenovirus expressing a targeting adapter molecule exhibits enhanced oncolytic potency on CAR-deficient tumors. Gene Ther 10:1982–1991
31. Hemminki A, Wang M, Hakkarainen T, Desmond RA, Wahlfors J, Curiel DT (2003) Production of an EGFR targeting molecule from a conditionally replicating adenovirus impairs its oncolytic potential. Cancer Gene Ther 10:583–588
32. He TC, Zhou S, da Costa LT, Yu J, Kinzler KW, Vogelstein B (1998) A simplified system for generating recombinant adenoviruses. Proc Natl Acad Sci U S A 95:2509–2514

Chapter 4

Development of Entry-Targeted Oncolytic Measles Viruses

Michael D. Mühlebach and Roberto Cattaneo

Abstract

This chapter describes the development of recombinant oncolytic measles viruses (MeV) that selectively enter and destroy tumor cells. The envelope of MeV is a favorable targeting substrate because receptor attachment and membrane fusion functions are separated on two proteins: the hemagglutinin (H) that binds receptors, and the fusion (F) protein that fuses the viral envelope with the cell membrane. The cell entry process, which depends on receptor recognition and occurs at the plasma membrane at neutral pH, results in the delivery of encapsidated genomes to the cytoplasm, where they replicate. Towards improving cancer specificity of oncolytic MeV, two types of cell entry targeting have been achieved. First, entry has been redirected through cancer-specific cell surface proteins. This was done by displaying specificity domains on H while also ablating binding to its natural receptors. Second, activation of the F protein was made dependent on secreted cancer proteases, while also interfering with F cleavage/activation by a ubiquitous intracellular protease. This chapter describes how entry-targeted MeV are produced: In short, gene cassettes with modified H or F coding regions are generated, and then introduced into the viral genome available on plasmid DNA. Such full-length genome plasmids are transfected in cell lines expressing, stably or transiently, the three viral proteins necessary for genome replication. Infectious centers form among these "rescue" cells, which allow isolation of clonal recombinant viruses. These are amplified, characterized in vitro, and then evaluated for their oncolytic activity in appropriate preclinical animal models.

Key words Recombinant measles virus, Receptor targeting, Protease targeting, Entry targeting, Rescue of *Morbillivirus*, *Paramyxoviridae*

1 Introduction

1.1 Background

Virus families have evolved specificities for different cell types, and this natural diversity is being used for their development as anticancer therapeutics [1]. Different properties of individual viruses are now being modified to improve infection of cancer tissue while reducing infection of normal tissue [2]. The MeV oncolytic platform stands on four premises. First, safety: live-attenuated MeV has been administered as a vaccine to at least one billion children with outstanding safety and efficacy records [3–5]. Second, natural oncolytic properties: clinical observations from different groups documented cancer regression following wild-type measles

Christine E. Engeland (ed.), *Oncolytic Viruses*, Methods in Molecular Biology, vol. 2058,
https://doi.org/10.1007/978-1-4939-9794-7_4,

infections [6]. Third, oncolytic efficacy of vaccine-lineage recombinant MeV has been demonstrated in different animal cancer models [7]. Fourth, ongoing phase I and II clinical trials confirmed safety and yielded promising efficacy indications [8]. Based on these premises, MeV has established itself as a preferred oncolytic platform.

The next step is to improve tumor targeting of MeV, thereby enhancing oncolytic efficacy. MeV is one of the best systems for the development of entry-targeted viruses owing to the plasticity of its glycoproteins, and the separation of receptor binding and membrane-fusion functions: the attachment protein hemagglutinin (H) binds receptors, while the fusion protein fuses the viral envelope with the cell membrane [7]. This chapter reveals how these two different but complementary strategies to modify each glycoprotein to increase tumor specificity can be applied and further developed.

1.2 Re-targeting Cell Entry Through Modifications of the Attachment Protein

MeV is a negative strand RNA virus of the genus Morbillivirus in the family *Paramyxoviridae* [9]. The particles of these viruses have similar pleomorphic morphologies and may contain more than one genome [10] (Fig. 1, top). This implies that additional genes can be inserted in the viral genome without negative impact on efficient

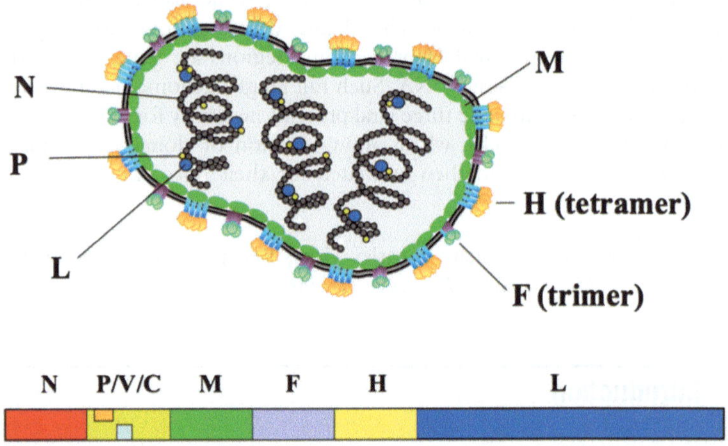

Fig. 1 Diagram of a *Morbillivirus* particle (top) and of the viral genome (bottom). Top: The nucleocapsid (N), phosphoprotein (P), polymerase (large, L), matrix (M), fusion (F), and hemagglutinin (H) proteins are indicated with different colors. Bottom: The corresponding open reading frames are indicated on the genome. The genome is one single molecule of RNA divided in six consecutive transcription units. Five units code for a single protein, but the second codes for three proteins, P, V, and C. While P is translated from the first start codon, C (depicted by small orange box) is translated from an alternative start codon of the native mRNA. In contrast, V is translated from mRNA, which is edited via transcriptional insertion of a nontemplated guanosine at a specific position, the editing site (depicted by small gray box). Thereby, V shares its amino-terminal half with P, but its carboxyl-terminal domain is distinct

particle formation, favoring the stability of recombinant viruses. All the morbillivirus genomes are organized into six contiguous, non-overlapping transcription units separated by three untranscribed nucleotides and coding for eight viral proteins in the order (positive strand): 5′-N-P/V/C-M-F-H-L-3′ (Fig. 1, bottom).

Cellular surface proteins used as entry receptors are one determinant of the tropism of MeV and the other morbilliviruses [11]. Since the pathogenic wild type viruses utilize sequentially two receptors, they are considered dual-tropic. After contagion the signaling lymphocyte activation molecule (SLAM) mediates virus entry and favors the extensive spread in immune tissues [12]. Infected immune cells then deliver MeV to epithelial cells that express nectin-4, a component of the adherens junction expressed preferentially in epithelia of the trachea [13]. Preferential MeV replication in SLAM-expressing transformed lymphocytes may, in retrospect, account for remissions from Burkitt lymphoma observed after wild type MeV infections [6, 14]. Nectin-4 is a cellular marker of several types of cancer, which has implications for tumor targeting [15–17]. In addition to the two "wild type" MeV receptors, the vaccine strain H protein also binds the complement membrane co-factor protein CD46, which is expressed preferentially in many cancer cell types, allowing for tropism expansion [7]. CD46-dependent entry has been acquired during the process of attenuation [11].

To completely re-target MeV tropism at the level of receptor recognition, cell entry through natural and acquired receptors must be ablated. Toward this aim, residues on the surface of H interacting specifically with each receptor were identified, and then mutagenized in combination, resulting in the generation of selectively receptor-blind viruses [18, 19]. The second step is to display on H a specificity domain binding a protein expressed preferentially on cancer cells. While small specificity domains can be inserted in exposed loops of the H protein [20], the carboxy-terminus of the H protein is the preferred insertion site because it tolerates the addition of large or modular specificity domains of 100 or more amino acids [21–24]. One focus of this chapter is therefore the H-carboxy-terminal extension methodology (Fig. 2, left).

1.3 Re-targeting Particle Activation Through Modifications of the Fusion Protein

A complementary cancer re-targeting strategy involves modification of the F protein cleavage/activation sequence (Fig. 2, right). The F precursor protein of *Paramyxoviridae*, F_0, requires cleavage into two subunits, F_2 and F_1, for activation of function. In particular, the MeV F protein is cleaved intracellularly by the ubiquitous trans-Golgi protease furin [25]. Since the F proteins in released MeV particles are proteolytically activated, their tropism is not restricted at the level of F activation. However, the tropism of the animal *Paramyxoviridae* Sendai Virus and Newcastle Disease Virus is restricted in most tissues because their F proteins require

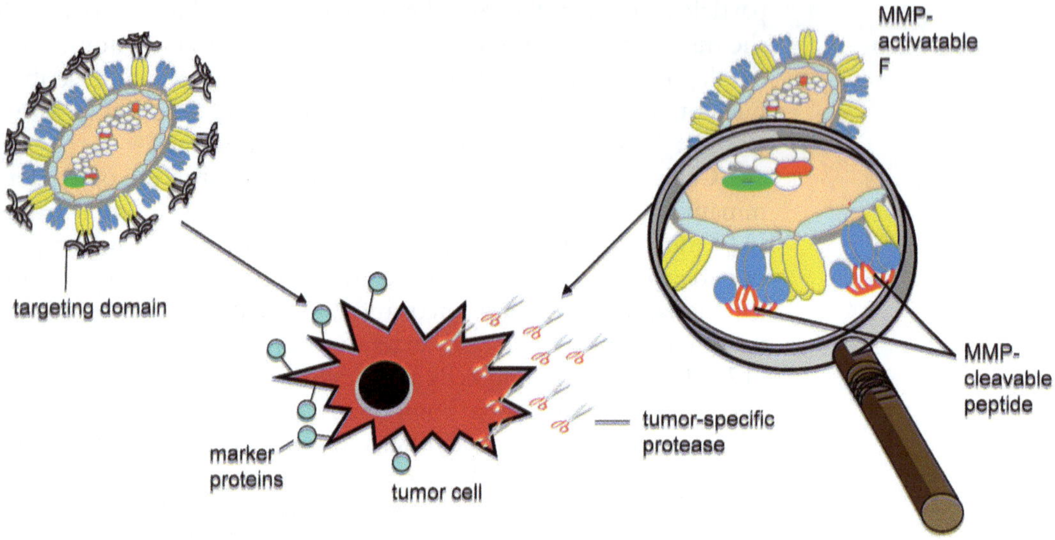

Fig. 2 Schematic depiction of targeting principles. Tumor cells can be characterized by the expression of specific surface marker proteins (turquois) and their invasive growth is facilitated by tumor specific proteases. Tumor targeting by oncolytic measles virus takes advantage of these properties by on the one hand rendering attachment of the viral hemagglutinin (yellow) specific for the marker proteins via fusion to respective ligands as targeting domains. On the other hand, the required proteolytic activation of the fusion protein (blue) can be rendered specific for tumor-associated proteases via a protease-cleavable peptide (red)

activation through secreted proteases that are expressed only in certain tissues [26, 27].

Based on this principle, it was shown, first, that the 5-amino acid sequence situated upstream of the MeV F_0 cleavage site can be mutagenized to inhibit cleavage through furin, while enabling cleavage and activation through an exogenous protease, trypsin [28]. Second, it was shown that MeV F_0 can be selectively activated by tumor-secreted matrix metalloproteinases (MMP), and that a recombinant MeV carrying an MMP-activatable F protein has enhanced safety and efficacy in a mouse xenograft model [29]. Third, based on protease activity profiles of cell line and primary liver tumors, a recombinant MeV with an optimized 6-amino acid MMP-activatable F protein was generated which replication was strongly restricted to human tumor tissue sections [30]. The methods in this chapter focus on how the F protein cleavage activation site can be modified.

We also note that, until now, re-targeting efforts have been focused on either H or F protein modification. In addition, with one notable exception involving grafting of the MeV envelope proteins on lentiviral vectors [31], re-targeting efforts have been focused exclusively on the vaccine strain envelope proteins. Time has come to combine the F and H targeting strategies. In addition, since multiple lines of evidence suggest that wild type MeV envelopes are more stable, and more discriminating, than those of the vaccine strain, time has come to base targeting strategies on a wild type H protein. This chapter should facilitate both lines of research.

2 Materials

2.1 Plasmid DNA and cDNA

1. Full-length MeV genome plasmids such as pBR-MV_{vac2}-GFP (H) [32] or p(+)PolII-MV_{NSe}-GFP(N) [22] (*see* **Note 1**). These genomes contain the measles virus genes organized in single transcription units as found in the virus genome, but single-cutter restriction endonuclease recognition sites have placed between certain genes during cloning to facilitate easy exchange of individual gene cassettes. In the abovementioned examples, the H ORF is flanked with *Pac*I and *Spe*I and the F ORF by *Nar*I and *Pac*I recognition sites (*see* **Note 2**).
2. T7-promoter driven expression plasmids for MeV proteins being components of the viral RNP complex (i.e., nucleocapsid protein N, phosphoprotein P, viral RNA-dependent RNA-Polymerase L), such as pEMC-Na, pEMC-Pa, or pEMC-La [33, 34], respectively.
3. Pol II-Promoter driven expression plasmids for MeV RNP complex proteins, such as pCA-MV-N, pCA-MV-L, and pCA-MV-P [35].
4. An expression plasmid for MeV-H, which is mutated in its receptor-binding surface (H_{mut}) [19]. H_{mut} then additionally harbors a 6×His-Tag at its carboxy-terminus, which is preceded by a short Gly-Ser-linker and *Sfi*I/*Not*I restriction sites, in between, that allow insertion of a targeting domain of choice [36].
5. cDNA of the targeting domain (TD) to be fused to the ectodomain of H. The targeting domain sequence has to be flanked 5′ by *Sfi*I and 3′ by *Not*I restriction sites (or alternative sites being spaced between the linker and the His-Tag) such that the genome length of the putative recombinant virus still obeys to the "rule-of-six" [37] (*see* **Note 3**).
6. An expression plasmid for MeV-F. The F ORF is flanked by restriction endonuclease restriction sites that allow easy shuttling of the F ORF between the expression plasmid and MeV genome plasmids.

2.2 PCR Components

1. Expand High-Fidelity PCR System (Roche) or standard PCR protocol.
2. Primers to generate targeting domain cDNA to be cloned into the H carboxy-terminus of MeV H (*see* Fig. 3).

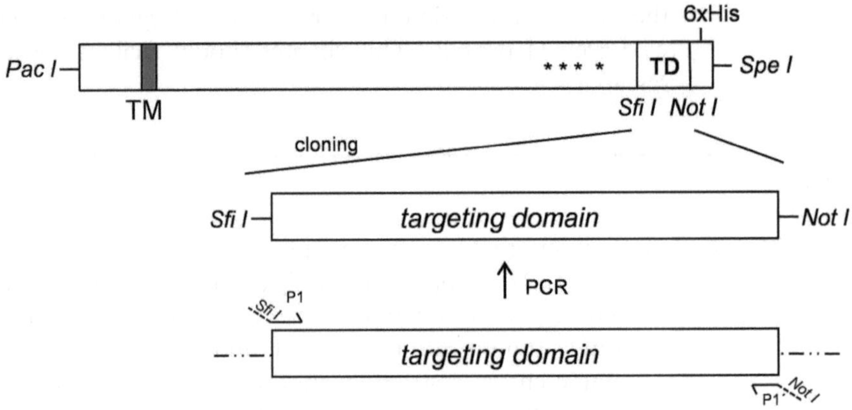

Fig. 3 Generation of tumor marker-targeted H protein. A tumor-specific targeting domain is amplified by PCR using flanking primers P1 and P1′ to generate terminal restriction sites (e.g., *SfiI* and *NotI*). These facilitate cloning into the mutated, receptor-blind H-ORF (blinding mutations [18, 19] indicated by asterisks) near to the carboxy-terminus of the ectodomain of H. The targeting domain is followed by a carboxy-terminal 6×His-Tag to allow amplification on Vero-αHis cells [24] bearing a recombinant receptor recognized by the His-Tag. The targeting domain is usually cloned into an H-ORF on an expression plasmid to facilitate in vitro generation of the recombinant protein. Shuttling into full-length MeV genomes is facilitated via flanking restriction sites (e.g., *PacI* and *SpeI*). Dark gray, transmembrane domain (TM)

- Sense primers have to encompass the 5′ restriction site preceding the targeting domain and the approx. 20–25 following nucleotides of the targeting domain sequence.
- Antisense primers have to encompass the 3′ restriction site following the targeting domain and the approx. 20–25 preceding nucleotides of the targeting domain sequence. It may be necessary to include extra nucleotides between restriction site and 6×His-Tag to obey the rule-of-six (*see* **Note 3**).

3. Primers to generate an F_1–F_2 fragment separated by the protease cleavage peptide to substitute the furin cleavage site by primer extension (fusion) PCR (*see* Fig. 4).
   - Sense primers for fragment 1 encompass a short oligonucleotide, which is binding in the F gene up-stream of a single-cutting restriction site, while the antisense primer binds just at the furin site and is 5′ elongated by the antisense coding sequence of the protease motif of interest.
   - Sense primers for fragment 2 bind at the furin site and are 3′ elongated by the coding sequence of the protease motif of interest, while the antisense primer encompasses a short oligonucleotide, which is binding in the F gene downstream of a single-cutting restriction.

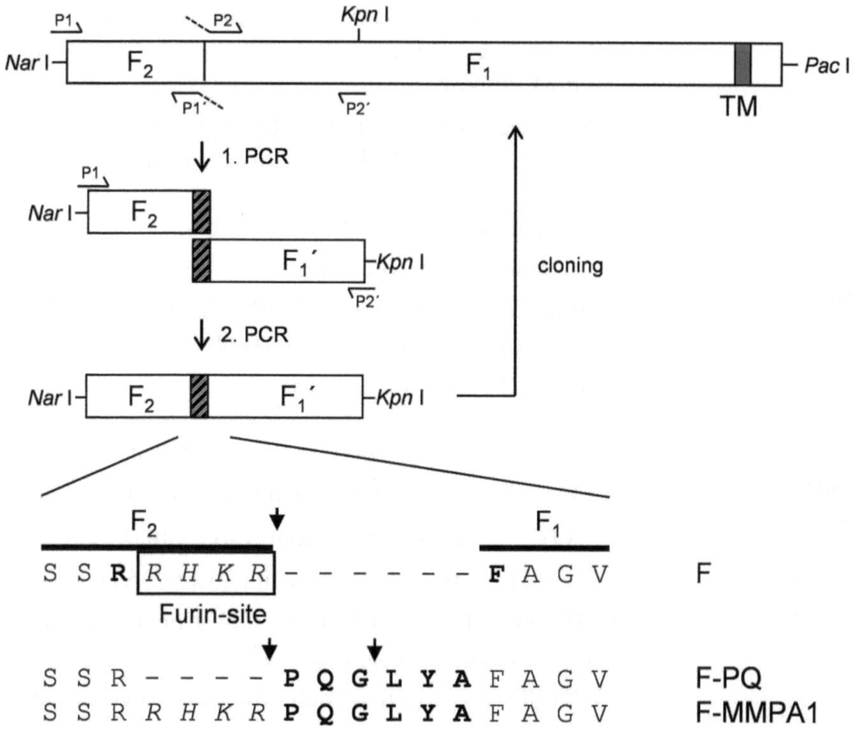

Fig. 4 Generation of protease-targeted F protein. Both flanks of the F_1/F_2 border region are amplified by PCR using primer pairs P1/P1′ and P2/P2′, respectively, to generate F-ORF-fragments F2 and F1′. In these fragments, the furin protease cleavage motif (italics, single amino acid code) has been substituted [30] or expanded [29] by an additional cleavage motif (bold/hashed boxes) recognized by tumor-specific protease (s) (hashed boxes). These fragments are fused in a second step by primer-extension PCR and then cloned back into the F-ORF using single-cutting restriction sites, (e.g., *NarI* and *KpnI*). Shuttling of the modified F-ORF into full-length MeV genomes is facilitated via flanking restriction sites (e.g., *PacI* and *NarI*) for MV_{NSe}. Dark gray, transmembrane domain (TM); arrowheads indicate proteolytic cleavage sites

2.3 Enzymatic Restriction Reaction Components

1. Commercially available restriction endonucleases and respective buffers.
2. Nuclease-free H_2O.
3. DNA to be digested.

2.4 Ligation Reaction Components

1. Commercially available standard ligation kit.
2. Purified DNA fragments with compatible ends after restriction digest.

2.5 Bacteria Culture Components

1. Bacteria: *E. coli* (e.g., Top10 F′ (F′{*lac*I^q Tn10 (Tet^R)}, mcrA, Δ(*mrr-hsd*RMS-*mcr*BC), Φ 80 *lac*ZΔM15, Δ*lac*X74, *deo*R, *rec*A1, *ara*D139, Δ(*ara-leu*)7697, *gal*U, *gal*K, *rps*L (Str^R), *end*A1, *nup*G)).
2. Luria–Bertani (LB) medium: 1% (v/w) bacto-tryptone, 0.5% (w/v) yeast extract, 1% (w/v) NaCl, pH 7.0.

3. S.O.C. medium: 2% (w/v) tryptone, 0.5% (w/v) yeast extract, 10 mM NaCl, 2.5 mM KCl, 10 mM $MgCl_2$, 10 mM $MgSO_4$, 20 mM glucose.
4. DNA isolation kits suited for DNA isolation from small (5 mL) or medium sized (200 mL) culture volumes.

2.6 Plasmid Transfection Components

1. $CaPO_4$ transfection: commercial kit.
2. 5 mL polystyrene round-bottom tube.
3. Lipofection: Lipofectamine® 2000 Transfection Reagent (Invitrogen) or other alternative transfection reagents.
4. Opti-MEM medium (Invitrogen).

2.7 Eukaryotic Cell Culture Components

1. Vero cells (African green monkey kidney): ATCC CCL-81.
2. 293T cells (human embryonal kidney): ATCC CRL-3216.
3. SK-OV-3 cells (human ovarian carcinoma): ATCC HTB-77.
4. CHO-K1 cells (Chinese hamster ovary): ATCC CCL-61.
5. CHO-SLAM, CHO-CD46, CHO-nectin-4 (MeV-receptor transgenic CHO cells) [12, 15].
6. HT1080 cells (MMP-rich human osteosarcoma): ATCC CCL-121.
7. 293-3-46 (transgenic HEK cells) [33].
8. 100 mg/mL Geneticin (G418) solution.
9. MMP inhibitor GM6001 (Merck, Darmstadt, Germany); final concentration: 10 μM.
10. Complete DMEM: Dulbecco's Modified Eagle's Medium supplemented with 10% fetal bovine serum and 2 mM L-Gln with additional 1.2 mg/mL of Geneticin, when rescued as described in [33]. All cells are cultured at 37 °C in a humidified atmosphere containing 6% CO_2 for no longer than 6 months after thawing of the original stock.

2.8 Western Blot

1. RIPA lysis buffer: 50 mM Tris, 150 mM NaCl, 1% (w/v) NP-40, 0.5% (w/v) sodium-deoxycholate, 0.1% (w/v) sodium dodecyl sulfate (SDS), pH 8.0. Supplemented with Protease Inhibitor Cocktail Complete (Roche Diagnostics, Mannheim, Germany).
2. Antibody/serum recognizing MeV H or MeV F, or MeV N for normalization of infection.
3. Standard SDS-PAGE and Western Blot equipment and material.

2.9 Animal Experiments

1. Mice deficient in adaptive immunity allowing transplantation of xenograft tumors (e.g., NOD-Scid [38]).
2. Human tumor cell lines, positive for the targeted receptor or expressing the targeted proteases and growing as xenograft

tumors in immunodeficient mice, for example, SK-OV-3 or HT1080 cells, respectively.

3. 30 G needles, one for each tumor cell line or oncolytic virus to be injected.
4. 2 mL syringes with fine scale, one for each virus to be injected.

3 Methods

3.1 Cloning of Tumor-Associated Antigen Re-targeted MeV-H

1. Generate gene segments encoding the desired targeting domains binding to tumor structures of choice by gene synthesis (Fig. 3). The ORF encompassing targeting domain codons is flanked by the respective restriction sites, which allow direct cloning of the gene segment into the carboxy-terminus of mutated H encoded on an expression plasmid. It may be necessary to include a specific number of extra nucleotides between the last codon of the TD and 3′ *Not*I restriction site to obey the rule-of-six (*see* **Note 3**).

   Alternatively, amplify the desired sequence on the basis of plasmid cDNA by PCR or cDNA after reverse transcription from the directly isolated mRNA of cells expressing the respective receptor ligand using RT-PCR. For this purpose, specific primers have to be designed as outlined above, depending on the exact structure of targeting domain/ligand sequence to be amplified.
2. Treat the resulting cDNA segments as well as an H encoding expression plasmid (e.g., pCG-H [39]), with the respective restriction endonucleases using the following standard reaction mix:

   1–10 μg DNA.

   5–10 units per restriction enzyme.

   5 μL 10× NEB buffer (depending on the enzyme used).

   5 μL 10× BSA (if required by the applied enzyme).

   Fill up to 50 μL with nuclease-free H_2O.

   Perform DNA digestions at the temperature required by the respective enzymes to allow thorough digestion of the DNA, optimally overnight.
3. Purify the desired segments after restriction digestion by standard agarose gel electrophoresis (*see* **Note 4**). Ligate the purified plasmid backbone encoding mutated H and the targeting domain-sequence containing insert using standard ligation conditions with at least tenfold molar excess of the insert:

   50 ng vector DNA.

   150 ng insert DNA.

   2 μL 5× DNA dilution buffer.

Fill up to 10 μL with nuclease-free water and mix.

10 μL 2× T4 DNA ligation buffer.

1 μL (5 U) T4 DNA ligase.

for 10–15 min at room temperature using, for example, commercially available ligation kits.

4. Directly transform ligated DNA into chemically competent standard *E. coli* strains such as DH5α, DH10b, or Top10F′ and cultured on LB-agar plates containing the selection antibiotic at 37 °C.
5. After overnight or 48 h of cultivation, pick single bacterial colonies into 5 mL LB medium, each, containing the respective selection antibiotic.
6. After culture at 37 °C employing permanent shaking, draw a 500 μL probe of each sample the next morning or 48 h after picking the colonies, and store for few hours at 4 °C during isolation of plasmid DNA using standard procedures for DNA preparation from small culture volumes.
7. Analyze the isolated DNA by treatment with appropriate restriction enzymes (e.g., *Bgl*II or *Ava*I for Edmonston B-derived H [GenBank Acc. No. AB583749.1]).
8. Use probes of clones displaying the expected DNA pattern after agarose gel electrophoresis to inoculate 200 mL LB medium, each, containing the respective selection antibiotic. Pellet bacteria 24 or 48 h after inoculation (the bacteria optimally being still in the logarithmic growth phase) and isolate plasmid DNA using standard procedures for DNA isolation from medium sized bacterial cultures.
9. Check the integrity and identity of the isolated plasmid DNA by enzymatic restriction nuclease treatment and agarose gel electrophoresis. The plasmids containing the mutated H fused to a targeting domain of choice are ready for analysis of expression and fusion-helper function, now.

3.2 Analysis of Expression and Fusion-Helper Function of Receptor-Targeted H

The integrity and function of the fused H-targeting domain (H-TD) constructs can be tested before insertion of the respective cassette into MeV genomes by transient transfection together with an expression plasmid for the viral fusion protein F into susceptible cell lines. To this end, purified expression plasmids for H_{mut}-TD and F are co-transfected into standard cell lines such as 293T cells using standard transfection procedures in a 1:1 ratio.

1. Seed 8×10^5 293T or 1×10^5 Vero-αHis or tumor cells positive for the targeted receptor per well in 2 mL complete DMEM into 6-well plates and incubate overnight.

2. The next day, ensure cells are approximately 80% confluent bevor starting transfection.
3. Transfection: Mix 4 μg of the respective H_{mut}-TD plasmid with 4 μg pCG-F (F protein expression plasmid) followed by addition of 250 μL Opti-MEM to this mixture (*see* **Note 5**). In parallel mix 12.5 μL Lipofectamine 2000 and fill up to 250 μL with Opti-MEM. Incubate both mixtures for 5 min. Mix both solutions and gently mix by vortexing at level 4. Incubate for 20 min to build Lipid–DNA complexes. During the incubation period change medium on the prepared 6-well plates to 1.5 mL Opti-MEM. After incubation add the total of the Lipid–DNA complex solution (500 μL) dropwise to the 6-well. Incubate for 4 h. Then, change medium again to complete DMEM and incubate for 2 days at 37 °C.
4. Lyse the transfected 293T cells and use lysates for determination of H_{mut}-TD expression as well as protein integrity and stability by Western blot analysis: Wash the cells with 1 mL PBS at 4 °C for 5 min (*see* **Note 6**).
5. Then incubate the cells with 500 μL RIPA buffer for 10 min on ice (*see* **Note 7**).
6. Transfer the cell suspension into pre-cooled 1.5 mL reaction tubes and centrifuge at 17,000 × *g*, 15 min, 4 °C to remove the cell debris.
7. Transfer the protein-containing supernatant into a fresh pre-cooled 1.5 mL reaction tube and store at −80 °C. Thaw frozen cell lysates on ice for further Western Blot applications according to standard conditions using antibodies recognizing MeV glycoproteins H or F, (e.g., α-H_{cyt} or α-F_{cyt} [40]) to allow assessment of amount, size, and stability of the glycoprotein complexes in cells, but especially of the modified H protein.
8. Check the transfected Vero-αHis or receptor-positive tumor cell cultures for MeV glycoprotein-mediated cell-to-cell fusion resulting in multi-nucleated cells, the so-called syncytia. Syncytia formation and the extent of the same reveal fusion-helper activity of H_{mut}-TD protein. These cell cultures can also be lysed afterwards to be checked for protein expression as outlined for the transfected 293T cells.

3.3 Cloning of Protease-Activatable MeV-F

1. Generate the gene encoding a modified protease-activatable F protein, $F_{protease}$, by a two-step primer-extension PCR protocol (Fig. 4). The desired sequence is introduced by amplification of F sequences surrounding the F_1/F_2 border on the basis of plasmid DNA using two sets of primers and a two-step protocol. For this purpose, specific primers have to be designed as outlined above. It may be necessary to include a codon-triplet for an extra amino acid to obey the rule-of-six (*see* **Note 3**). In

the first step, fragments 1 and 2 are amplified separately, and the coding sequence for the protease recognition motif of interest is added at the amino or carboxy-terminus, respectively, due to primer design. In the second step, purified fragments are used as template in an equimolar mixture, and fused and amplified by PCR using the outer flanking primers.

2. Treat the resulting cDNA segments as well as an F-encoding expression plasmid (e.g., pCG-F) [39], with the respective restriction endonucleases using the following standard reaction mix:

   1–10 μg DNA.

   5–10 units per restriction enzyme.

   5 μL 10× NEB buffer (depending on the enzyme used).

   5 μL 10× BSA (if required by the applied enzyme).

   Fill up to 50 μL with nuclease-free H_2O.

   Perform DNA digestions at the temperature required by the respective enzymes to allow thorough digestion of the DNA, optimally overnight.

3. After restriction digestion purify the desired segments by standard agarose gel electrophoresis (*see* **Note 4**). Ligate the purified plasmid backbone encoding F and the protease recognition motif-sequence containing insert using standard ligation conditions with at least tenfold molar excess of the insert:

   50 ng vector DNA.

   150 ng insert DNA.

   2 μL 5× DNA dilution buffer.

   Fill up to 10 μL with nuclease-free water and mix.

   10 μL 2× T4 DNA ligation buffer.

   1 μL (5 U) T4 DNA ligase.

   for 10–15 min at room temperature using, e.g., commercially available ligation kits.

4. Directly transform ligated DNA into chemically competent standard *E. coli* strains such as DH5α, DH10b, or Top10F′ and culture on LB-agar plates containing the selection antibiotic at 37 °C.

5. After overnight or 48 h of cultivation, pick single bacterial colonies into 5 mL LB medium, each, containing the respective selection antibiotic.

6. After culture at 37 °C employing permanent shaking, draw a 500 μL probe of each sample the next morning or 48 h after picking the colonies, and store for few hours at 4 °C during

isolation of plasmid DNA using standard procedures for DNA preparation from small culture volumes.

7. Analyze the isolated DNA by treatment with appropriate restriction enzymes (e.g., *Age*I or *Eco*RV for Edmonston B-derived F [GenBank Acc. No. KM054581.1]).
8. Use probes of clones displaying the expected DNA pattern after agarose gel electrophoresis to inoculate 200 mL LB medium, each, containing the respective selection antibiotic. Pellet bacteria 24 or 48 h after inoculation (the bacteria optimally being still in the logarithmic growth phase) and isolate plasmid DNA using standard procedures for DNA isolation from medium-sized bacterial cultures.
9. Check the integrity and identity of the isolated plasmid DNA by enzymatic restriction nuclease treatment and agarose gel electrophoresis. The plasmids now containing the modified F containing a protease site of choice are ready for analysis of expression and protease-dependent fusion function.

3.4 Analysis of Expression and Fusion Function of Protease-Targeted $F_{protease}$

The integrity and function of the $F_{protease}$ constructs can be tested before insertion of the respective cassette into MeV genomes by transient transfection together with an expression plasmid for the viral fusion-helper protein H into protease-positive cell lines. Purified expression plasmids for H and $F_{protease}$ are co-transfected into protease-positive cell lines such as HT 1080 cells using standard transfection procedures in a 1:1 ratio.

1. Seed 1×10^5 HT1080 or other tumor cells positive for the targeted protease per well in 2 mL complete DMEM into 6-well plates and incubate overnight.
2. The next day, ensure cells are approximately 80% confluent bevor starting transfection.
3. Transfection: Mix 4 μg of the respective $F_{protease}$ plasmid with 4 μg pCG-H followed by addition of 250 μL Opti-MEM to this mixture (*see* **Note 5**). In parallel mix 12.5 μL Lipofectamine 2000 and fill up to 250 μL with Opti-MEM. Incubate both mixtures for 5 min. Mix both solutions and gently mix by vortexing at level 4. Incubate for 20 min to build Lipid–DNA complexes. During the incubation period change medium on the prepared 6-well plates to 1.5 mL Opti-MEM. After incubation add the total of the Lipid–DNA complex solution (500 μL) dropwise to the 6-well. Incubate for 4 h. Then, change medium again to complete DMEM and incubate for 2 days at 37 °C.
4. Check the transfected HT1080 cells or other receptor-positive tumor cell cultures before lysis for MeV glycoprotein-mediated cell-to-cell fusion resulting in multi-nucleated cells, the

so-called syncytia. Syncytia formation and the extent of the same reveal protease-dependent fusion activity of $F_{protease}$ protein.

5. Lyse the transfected HT1080 cells and use lysates for determination of $F_{protease}$ expression as well as protein cleavage by Western Blot analysis: Wash the cells with 1 mL PBS at 4 °C for 5 min (*see* **Note 6**).
6. Then incubate the cells with 500 μL RIPA buffer for 10 min on ice (*see* **Note 7**).
7. Transfer the cell suspension into pre-cooled 1.5 mL reaction tubes and centrifuge at 17,000 × *g*, 15 min, 4 °C to remove the cell debris.
8. Transfer the protein-containing supernatant into a fresh pre-cooled 1.5 mL reaction tube and store at −80 °C. Thaw frozen cell lysates on ice for further Western Blot applications according to standard conditions using antibodies recognizing MeV glycoproteins H or F, (e.g., α-H_{cyt} or α-F_{cyt} [40]) to allow assessment of amount, size, and degradation of the glycoprotein complexes in cells, but especially of the modified F protein.

3.5 Cloning of Full-Length MeV Genomes Encompassing Targeted H or F

The functional H_{mut}-TD or $F_{protease}$ gene cassette can then be inserted via the restriction sites flanking the H or F gene cassettes in both genome and expression plasmid. For this purpose, both H_{mut}-TD or $F_{protease}$ and MeV genome plasmids have to be digested using matching restriction endonucleases. Afterwards, inserts containing the re-targeted H or F proteins can be ligated into the corresponding positon of recombinant MeV genomes.

1. Treat the H_{mut} or $F_{protease}$ as well as MeV full-length genome plasmids with the respective restriction endonucleases (e.g., *Pac*I and *Spe*I) using the following standard reaction mix:

   1–10 μg DNA.

   5–10 units per restriction enzyme.

   5 μL 10× NEB buffer (depending on the enzyme used).

   5 μL 10× BSA (if required by the applied enzyme).

   Fill up to 50 μL with nuclease-free H_2O.

   Perform DNA digestions at the temperature required by the respective enzymes to allow thorough digestion of the DNA, optimally overnight.

2. Purify the desired segments after restriction digestion by standard agarose gel electrophoresis (*see* **Note 4**). Ligate the purified genome-containing plasmid backbone and the re-targeted H_{mut}-TD or $F_{protease}$ containing insert using standard ligation conditions with at least tenfold molar excess of the insert:

50 ng vector DNA.

150 ng insert DNA.

2 μL 5× DNA dilution buffer.

Fill up to 10 μL with nuclease-free water and mix.

10 μL 2× T4 DNA ligation buffer.

1 μL (5 U) T4 DNA ligase.

for 10–15 min at room temperature using, e.g., commercially available ligation kits.

3. Directly transform ligated DNA into chemically competent *E. coli* strains such as Top10F′ being able to amplify large plasmids and culture on LB-agar plates containing the selection antibiotic at the appropriate temperature (*see* **Note 1**).
4. After overnight or 48 h of cultivation, pick single bacterial colonies into 5 mL LB medium, each, containing the respective selection antibiotic.
5. After culture at the appropriate temperature (*see* **Note 1**) employing permanent shaking, draw a 500 μL probe of each sample the next morning or 48 h after picking the colonies, and store for few hours at 4 °C during isolation of plasmid DNA using standard procedures for DNA preparation from small culture volumes.
6. Analyze the isolated DNA by treatment with appropriate restriction enzymes (e.g., *Hind*III for Edmonston B-derived MeV).
7. Use probes of clones displaying the expected DNA pattern after agarose gel electrophoresis to inoculate 200 mL LB medium, each, containing the respective selection antibiotic. Pellet bacteria 24 or 48 h after inoculation (the bacteria optimally being still in the logarithmic growth phase) and isolate plasmid DNA using standard procedures for DNA isolation from medium-sized bacterial cultures.
8. Check the integrity and identity of the isolated plasmid DNA by enzymatic restriction nuclease treatment and agarose gel electrophoresis. The genome plasmids now containing the modified MeV genomes are ready for the rescue of recombinant viruses.

3.6 Introduction to Rescue of Recombinant MeV

At least three protocols for the rescue of recombinant MeV are available [33–35] (*see* **Note 8**), which differ mainly in the cell lines, plasmids, as well as the transfection protocol used for the transfection of recombinant MeV DNA into the rescue cell line.

Carry out all procedures aseptically under laminar flow and at room temperature unless otherwise specified. Eukaryotic cells are generally cultured at 37 °C, 5% CO_2 and 95% humidity in an incubator.

3.7 Rescue by Transfection of the 293-3-46 Helper Cell Line [33]

1. Seed 8 × 10^5 293-3-46 cells/well in 2 mL complete DMEM +1.2 mg/mL G418 into 6-well plates and incubate overnight (*see* **Note 9**).
2. The next day, ensure cells are approximately 80% confluent before starting with transfection and replace the medium of the cells by complete DMEM without G418 approximately 3 h before transfection.
3. Transfection of 293-3-46 helper cells (calcium phosphate transfection): Use the ProFection Mammalian Transfection System for transfection. In detail, add 5 μg of MeV genome plasmid and 25 ng pEMC.La to 25 μL $CaCl_2$ [2 M] in a 1.5 mL reaction tube. Fill the reaction volume up to 200 μL with nuclease-free H_2O. Fill 200 μL 2× HBS buffer into a 5 mL polystyrene round-bottom tube and strongly vortex, while carefully adding the DNA-containing solution dropwise to the 2× HBS buffer. Then incubate the transfection mix for 30 min at RT, add dropwise to the medium of the cells, and incubate cells with transfection mix for 24 h at 37 °C. After incubation, perform a heat shock of the transfected 293-3-46 cells for 3 h at 42 °C in a water bath (*see* **Note 5**).
4. In parallel, prepare overlay cell cultures by seeding 8 × 10^5 Vero-αHis or HT1080 cells/10 cm cell culture dish in 10 mL complete DMEM. After the heat shock, cultivate 293-3-46 cells for further 48 h at 37 °C, before overlaying 293-3-46 cells onto prepared overlay cells. During the incubation period examine for syncytia formation, daily (*see* **Note 10**).

3.8 Transfection Protocol Using T7 Polymerase Based Rescue System and a T7 Pol Expressing Vaccinia Virus [34]

1. Seed 5 × 10^5 Vero-αHis or HT1080 cells/well in 2 mL complete DMEM into 6-well plates and incubate overnight (*see* **Note 9**).
2. The next day, ensure cells are approx. 80% confluent before starting transfection.
3. Infection: Infect cells with a T7-encoding vaccinia virus such as MVA-T7 at an MOI of 1–5 (*see* **Note 11**). Forty-five minutes after infection, replace the medium and transfect the cells.
4. Transfection: Mix 1.5 μg of the respective MeV genomic cDNA plasmid with 1.5 μg pEMC-Na, 1.5 μg pEMC-Pa, and 0.5 μg EMC.La (helper plasmids) and transfect using a commercially available lipofection method as described below in the protocol using the Pol II-based rescue system. Culture transfected cells and examine for syncytia formation, daily (*see* **Note 9**).

3.9 Transfection Protocol Using a Pol II Polymerase-Based Rescue System [35]

1. Seed 8 × 10^5 293T cells/well in 2 mL complete DMEM into 6-well plates and incubate overnight (*see* **Note 9**).
2. The next day, ensure cells are approximately 80% confluent before starting transfection.

3. Transfection: Mix 4 μg of the respective MeV genomic cDNA plasmid with 0.4 μg pCA-MV-N, 0.4 μg pCA-MV-L, and 0.1 μg pCA-MV-P (helper plasmids); add 250 μL Opti-MEM to this mixture (*see* **Note 12**). In parallel mix 12.5 μL Lipofectamine 2000 and fill up to 250 μL with Opti-MEM. Incubate both mixtures for 5 min. Mix both solutions and gently mix by vortexing at level 4. Incubate for 20 min to build Lipid–DNA complexes. During the incubation period change medium on the prepared 6-well plates to 1.5 mL Opti-MEM. After incubation add the total of the Lipid–DNA complex solution (500 μL) dropwise to the 6-well. Incubate for 4 h. Then, change medium again to complete DMEM and incubate for 2 days.
4. Seed 3×10^6 Vero-αHis or HT1080 cells/10 cm cell culture dish in 10 mL complete DMEM and incubate for 4 h. Completely suspend transfected 293T cells by pipetting up and down and spread 1 mL of the solution evenly onto the prepared 10 cm cell culture dish (overlay; *see* Subheading 3.10). Incubate for 3 days. During the incubation period examine for syncytia formation, daily (*see* **Note 10**).

3.10 Isolation of Single Infectious MeV Clones

1. If you find syncytia, pick them as outlined below (**step 4**).
2. If no syncytia are visible on day 4 post overlay, split the overlay culture by passaging the culture 1:4. For that purpose, seed 8×10^5 Vero-αHis or HT1080 cells/10 cm cell culture dish in 10 mL complete DMEM and incubate for 4 h. Aspirate the medium from the 10 cm cell culture dish of the overlay and wash once with 5 mL PBS. Detach the cells by incubating with 1.5 mL trypsin/EDTA for 5 min. After complete detaching (check by microscope) stop trypsin incubation by adding 2.5 mL complete DMEM and suspend the cells by pipetting up and down. Seed 25% of the cell suspension to the prepared 10 cm cell culture dish. Transfer the remaining 75% of the cell suspension into a 15 mL tube and snap-freeze in liquid nitrogen for 5 min. Thaw the frozen cell suspension at 37 °C in a water bath and pellet the resulting cell debris by centrifugation at $2{,}500 \times g$ for 5 min at 4 °C. Transfer the supernatant to the 10 cm cell culture dish dropwise and incubate overnight.
3. Again check for syncytia formation (*see* **Note 10**). If there is no syncytia formation wait for one additional day, and if there is still no syncytia formation, repeat the rescue.
4. If there is syncytia formation, seed one 6-well plate/virus with 3×10^5 Vero-αHis cells/well in 2 mL complete DMEM per well and incubate for 4 h. Identify six different syncytia (*see* **Note 10**) and mark them by surrounding the syncytia at the bottom side of the 10 cm cell culture dish. Harvest one syncytium by scratching the marked area with a 200 μL pipette tip

and aspirating in parallel 60 μL. Transfer the aspirated 60 μL to one of the prepared 6-wells. Repeat the procedure to obtain six different virus clones (one 6-well plate/virus).

5. Incubate the 6-well plate and check regularly for the level of infection indicated, for example, by formation of syncytia or expression of GFP marker protein (at least twice a day). If a well is near complete infection, aspirate 1 mL medium and detach the infected cells by scratching with a cell scraper. Transfer the remaining 1 mL including the virus to a 15 mL tube and snap-freeze in liquid nitrogen for 5 min (*see* **Note 13**). Thaw the frozen virus solution at 37 °C in a water bath (*see* **Note 14**) and pellet the remaining cell debris by centrifugation at 2,500 × *g* for 5 min at 4 °C. Aliquot the so-called "passage 0" (P0) virus supernatant at 300 μL each and store at −80 °C.
6. For infection of virus in passage 1 (P1), seed 5 × 10^6 Vero-αHis or HT1080 cells/15 cm cell culture dish in 20 mL complete DMEM and incubate for 4 h. Thaw one of the P0 virus aliquots at room temperature and distribute the virus suspension across the 15 cm cell culture dish. Slew the 15 cm cell culture dish immediately for a few seconds (*see* **Note 15**). Incubate the 15 cm cell culture dish and check regularly for the level of infection (*see* **Note 16**).
7. If the 15 cm cell culture dish is completely infected aspirate the medium completely and add 1 mL Opti-MEM. Detach the infected cells by scratching with a cell scraper. Transfer the medium including all cell debris using a 1 mL pipette (*see* **Note 17**) to a 15 mL falcon and snap-freeze in liquid nitrogen for 5 min (*see* **Note 13**). Thaw the frozen virus solution at 37 °C in a water bath (*see* **Note 14**) and pellet the remaining cell debris by centrifugation at 2,500 × *g* for 5 min at 4 °C. Distribute the P0 virus supernatant into 300 μL aliquots and store at −80 °C. To obtain the virus titer measured as 50% Tissue Culture Infective Dose ($TCID_{50}$), perform an endpoint dilution assay with one aliquot according to Spearman-Karber [41].

3.11 Amplification of Recombinant MeV (See Note 18)

1. For generation of P1 virus stocks seed four times 5 × 10^6 Vero-αHis or HT1080 cells in 15 cm dishes and cultivate for around 4 h at 37 °C.
2. Infect the cells with 300 μL of P0 stock. Cultivate Vero-αHis or HT1080 cells infected with re-targeted oncolytic MeV at 32 or 37 °C, depending on the desired speed of virus amplification.
3. When almost all cells are infected (*see* **Notes 10** and **16**), harvest the P1 culture by freezing and thawing of cells. Remove the medium of the infected cells completely (*see* **Note 19**) and add 1 mL Opti-MEM to the infected cells. Carefully detach the

cells using a cell scraper and transfer the cell suspension into a 15 mL tube, and snap-freeze in liquid nitrogen (*see* **Note 13**). Then, thaw the suspension at 37 °C (*see* **Note 14**) and subsequently centrifuge at approx. 3,000 × *g*, 5 min, 4 °C. Aliquot the supernatant into 300 μL aliquots in cryotubes and store at −80 °C (*see* **Note 20**). Titrate the P1 virus stocks on Vero-αHis cells according to Kaerber and Spaerman [41].

4. To generate P2 and subsequent MeV cultures (*see* **Note 21**), seed 5 × 10^6 Vero-αHis or HT1080 cells in quadruplicates in 15 cm dishes and cultivate for around 4 h at 37 °C. Then, infect the cells with a multiplicity of infection (MOI) of 0.03 of the titrated parental virus stock and cultivate at 32 or 37 °C. When almost all cells are infected, harvest the P2 culture as described above and titrate on Vero-αHis or HT1080 cells [41].

3.12 Characterization of Re-targeted Oncolytic Measles Virus by Western Blot Analysis

1. For the preparation of cell lysates incubate 3 × 10^5 Vero-αHis or HT1080 cells/well in 2 mL complete DMEM per 6-well for around 4 h.
2. Infect the cells with MOI of 0.03 of the respective virus.
3. When almost all cells are infected, remove medium and wash the cells with 1 mL PBS at 4 °C for 5 min (*see* **Note 6**).
4. Then, incubate the cells with 500 μL RIPA buffer for 10 min on ice (*see* **Note 7**).
5. Transfer the cell suspension into pre-cooled 1.5 mL reaction tubes and centrifuge at 17,000 × *g*, 15 min, 4 °C to remove the cell debris.
6. Transfer the protein-containing supernatant into a fresh pre-cooled 1.5 mL reaction tube and store at −80 °C. Thaw frozen cell lysates on ice for further Western Blot applications according to standard conditions using antibodies recognizing the MeV nucleocapsid protein, to standardize for infection. Other antibodies recognizing the viral glycoproteins allow assessment of glycoprotein complex expression by the recombinant viruses as well as glycoprotein integrity and stability/processing by, for example, the appearance of degradation products.

3.13 Characterization of Receptor Targeting In Vitro

For the characterization of the receptor usage of re-targeted MeV in vitro, a panel of CHO cell lines transgenically expressing either one of the natural MeV receptors (SLAM, nectin-4, or CD46), the parental CHO-K1 cells expressing no receptor for MeV, or CHO cells transgenically expressing the targeted tumor-associated antigen as receptor can be used.

1. To seed cells, seed 1 × 10^5 of each CHO cell line per well of a 12-well plate. After 24 h, infect the cells with the different targeted viruses of interest and in parallel with a parental, non-targeted MeV with an MOI of 1.

2. Assess infection of individual CHO cell cultures by expression of GFP, if this protein is encoded as a marker gene by the targeted MeV, or by cytopathic effect (CPE), that is, induction of syncytia.
3. The infection profile of each targeted virus reveals the targeting specificity of each virus.
4. As an alternative, mix target cells with non-target cells marked by a recombinantly expressed fluorescent protein (e.g., dsRed2) and infect the mixed culture 24 h after seeding with a green-fluorescent, targeted MeV or with green-fluorescent, non-targeted parental MeV.
5. Specificity of targeting can be assessed by separation of virus-encoded green and non-target cell-expressed red fluorescence and subsequent depletion of non-fluorescent target cells from the culture, over time.
6. Absence of mixed fluorescence in infected mixed cultures over time demonstrates stability of targeting/de-targeting of oncolytic MeV.

3.14 Characterization of Protease Targeting In Vitro

For the characterization of activation of protease-targeted MeV in vitro, a panel of cell lines expressing targeted proteases or not are used.

1. To seed cells, seed 1×10^5 of each cell line per well of a 12-well plate. After 24 h, infect the cells with the different targeted viruses of interest and in parallel with a parental, non-targeted MeV with an MOI of 1.
2. In addition, incubate protease-positive cell lines infected with the re-targeted MeV with a protease inhibitor such as GM6001 few hours after infection to specifically inhibit protease activity in the given culture.
3. Assess spread of re-targeted MeV in individual cultures by expression of GFP, if this protein is encoded as a marker gene by the targeted MeV, or by CPE, that is, induction of syncytia.
4. The infection profile of each targeted virus reveals the targeting specificity of each virus.

3.15 In Vivo Characterization of Targeted Oncolytic MeV

1. Immunodeficient mice (e.g., NOD-Scid (*see* **Note 22**) aged 6–8 weeks) receive a cell type-specific number of viable tumor cells expressing the targeted antigen (e.g., 5×10^6 SK-OV-3 cells for testing HER2-targeted MeV or HT1080 cells for testing protease-targeted MeV) by subcutaneous injection in a final volume of 100 μL PBS into the shaved flanks to implant tumors.
2. Check mice at the injection site for the appearance of palpable tumors.
3. When the tumors reach a mean volume of 50 mm^3 (measured by a caliper using the formula $a \times b^2/2$, where a is the longest

tumor diameter and *b* the shortest), stratify mice into different groups for a similar mean tumor volume.

4. Then, mice receive 1×10^6 $TCID_{50}$ of targeted MeV or non-targeted parental MeV diluted in Opti-MEM, or Opti-MEM alone to a final volume of 100 μL by direct intra-tumoral injections on 5 consecutive days, each.
5. Check mice daily for tumor volume using a caliper as described before, and for other symptoms of morbidity.
6. Sacrifice mice when reaching predefined humane endpoints, mainly a tumor volume of 1,000 mm^3.
7. Tumor growth kinetics and survival of mice in different treatment cohorts indicate therapeutic efficacy of targeted oncolytic MeV in direct comparison to non-targeted, parental oncolytic MeV.
8. Prepare tumors of mice, resuspend the tumor cells and directly check for expression of targeted receptor or take tumor cells into culture.
9. Downregulation of targeted receptors or protease activity determined by gel zymography of suspended tumor tissue could indicate resistance development against targeted OV, while tumor cell cultures can reveal persistence of oncolytic virus in tumor tissue and infiltration of other cell types (e.g., NK cells).

4 Notes

1. Both examples pBR-MV_{vac2}-GFP(H) and p(+)PolII-MV_{NSe}-GFP(N) contain, as all genome plasmids for the rescue of recombinant MeV, a full-length MeV genome. pBR-MV_{vac2}-GFP(H) is derived from plasmid backbone pBR322 (low copy) and expresses the viral RNA antigenome under the control of a T7-promoter, whereas p(+)PolII-MV_{NSe}-GFP(N) has a pBluescript (high copy) backbone and is Pol II-driven. Plasmids containing full-length MeV genomes tend to be a bit delicate to handling procedures. Therefore, plasmids with a high-copy plasmid backbone should be amplified in *E. coli* at 30 °C, whereas low-copy plasmids can be amplified at 37 °C. In addition, it pays to directly pick clones or isolate plasmids from growing cultures instead of storing liquid bacteria cultures or colony plates with MeV genome-containing plasmids at 4 °C for more than few hours (never overnight!). As an alternative, bacteria pellets can be stored frozen at −20 °C before plasmid isolation.
2. Each MeV gene cassette is flanked by conserved genetic elements representing start and stop-signals for the viral

polymerase complex. To separate single transcription units encoding the different MeV proteins, single-cutting restriction enzyme recognition sites are inserted in some genome plasmids such as those mentioned above (*see* **Note 1**). Thereby, single virus genes can be easily cloned out of and back into the virus genome after genetic manipulation such as mutation of the natural receptor binding surface or insertion of a targeting domain.

3. The number of nucleotides in MeV genomes can be exactly divided by 6, presumably due to one N protein binding to six nucleotides. Also recombinant genomes have to obey this rule; otherwise no recombinant virus can be rescued. Therefore, it has to be considered that the length of the inserted targeting domain coding region can be divided with the same rest by 6 as the length of the segment removed from the genome during cloning, if cassettes are inserted into the glycoprotein open reading frame. Thereby, one makes sure that the length of the resulting full-length genome is multiple of six, again.
4. Due to the size of the genome-containing plasmid (approximately 20 kDa), voltage during agarose gel electrophoresis should be restricted to max. 70 V.
5. Seal 6-well plate properly to avoid contamination of the cells.
6. Shake the plate gently a few times during incubation.
7. Shake the plate vigorously a few times during incubation.
8. While T7-based rescue of MeV, especially using the helper cell line 293-3-46, is the standard rescue system guaranteeing precise start and stop of the transcribed antigenomic viral RNA due to the precise start of T7-driven transcription and stop due to specific termination signals in conjunction with the ribozyme flanking the viral genome sequences [33], efficiency of the rescue is quite variable and depends considerably on the status of the rescue cells. Moreover, syncytia formation using 293-3-46 cells after overlay is not too efficient [33], further limiting rescue efficiency especially of virus variants with limited fusion activity (unpublished own observation). Vaccinia Virus-driven rescue allows usage of cell lines other than 293, which may be better suited for propagation of one specific MeV variant, but depends on the quality of the used plasmids. Pol II-driven virus rescue is very efficient and usually results in recombinant MeV with precise genomes, but the lower precision of start and stop of Poll II transcription may allow completion of virus from genomes not being in line with the rule of six [35].
9. Prepare one more 6-well plate than required for the rescue of the different MeV. Use this additional well as transfection control, as explained in **Note 12**.

10. If the recombinant MeV will additionally express GFP or other fluorescent marker proteins, check for GFP-expression with the help of a fluorescence microscope to identify syncytia. If no marker protein for easy evaluation is encoded, check for syncytia formation via light microscope in phase contrast.
11. Using another Vaccinia Virus than MVA or a similarly attenuated virus for T7 Pol expression, one has to make sure that the rescued recombinant MeV can be separated from co-replicating Vaccinia Virus, which may be a tedious process due to the laborious methods needed to physically separate the different virus populations completely.
12. For transfection control, use 4 μg of, for example, EGFP expression plasmids such as pEGFP-N1 without any MeV cDNA or helper plasmids and transfect as described for MeV cDNA.
13. After quick-freezing in liquid nitrogen, the tube containing the virus suspension can be stored at −80 °C for several days.
14. Check thawing of the virus regularly for liquefied content of the tube; the virus is heat sensitive and incubating the virus at 37 °C will lead to significantly reduced virus titers.
15. It is essential to shake the cell culture dish immediately as well as properly to avoid erratic infection of Vero cells, which will lead to lower virus titers.
16. The growth rate is depending on the virus strain and might be modulated by the targeting domain. Therefore, check the virus growth regularly to develop a feeling for the growth rate.
17. Scratch the cells to one edge of the cell culture dish and hold the plate inclined while aspirating the medium including all cell debris.
18. While MeV vaccine strains authorized to be used as human vaccines are usually regarded as safe and accordingly being handled under BSL-1 conditions, recombinant viruses—even those bearing identical vaccine strain genomes—may be handled under BSL-1 or BSL-2 conditions depending on the locally responsible national or regional regulatory authorities.
19. Try to remove the medium completely by following **Note 17**. This will concentrate virus.
20. The number of aliquots can be modified, depending on freezer-space and experimental plans.
21. For passages >P5 it is sufficient to infect only one 15 cm dish and store around five aliquots because those viruses are not used for experiments, normally.
22. Experimental mouse work has to be carried out in compliance with the regulations of the respective animal protection law.

Acknowledgments

The authors thank U. Schneider for providing the Pol II Rescue System.

References

1. Cattaneo R, Miest T, Shashkova EV et al (2008) Reprogrammed viruses as cancer therapeutics: targeted, armed and shielded. Nat Rev Microbiol 6(7):529–540. https://doi.org/10.1038/nrmicro1927
2. Miest TS, Cattaneo R (2014) New viruses for cancer therapy: meeting clinical needs. Nat Rev Microbiol 12(1):23–34. https://doi.org/10.1038/nrmicro3140
3. Griffin DE (2002) Measles virus. Wiley encyclopedia of molecular medicine. John Wiley & Sons, Inc, Hoboken, NJ
4. Davidkin I, Jokinen S, Broman M et al (2008) Persistence of measles, mumps, and rubella antibodies in an MMR-vaccinated cohort: a 20-year follow-up. J Infect Dis 197(7):950–956. https://doi.org/10.1086/528993
5. Hilleman MR (2001) Current overview of the pathogenesis and prophylaxis of measles with focus on practical implications. Vaccine 20 (5–6):651–665
6. Kelly E, Russell SJ (2007) History of oncolytic viruses: genesis to genetic engineering. Mol Ther 15(4):651–659. https://doi.org/10.1038/sj.mt.6300108
7. Navaratnarajah CK, Leonard VHJ, Cattaneo R (2009) Measles virus glycoprotein complex assembly, receptor attachment, and cell entry. Curr Top Microbiol Immunol 329:59–76
8. Cattaneo R, Russell SJ (2017) How to develop viruses into anticancer weapons. PLoS Pathog 13(3):e1006190. https://doi.org/10.1371/journal.ppat.1006190
9. Lamb RA, Parks GD (2013) Paramyxoviridae: the viruses and their replication. In: Fields BN, Knipe DM (eds) Fields virology, vol 1, 6th edn. Wolters Kluwer Lippincott Williams & Wilkins, Philadelphia, PA, pp 957–995
10. Rager M, Vongpunsawad S, Duprex WP et al (2002) Polyploid measles virus with hexameric genome length. EMBO J 21(10):2364–2372. https://doi.org/10.1093/emboj/21.10.2364
11. Mateo M, Navaratnarajah CK, Cattaneo R (2014) Structural basis of efficient contagion: measles variations on a theme by parainfluenza viruses. Curr Opin Virol 5:16–23. https://doi.org/10.1016/j.coviro.2014.01.004
12. Tatsuo H, Ono N, Tanaka K et al (2000) SLAM (CDw150) is a cellular receptor for measles virus. Nature 406(6798):893–897. https://doi.org/10.1038/35022579
13. Mühlebach MD, Mateo M, Sinn PL et al (2011) Adherens junction protein nectin-4 is the epithelial receptor for measles virus. Nature 480(7378):530–533. https://doi.org/10.1038/nature10639
14. Bluming AZ, Ziegler JL (1971) Regression of Burkitt's lymphoma in association with measles infection. Lancet 2(7715):105–106
15. Fabre-Lafay S, Garrido-Urbani S, Reymond N et al (2005) Nectin-4, a new serological breast cancer marker, is a substrate for tumor necrosis factor-alpha-converting enzyme (TACE)/ADAM-17. J Biol Chem 280 (20):19543–19550. https://doi.org/10.1074/jbc.M410943200
16. Derycke MS, Pambuccian SE, Gilks CB et al (2010) Nectin 4 overexpression in ovarian cancer tissues and serum: potential role as a serum biomarker. Am J Clin Pathol 134(5):835–845. https://doi.org/10.1309/AJCPGXK0FR4MHIHB
17. Takano A, Ishikawa N, Nishino R et al (2009) Identification of nectin-4 oncoprotein as a diagnostic and therapeutic target for lung cancer. Cancer Res 69(16):6694–6703. https://doi.org/10.1158/0008-5472.CAN-09-0016
18. Leonard VHJ, Sinn PL, Hodge G et al (2008) Measles virus blind to its epithelial cell receptor remains virulent in rhesus monkeys but cannot cross the airway epithelium and is not shed. J Clin Invest 118(7):2448–2458. https://doi.org/10.1172/JCI35454
19. Vongpunsawad S, Oezgun N, Braun W et al (2004) Selectively receptor-blind measles viruses: identification of residues necessary for SLAM- or CD46-induced fusion and their localization on a new hemagglutinin structural model. J Virol 78(1):302–313
20. Navaratnarajah CK, Oezguen N, Rupp L et al (2011) The heads of the measles virus attachment protein move to transmit the fusion-triggering signal. Nat Struct Mol Biol 18 (2):128–134. https://doi.org/10.1038/nsmb.1967

21. Hanauer JR, Gottschlich L, Riehl D et al (2016) Enhanced lysis by bispecific oncolytic measles viruses simultaneously using HER2/neu or EpCAM as target receptors. Mol Ther Oncolytics 3:16003. https://doi.org/10.1038/mto.2016.3
22. Friedrich K, Hanauer JR, Prüfer S et al (2013) DARPin-targeting of measles virus: unique bispecificity, effective oncolysis, and enhanced safety. Mol Ther 21(4):849–859. https://doi.org/10.1038/mt.2013.16
23. Hammond AL, Plemper RK, Zhang J et al (2001) Single-chain antibody displayed on a recombinant measles virus confers entry through the tumor-associated carcinoembryonic antigen. J Virol 75(5):2087–2096. https://doi.org/10.1128/JVI.75.5.2087-2096.2001
24. Nakamura T, Peng K-W, Harvey M et al (2005) Rescue and propagation of fully retargeted oncolytic measles viruses. Nat Biotechnol 23(2):209–214. https://doi.org/10.1038/nbt1060
25. Watanabe M, Hirano A, Stenglein S et al (1995) Engineered serine protease inhibitor prevents furin-catalyzed activation of the fusion glycoprotein and production of infectious measles virus. J Virol 69(5):3206–3210
26. Shengqing Y, Kishida N, Ito H et al (2002) Generation of velogenic Newcastle disease viruses from a nonpathogenic waterfowl isolate by passaging in chickens. Virology 301(2):206–211
27. Nagai Y, Klenk HD, Rott R (1976) Proteolytic cleavage of the viral glycoproteins and its significance for the virulence of Newcastle disease virus. Virology 72(2):494–508
28. Maisner A, Mrkic B, Herrler G et al (2000) Recombinant measles virus requiring an exogenous protease for activation of infectivity. J Gen Virol 81(Pt 2):441–449. https://doi.org/10.1099/0022-1317-81-2-441
29. Springfeld C, von Messling V, Frenzke M et al (2006) Oncolytic efficacy and enhanced safety of measles virus activated by tumor-secreted matrix metalloproteinases. Cancer Res 66(15):7694–7700. https://doi.org/10.1158/0008-5472.CAN-06-0538
30. Mühlebach MD, Schaser T, Zimmermann M et al (2010) Liver cancer protease activity profiles support therapeutic options with matrix metalloproteinase-activatable oncolytic measles virus. Cancer Res 70(19):7620–7629. https://doi.org/10.1158/0008-5472.CAN-09-4650
31. Funke S, Schneider IC, Glaser S et al (2009) Pseudotyping lentiviral vectors with the wild-type measles virus glycoproteins improves titer and selectivity. Gene Ther 16(5):700–705. https://doi.org/10.1038/gt.2009.11
32. Reyes-del Valle J, Hodge G, McChesney MB et al (2009) Protective anti-hepatitis B virus responses in rhesus monkeys primed with a vectored measles virus and boosted with a single dose of hepatitis B surface antigen. J Virol 83(17):9013–9017. https://doi.org/10.1128/JVI.00906-09
33. Radecke F, Spielhofer P, Schneider H et al (1995) Rescue of measles viruses from cloned DNA. EMBO J 14(23):5773–5784
34. Schneider H, Spielhofer P, Kaelin K et al (1997) Rescue of measles virus using a replication-deficient vaccinia-T7 vector. J Virol Methods 64(1):57–64
35. Martin A, Staeheli P, Schneider U (2006) RNA polymerase II-controlled expression of antigenomic RNA enhances the rescue efficacies of two different members of the Mononegavirales independently of the site of viral genome replication. J Virol 80(12):5708–5715. https://doi.org/10.1128/JVI.02389-05
36. Nakamura T, Peng K-W, Vongpunsawad S et al (2004) Antibody-targeted cell fusion. Nat Biotechnol 22(3):331–336. https://doi.org/10.1038/nbt942
37. Calain P, Roux L (1993) The rule of six, a basic feature for efficient replication of Sendai virus defective interfering RNA. J Virol 67(8):4822–4830
38. Mrkic B, Pavlovic J, Rulicke T et al (1998) Measles virus spread and pathogenesis in genetically modified mice. J Virol 72(9):7420–7427
39. Cathomen T, Buchholz CJ, Spielhofer P et al (1995) Preferential initiation at the second AUG of the measles virus F mRNA: a role for the long untranslated region. Virology 214(2):628–632. https://doi.org/10.1006/viro.1995.0075
40. Cattaneo R, Rose JK (1993) Cell fusion by the envelope glycoproteins of persistent measles viruses which caused lethal human brain disease. J Virol 67(3):1493–1502
41. Kärber G (1931) Beitrag zur kollektiven Behandlung pharmakologischer Reihenversuche. Arch Exp Pathol Pharmakol 162(4):480–483. https://doi.org/10.1007/BF01863914

Chapter 5

Insert Stability and In Vivo Testing of MicroRNA-Detargeted Oncolytic Picornaviruses

Autumn J. Schulze

Abstract

Although a variety of oncolytic viruses under clinical investigation have proven to be safe, the overall efficacy of oncolytic viruses as monotherapies has been suboptimal. While responses to combination therapies are much more promising, the development of oncolytic virus monotherapies with enhanced potency is imperative. With this initiative comes the need for improved mechanisms of virus targeting to prevent off-target toxicities. MicroRNA-detargeting has emerged as an invaluable tool for preventing unwanted toxicities of oncolytic viruses, particularly for picornaviruses. Here we describe methods to test the genetic stability of microRNA response elements in vitro and to evaluate the detargeting efficiency and therapeutic index of a microRNA-detargeted picornavirus in vivo. Although the methods described herein are specific to picornaviruses, microRNA-detargeting and these assays can be adapted for different classes of oncolytic viruses.

Key words Oncolytic picornavirus, MicroRNA, Targeting, Virotherapy, Genetic stability

1 Introduction

The *Picornaviridae* is a family of icosahedral viruses with positive-sense, single-stranded RNA genomes roughly 7–9 kb in length. These viruses infect a wide range of species and cause a variety of infections ranging from mild febrile illness to severe diseases of the heart, muscle, liver, and central nervous system. Several picornaviruses have been developed as oncolytic viruses including Coxsackievirus A21 (CVA21) [1–3], Mengovirus [4], a chimeric poliorhinovirus (PVSRIPO) [5–7], Seneca Valley Virus (SVV-001) [8], and a nonpathogenic melanoma-adapted enteric cytopathic human orphan type 7 virus (Rigvir) [9]. While some of these viruses are nonpathogenic in humans and animals, others can manifest serious clinical disease particularly in immune compromised hosts. Not only can this hinder preclinical evaluation of these viruses, as some only replicate in human cells and therefore can only be analyzed in xenogeneic models, but it also increases the risk for

Christine E. Engeland (ed.), *Oncolytic Viruses*, Methods in Molecular Biology, vol. 2058,
https://doi.org/10.1007/978-1-4939-9794-7_5, © Springer Science+Business Media, LLC, part of Springer Nature 2020

immunocompromised patients receiving these treatments. To ensure human pathogens do not cause disease in patients, clinical protocols are often designed to boost antiviral immunity prior to treatment which can diminish the oncolytic capacity of the therapy. Therefore, targeting mechanisms to prevent toxicities in patients prior to the development of antiviral immunity or in immunocompromised hosts are needed to improve the potency and safety of these virotherapies.

MicroRNA-detargeting has been demonstrated as an exceptional method for improving the therapeutic index of oncolytic picornaviruses as well as other classes of oncolytic viruses [10–13]. MicroRNAs are small noncoding RNAs that bind complementary target sequences in messenger RNAs (mRNAs) resulting in transcript degradation or translational silencing and subsequent regulation of cellular gene expression profiles [14–18]. This machinery can be exploited by incorporating target sequences into viral genomes such that in normal tissues expressing the microRNAs viral replication is suppressed preventing virulence, while in tumor cells that have aberrant expression of the microRNA (s) viral replication is left unhindered [10, 11, 19, 20]. To regulate picornaviral tropism, perfectly complementary microRNA response elements corresponding to microRNAs enriched within tissues of interest are incorporated into the viral genome targeting them for destruction in those tissues. These response elements generally consist of tandem repeats of the target sequences separated by short 4–6 nucleotide linkers and are commonly inserted within the noncoding regions (NCR) of the viral genome. This technique can theoretically be applied to all virus classes; however, it is particularly useful for targeting picornaviruses because of their limited carrying capacities that prevents the use of other targeting methods. We and others have successfully detargeted several picornaviruses using this method that maintain potent oncolytic activity and/or immunogenic potential in both xenogeneic and syngeneic mouse tumor models [2–4, 20, 21]. The continued improvement of this targeting mechanism over the past decade has brought the clinical translation of microRNA-detargeted oncolytic picornaviruses within sight.

In the following sections, we describe the methods employed in our laboratory for evaluating the genetic stability of microRNA target sequences during serial passaging in vitro, and determining the capacity of the response element to regulate viral tropism and enhance viral safety in vivo. We begin with a brief description of things to consider when designing microRNA response elements. For a comprehensive description of construction, rescue, amplification, titration, and in vitro characterization of microRNA-detargeted picornaviruses the reader is referred to reference [22].

2 Materials

2.1 Evaluating Response Element Genetic Stability In Vitro

2.1.1 Serial Passaging of MicroRNA-Detargeted Virus

1. Standard tissue/cell culture equipment, materials, and facility.
2. Six-well tissue culture treated plates.
3. Complete growth media for cell lines of interest.
4. Virus stocks of microRNA-detargeted and unmodified viruses.
5. Opti-MEM I Reduced Serum Medium. Store medium at 4 °C.
6. Sterile cell scrapers.
7. Cryogenic storage tubes.
8. Liquid nitrogen for freeze–thaw cycling.
9. 0.22 μm syringe filters.
10. Certified RNase and DNase-free pyrogen-free microcentrifuge tubes.
11. Standard tabletop microcentrifuge.

2.1.2 Viral RNA Extraction

1. Viral RNA Isolation Kit.
2. D/RNase surface decontamination spray.
3. 96–100% ethanol.
4. Certified RNase- and DNase-free pyrogen-free microcentrifuge tubes.
5. Nuclease free sterile water.

2.1.3 Amplification of MicroRNA Response Element Insert Region

1. Oligonucleotide primers flanking the insert region for amplification of the response element. Standard primer design guidelines for RT-PCR should be followed.
2. High Fidelity One-Step RT-PCR System.
3. Optional: Standard High-Fidelity polymerase chain reaction (PCR) reagents for nested PCR.
4. Thin-walled 0.2 mL PCR tubes.
5. Standard thermocycler.
6. Standard reagents and apparatus for agarose gel electrophoresis and transilluminator.

2.1.4 Gel Purification of RT-PCR Amplicons

1. Gel Extraction kit.
2. Sterile disposable razor blades.
3. Standard analytical weighing balance and heating block for processing agarose gel slices.
4. Optional: 3 M sodium acetate, pH 5.0.
5. Isopropanol.
6. DNA sequencer or sequencing facility.

2.2 Evaluating the Capacity of a MicroRNA Response Element to Control Tissue Tropism

1. Virus stocks of microRNA-detargeted virus.
2. Syringes.
3. Sterile dissection kit.
4. Standard reagents and equipment for humane animal euthanasia that adhere to all institutional animal care and use committee regulations.
5. 70% ethanol.
6. 5 mL polystyrene round bottom tubes.
7. PBS or Opti-MEM.
8. Tissue homogenizer or disposable tissue pellet pestles with microtubes.
9. Liquid nitrogen.
10. 96-well tissue culture treated plates for virus titration.

2.3 Evaluating the Therapeutic Index of a MicroRNA-Detargeted Virus In Vivo

1. Virus stocks of unmodified and microRNA-detargeted viruses.
2. Opti-MEM.
3. Handheld calipers.
4. Electronic scale and weighing container.
5. Blood collection tubes.
6. Lancets.
7. 70% ethanol.
8. Sterile dissection equipment.
9. Liquid nitrogen.

3 Methods

The design of the microRNA response element and its configuration within the viral genome is critical not only for successful targeting, but also for maintaining oncolytic potency. While some studies have shown that incorporating microRNA targets within the coding sequence of viruses can improve genetic stability, this is not always feasible [23]. The majority of microRNA-detargeted picornaviruses have response elements incorporated into the NCRs [4, 10, 20, 21]. MicroRNA targets are most commonly found within the 3′ NCRs of mRNAs. Picornaviruses can be successfully detargeted by encoding response elements within the 3′ NCR, however we have found that incorporation within the 5′ NCR can result in more efficient targeting and enhanced genetic stability while minimizing effects on viral replication [2, 24]. A commonality among RNA viruses is their dependence upon specific 3D fold(s) throughout the viral genome for functionality and optimized replication. The 3′ NCRs of picornaviruses are predicted

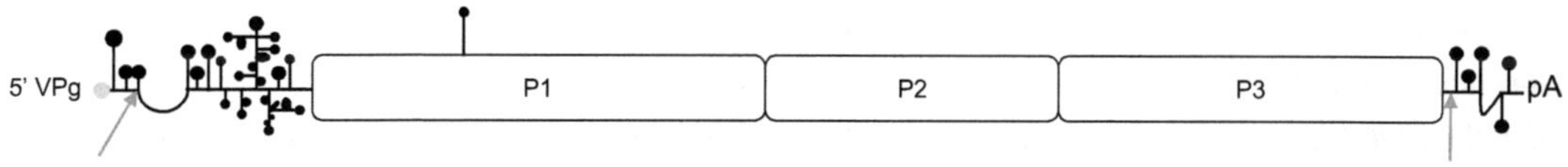

5′ NCR Insert:
miRT-124(2x) GGCATTCACCGCGTGCCTTAgcggccgcGGCATTCACCGCGTGCCTT

3′ NCR Insert:
miRT-133/208(2x) ACAGCTGGTTGAAGGGGACCAAAcgatACAGCTGGTTGAAGGGGACCAAActggagACAAGCTTTTTGCTCGTCTTATtcacACAAGCTTTTTGCTCGTCTTAT

Fig. 1 Schematic representation of predicted RNA secondary structures within the 5′ and 3′ NCRs of Mengovirus. MicroRNA response elements incorporated into a successfully detargeted virus (vMC$_{24}$NC) are listed. MicroRNA response element insertion sites are denoted by gray arrows. Modified from ref. 4

to contain folds that act as *cis*-replication elements. Our studies with Mengovirus and CVA21 have shown that insertion within the 3′ NCR can significantly alter the formation and/or functionality of these elements, particularly during the rescue of these viruses from in vitro-derived RNA transcripts [2]. Therefore, when determining response element insertion sites we recommend compiling as much information on RNA structural elements within the viral genome as possible. Since experimental determination of RNA structure is costly and often unfeasible we recommend performing RNA folding predictions using publicly available software to inform the design. A schematic representation of the secondary RNA structural elements within the NCRs of Mengovirus are shown in Fig. 1. Our studies have shown that incorporation of response elements prior to the poly(C) region in the 5′ NCR and upstream of the first stem loop within the 3′ NCR of Mengovirus have resulted in the least disruption of virus recovery from in vitro-derived RNA transcripts encoding the viral genome [4].

Previous studies have demonstrated that decreasing the homology of the direct repeats that flank a foreign insert improve its genetic stability [25]. This principle is likely applicable to microRNA response elements as well. Seamless cloning of the microRNA response element may increase the stability compared to insertion using a single restriction enzyme site. Additionally, using different microRNA targets in the response element as opposed to direct repeats of one or two sequences may elicit even better stability, however we have not formally tested this hypothesis. While trial and error is essential when optimizing microRNA-detargeting of a viral genome, rational design of the response element and its configuration within the genome are invaluable for saving time and effort.

3.1 Evaluating Response Element Genetic Stability In Vitro

3.1.1 Serial Passaging of MicroRNA-Detargeted Virus

1. Controls include mock infection and unmodified virus infection (*see* **Notes 1–2**).
2. Identify a cell line expressing microRNAs at levels similar to the tissue of interest (*see* **Notes 3–5**).
3. Plate cells in a 6-well plate such that they are 80% confluent at the time of infection (~24 h postseeding). Plate cells in 2 mL per well of complete growth media.
4. Dilute virus stocks in Opti-MEM to a concentration of 1 multiplicity of infection (MOI) per mL.
5. Remove the complete media from plated cells and wash wells two times by adding 0.5 mL of Opti-MEM per well, rocking gently and then aspirating. Take care not to disrupt the cells while aspirating.
6. Add 1 mL of virus dilution per well and incubate at 37 °C for 2 h.
7. Remove unincorporated virus by aspirating the media from all wells and wash two times by adding 0.5 mL complete growth media per well, rocking gently and then aspirating. Be sure to switch tips on aspirator between wells to avoid cross contamination.
8. Add 1.5 mL of complete growth media to each well and incubate at 37 °C for 24 h (*see* **Note 6**).
9. Transfer 1 mL of the media in each well to a cryogenic storage tube. Scrape the cells into the remaining supernatant by gently moving a rubber scraper across the entire well. Switch scrapers between wells to avoid cross contamination. Transfer the supernatant mixture to the corresponding cryogenic storage tube containing the original 1 mL of supernatant. Make sure not to splash supernatant from one well into another during scraping resulting in cross contamination.
10. Subject the samples to three freeze–thaw cycles in liquid nitrogen and remove the cellular debris by centrifuging the samples at 1500 × *g* for 10 min at 4 °C and passing the clarified lysate through a 0.22 μm syringe filter.
11. Transfer the cleared supernatant to a clean, sterile cryogenic tube. Samples can be stored at −80 °C at this point.
12. Mix 40% of the clarified lysate (0.6 mL) with 1.2 mL of Opti-MEM (1 volume of lysate to 2 volumes of media). Reserve 140 μL of clarified lysate for RNA analysis. Store clarified lysates at −80 °C until all samples have been collected.
13. Remove media from fresh cells plated in a 6-well plate at 80% confluence and wash two times by adding 0.5 mL Opti-MEM per well, rocking gently and then aspirating.

14. Add the 1.8 mL of diluted clarified lysate per well and incubate at 37 °C for 2 h.
15. Remove unincorporated virus by aspirating the media from all wells and wash two times by adding 0.5 mL complete growth media per well, rocking gently and then aspirating.
16. Add 1.5 mL of complete growth media per well and incubate at 37 °C for 24 h (*see* **Note 6**).
17. Continue passaging of virus for ten passages. For cytopathic viruses that are efficiently detargeted, passaging can generally be discontinued if complete cytopathic effects arise and samples tested prior to passage 10.

3.1.2 Viral RNA Extraction

1. Isolate viral RNA from cleared lysates using the Viral RNA isolation kit according to the manufacturer's instructions (*see* **Notes 7–10**). Standard practices for generating and maintaining a nuclease-free environment should be used for all steps.
2. Elute viral RNA with sterile nuclease-free water.
3. Purified viral RNA can be stored at −15 to −30 °C or −80 °C for up to 1 year.

3.1.3 Amplification of MicroRNA Response Element Insert Region

The methods described for amplification of response element regions use a one-step RT-PCR protocol to minimize pipetting error and potential for contamination (*see* **Note 11**). A two-step procedure can be used as well, particularly if the PCR steps need to be optimized.

1. Prepare a master mix for each insert region. The volume of each master mix should be roughly 110% of the volume required for all samples including no template controls. The reagents should be added to the master mix in the order listed from top to bottom. The master mix includes the following per reaction:

Component	Volume	Final concentration
Nuclease-free H_2O	Up to 45 μL	
10× PCR buffer	5 μL	1×
25 mM $MgCl_2$	3 μL	3 mM
Deoxynucleotide mix	2 μL	400 μM each dNTP
RNase inhibitor	1 μL	0.4 U/μL
Sense primer (10 μM)	2 μL	0.4 μM
Anti-sense primer (10 μM)	2 μL	0.4 μM

(continued)

Component	Volume	Final concentration
eAMV-RT	1 μL	0.4 U/μL
JumpStart AccuTaq LA DNA polymerase	1 μL	0.05 U/μL
Total volume	45 μL	

2. Add 5 μL of viral RNA to a thin-walled 0.2 mL PCR tube (*see* **Note 12**). If the volume of viral RNA added is altered, adjust the volume of water accordingly for a total volume of 50 μL per reaction.
3. Add 45 μL of master mix per tube making sure to switch tips between tubes to avoid cross contamination (*see* **Note 13**).
4. Close the caps on the PCR tubes and place tubes in the thermocycler.
5. The following thermocycling parameters are recommended as a starting condition, however like all PCRs these may need to be optimized for the specific primers used.

Step	Temperature, °C	Time
First strand synthesis	50	50 min
RT inactivation and initial denaturation	94	2 min
35 Cycles:		
Denaturation Annealing Elongation	94 55[a] 68	15 s 30 s 1 min
Final extension	68	5 min
Storage	4	Hold

[a]Dependent upon primer Tm

6. Separate the amplicons by agarose gel electrophoresis (Fig. 2) using an appropriately sized DNA marker (*see* **Note 14**).

3.1.4 Gel Purification of RT-PCR Amplicons

1. Visualize the amplicons using a transilluminator and excise the DNA band using a sterile, disposable blade. Be sure to minimize the size of the gel by removing extra agarose (*see* **Note 15**). Transfer the gel slice into a preweighed clean microcentrifuge tube.
2. Purify the PCR products using a gel extraction kit (*see* **Notes 16** and **17**).
3. Elute amplicons in sterile, nuclease-free water.
4. Eluted RT-PCR amplicons can be stored at −20 °C.

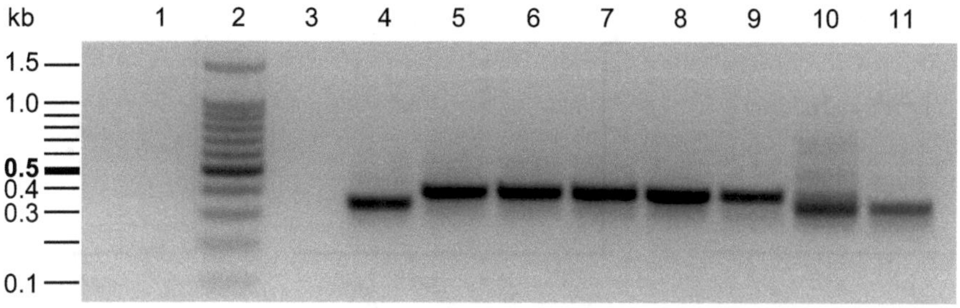

Fig. 2 RT-PCR analysis of viral RNA amplified from differentiated TE-671 (dTE-671) cells infected with a microRNA-detargeted CVA21. miRT-CVA21 was passaged in dTE-671 cells for a total of seven passages. A 394 nt-long fragment, spanning the microRNA response element insertion site, was amplified by nested PCR. Ten microliters of each reaction was separated on a 1.5% TAE agarose gel containing ethidium bromide and visualized with a UV-transilluminator. Lane 1: no template control. Lane 2: 100 bp DNA marker. Lane 3: mock infection negative control. Lane 4: unmodified CVA21 infection. Lanes 5–11: miRT-CVA21 passage 1–7. Reversion mutants (confirmed by sequencing) are observed in lanes 10 and 11

5. Sequence amplicons using the same primers used for amplification (or nested primers) and standard Sanger Sequencing methods (*see* **Note 18**).

3.2 Evaluating the Capacity of a MicroRNA Response Element to Control Tissue Tropism

The ability of the microRNA-detargeting to restrict virus replication in specific tissues can be analyzed by injecting the virus directly into the tissue of interest or via a clinically relevant route. If the desired effect is to eliminate unwanted toxicities, the model and dose used should recapitulate these toxicities when administered unmodified virus. Clinical signs of toxicity can be observed following intracranial injection of an attenuated, poly(C)-truncated Mengovirus (vMC_{24}) in 4–5 week old C57BL/6 mice, with peak virus replication 4–5 days postinjection. Insertion of a microRNA response element comprised of two repeat sequences complementary to miR-124, known to be enriched in the brain, can reduce neurovirulence. Therefore, to test the targeting efficacy, unmodified, control microRNA-targeted and miR-124-targeted viruses were administered through intracranial injection and viral titers within the brain, spinal cord and heart determined 4 days postinjection (Fig. 3).

1. Inject microRNA-detargeted virus at a dose known to cause toxicity for the unmodified virus or at a therapeutic dose. If necessary determine the LD_{50} of the unmodified virus via the specified route by conducting a dose escalation study.
2. Harvest tissues when peak virus replication is expected (generally 4–6 days post-virus injection). Tissues collected should include both target and off-target tissues for comparative analysis of titer (*see* **Note 19**).
3. Euthanize mice according to guidelines set forth and approved by your institutional animal care and use committee. The

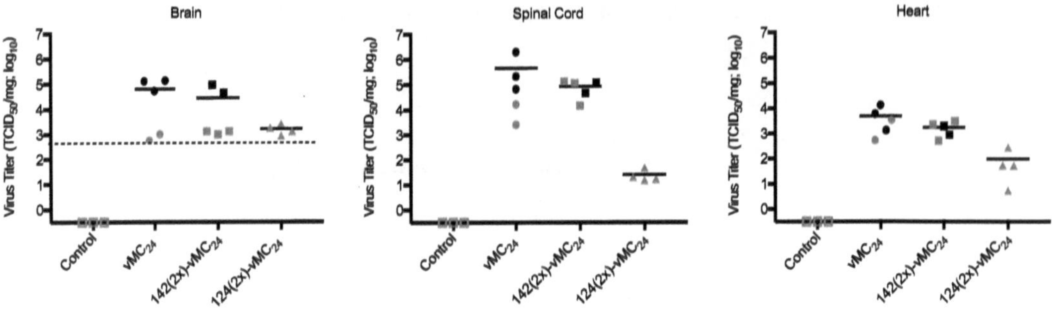

Fig. 3 Analysis of microRNA response element regulation of tissue tropism. Modified from ref. 4. C57BL/6 mice were injected intracranial (IC) with Opti-MEM (control $n = 3$) or 1×10^5 $TCID_{50}$ of vMC_{24} (unmodified virus, $n = 5$), 142(2×)-vMC_{24} (irrelevant control targeted virus, $n = 5$), or 124(2×)-vMC_{24} (neuro-detargeted virus, $n = 5$). Four days postinjection, all mice were euthanized and the infectious viral titers in the brain, spinal cord and heart determined. Black symbols denote mice that developed hind-limb paralysis. The dotted line in brain tissue graph represents the theoretical maximum input virus recovery and is based on the average brain weight of all mice analyzed. Individual viral titers are plotted along with the mean viral load (solid line). Reduced viral titers were observed in all tissues from mice treated with neuro-detargeted 124(2×)-vMC_{24}

reader is referred to references [26–28] for guidance on euthanasia recommended by the International Council for Laboratory Animal Science.

4. Drench mice in 70% ethanol prior to dissection to minimize the risk of contamination.
5. Dissect and weigh whole organs using sterilized dissection equipment. Tissues can be flash frozen in liquid nitrogen and stored at −80 °C for titration at a later date.
6. Place the tissue in a 5 mL polystyrene round-bottom falcon tube and add 1 mL of PBS or Opti-MEM media. Homogenize the tissue using a tissue homogenizer. If a tissue homogenizer is unavailable, homogenization using a mortar and pestle can also be done.
7. Subject the homogenates to three freeze–thaw cycles in liquid nitrogen.
8. Pellet cell debris by centrifuging at 10,000 × *g* for 15–30 min at 4 °C.
9. Titrate virus in cleared supernatant on susceptible cell line (virus producer cell line not expressing cognate microRNAs). For a full protocol of virus titration the reader is referred to ref. 22. For Mengovirus and CVA21 our lab quantifies infectious virus particles by determining the 50% tissue culture infectious dose ($TCID_{50}$) in H1-HeLa cells. Viruses can also be quantified by viral plaque assay.

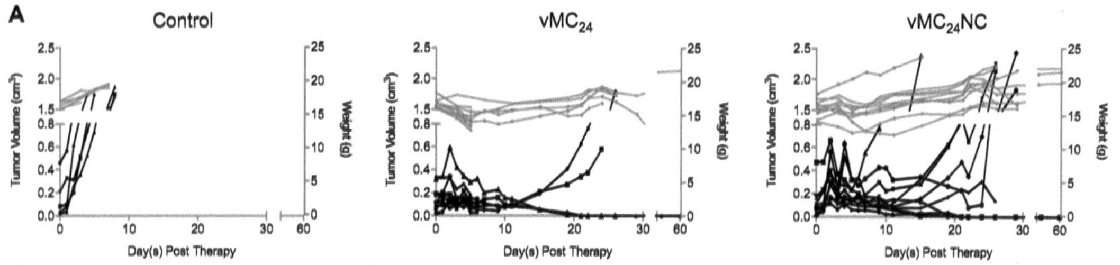

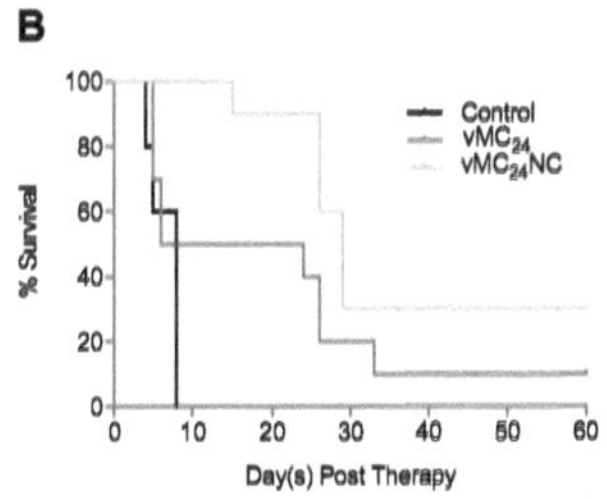

C

	Control	vMC_{24}	$vMC_{24}NC$
HLP	0/5	7/10	2/10
Median Survival	8 d	15 d	29 d
p vs Ctrl (overall)		0.1709	<0.0001
p vs vMC_{24} (1st 10 d)	0.1709		0.0116

Fig. 4 Comparative analysis of therapeutic index between unmodified and microRNA-detargeted virus. Modified from ref. 4. BALB/c mice bearing subcutaneous (SC) MPC-11 tumors were treated with a single intratumoral (IT) injection of 1×10^6 $TCID_{50}$ of vMC_{24} or $vMC_{24}NC$. (**a**) Disease burden was monitored by calculating tumor volume versus time using repeat caliper measurements (black lines). Weight was monitored for the duration of the experiment (gray lines). (**b**) Overall survival of control Opti-MEM-treated ($n = 5$), vMC_{24}-treated ($n = 10$), and $vMC_{24}NC$-treated ($n = 10$) mice was assessed using Kaplan-Meier survival curves. (**c**) Tabulation of mice that developed hind-limb paralysis (HLP), median survival, significance of overall survival benefit in treated versus control mice, and significance of survival benefit during the first 10 days following treatment in mice versus vMC_{24} treated mice. Virus-mediated toxicity is generally observed during the first 10 days. Significance in survival benefit was based on log-rank analyses

3.3 Evaluating the Therapeutic Index of a MicroRNA-Detargeted Virus In Vivo

Ideally, an oncolytic virus will have the ability to replicate in both animal and human cells so as to facilitate significantly relevant preclinical analysis of the therapeutic index of the virus. In these cases, studies should be conducted both in syngeneic and xenogeneic tumor models. In animals with intact immune systems, the safety of the microRNA-detargeted virus in the presence of a tumor may be skewed by the additional support of the antiviral immune response. Therefore, conducting a similar analysis in an immunocompromised host will provide the most in-depth evaluation of the therapeutic index (*see* **Note 20**). We have shown that dual-detargeting of an attenuated, poly(C)-truncated Mengovirus (vMC_{24}) by incorporating response elements complementary to miR-124 (neuro-enriched), miR-133 and miR-208 (muscle/cardiac-enriched) enhances the therapeutic index of the virus (Fig. 4).

3.3.1 Tumor Implantation, Virus Treatment, and Data Collection

1. Mice can be injected subcutaneously in the flanks with 1×10^5 to 1×10^7 cells (>95% viable) resuspended in Opti-MEM media. The number of cells is dependent upon the tumor cell line. Our lab routinely injects 5×10^6 MPC-11 (syngeneic mouse myeloma model) and Mel-624 (xenogeneic human

melanoma model) cells to establish tumors for Mengovirus and CVA21 studies, respectively.

2. Once tumors have reached a mean diameter of 0.5 cm, virus can be injected either IT or IV.
3. Tumor volumes should be measured at regular intervals with handheld calipers and the volume determined using the formula volume = (length × (width)^2)/2.
4. Weights should also be determined at regular intervals and observations on clinical manifestations of toxicity recorded. Comparisons in overall survival can be determined using log-rank statistics (*see* **Note 21**).

3.3.2 Evaluating Viremia

Intravenous administration of 1 × 10^7 $TCID_{50}$ of a dual-detargeted, poly(C)-truncated Mengovirus ($vMC_{24}NC$) to MPC-11 tumor bearing mice results in high viremia 2 days post-treatment (Fig. 5).

1. Collect blood from mice 1–2 days post-virus injection in a blood collection tube. Our lab routinely uses the facial vein and submandibular vein bleed techniques to obtain blood for titration of virus in both plasma and serum [29]. Anticoagulants other than heparin are recommended if RNA isolation will also be conducted.
2. Invert plasma collection tube 8–10 times to mix with anticoagulant or allow blood to clot for 30 min at room temperature for serum tubes.
3. Centrifuge the tube at 7000 × *g* for 10 min to separate plasma/serum.
4. Pipet plasma/serum into a sterile cryogenic storage tube. Samples can be stored at −80 °C for titration at a later date.

3.3.3 Tissue Analysis

Intravenous administration of 1 × 10^7 $TCID_{50}$ of a dual-detargeted, poly(C)-truncated Mengovirus ($vMC_{24}NC$) to MPC-11 tumor bearing mice results in high viral loads in tumor and significantly less virus in other tissues 4 days posttreatment (Fig. 6). Sequence analysis of virus within tissues/blood will help determine whether observed toxicities are due to the development of reversion mutants or if the microRNA-detargeting is insufficient to prevent toxicity.

1. At the time of euthanasia, tumor (if available) and other organs should be collected for virus titration and viral sequence analysis. Drench mice in 70% ethanol prior to dissection of tissues. Harvested tissues should be sectioned using sterilized dissection equipment and disposable blades. Clean instruments in between each tissue harvested and use a new blade for each sectioning. Sectioning of tissues should be kept consistent for

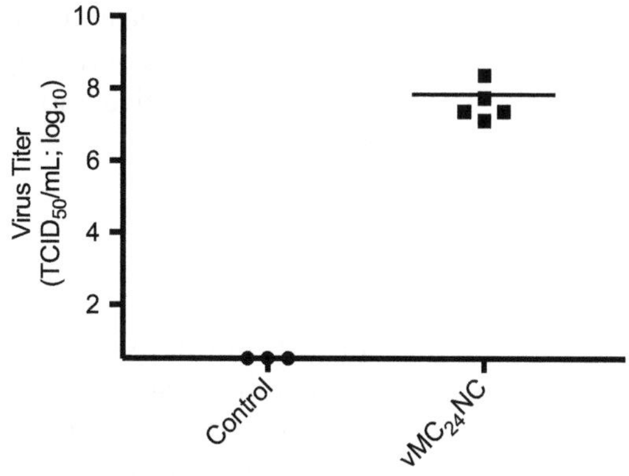

Fig. 5 Determination of viremia following IV treatment with microRNA-detargeted virus. Modified from ref. 4. BALB/c mice bearing s.c. MPC-11 tumors were treated with a single IV injection via the tail vein of Opti-MEM control or 1 × 10^7 $TCID_{50}$ $vMC_{24}NC$. Plasma viral loads in all mice were determined on day 2 posttherapy

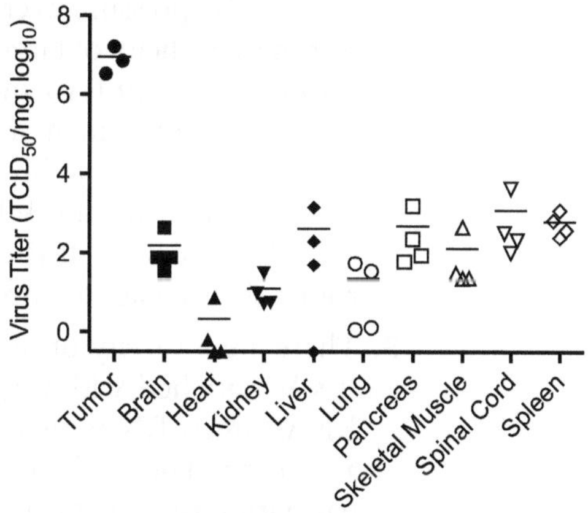

Fig. 6 Viral load in tissues following IV treatment with microRNA-detargeted virus. Modified from ref. 4. BALB/c mice bearing s.c. MPC-11 tumors were treated with a single IV injection via the tail vein of Opti-MEM control or 1 × 10^7 $TCID_{50}$ $vMC_{24}NC$. Four days post-virus treatment, four mice from each group were euthanized and viral loads within various tissues were determined. All tissues from control-treated mice were negative (data not shown). Note that one mouse per group did not have tumors at the time of harvest

all animals (i.e., the same regions of the tissue analyzed for virus titer among different animals). Tissue sections can be flash frozen in liquid nitrogen and stored at −80 °C for processing at a later date.

2. Tissues should be processed as described in Subheading 3.2, **steps 5–8**. Virus titer in the tissue should be determined as described in Subheading 3.2, **step 9** [22]. Viral RNA should be

extracted from cleared supernatant as described in Subheading 3.1.2. Regions containing the microRNA response elements should be amplified and sequenced as described in Subheading 3.1.3.

4 Notes

1. All institutional safety guidelines should be followed for the safe handling, storage, decontamination and disposal of materials that come into contact with the virus. For CVA21 and Mengovirus this includes working under biosafety level 2 conditions with appropriate personal protective equipment, disinfecting materials with a virucidal agent (e.g., Wescodyne) and disposal in biohazardous waste containers.
2. Conducting genetic stability evaluation in a cell line that does not express cognate microRNAs will allow for a comparative analysis of selective pressure exerted by microRNAs versus nonspecific pressure exerted by insertion. Another method to test the specificity of targeting and its role on the stability of the response element is to include a virus with a nonspecific microRNA response element in the same position as the experimental virus. miR-142 is highly expressed in hematopoietic cells and has been used routinely by our lab as an off-target control. A scrambled version of the microRNA response element is also routinely used as a control by many labs.
3. There are a variety of online databases to help identify microRNAs enriched within specific tissues, target genes of microRNAs, microRNAs that are dysregulated in different diseases, and so on. The reader is referred to reference [30] as a tool for navigating these databases to best fit your needs when designing microRNA response elements.
4. If a cell line expressing the desired microRNAs is not known the microRNA content can be quantified by qRT-PCR. This will require extraction of mature microRNAs from a panel of cells, reverse transcription of the microRNAs, and real-time quantification. Our lab routinely uses MirVana microRNA Isolation Kits (Ambion), Taqman microRNA Reverse Transcription Kits (Applied Biosystems), and TaqMan MicroRNA Assays (Applied Biosystems) for these steps, respectively. Generally, the expression levels can be standardized to miR-Let7a expressed in H1-HeLa cells as this concentration of microRNA is known to be sufficient for targeting oncolytic picornaviruses encoding miR-Let7a response elements.
5. If a cell line expressing the desired microRNAs is not available or difficult to culture (e.g., neuronal and cardiomyocyte cell

lines) it is suggested to test the stability in vivo. If in vitro analysis is desired there are protocols available to generate stable cell lines for constitutive microRNA expression. However, the reader is cautioned that artificial expression of a microRNA in a cell line will not always recapitulate the infection of normal cells as the global microRNA signatures are likely much different.

6. A single round of replication for a picornavirus is generally between 8 and 12 h. If the exact round of replication when response element mutation/reversion occurs is to be determined, samples should be collected around ~9 h postinfection.
7. Designation of a biological hood, pipets, tips, centrifuges and plastics that are for nuclease-free RNA work is preferred. Sample preparation and preparation for amplification and detection should be conducted in different areas. Make sure to use standard practices for generating a nuclease-free environment. Our laboratory routinely uses DNase/RNase-free surface decontaminant (Argos Technologies) for this purpose.
8. Our lab routinely uses the QIAamp Viral RNA kit for isolating viral RNA. For extremely low viral titer samples, up to 560 μL of clarified lysate can be processed using this kit by increasing the initial buffer volumes proportionally (AW1 and AW2 wash volumes remain the same) and loading the QIAamp Mini column multiple times. For even lower viral titers, virus can be concentrated through ultracentrifugation and resuspended in 140 μL of 1× PBS for processing.
9. Do not use denatured ethanol for RNA isolation as this can reduce RNA yield and purity.
10. The use of aerosol-barrier low binding tips is recommended for all steps.
11. Our lab routinely uses the Enhanced Avian HS RT-PCR kit (Sigma-Aldrich) for amplification of microRNA response element insert regions for sequencing since it contains the components necessary for both one-step and two-step procedures. However, for certain types of sequences, avian reverse transcriptase may lead to mispriming. One example of this is amplifying long poly(C) tracts. In these cases, we recommend using a murine reverse transcriptase.
12. The final concentration of template RNA in the one-step RT-PCR is recommended to be 2 pg/μL to 20 ng/μL. The majority of the RNA isolated will be carrier RNA (5.6 μg). The yield of viral RNA is normally less than 1 μg. In a volume of 60 μL this is 0.017 μg/μL. Therefore adding 5 μL of the isolated viral RNA prep equates to ~1.7 ng/μL of viral RNA and 11 ng/μL total RNA.

13. Preparing the no template control last is the best method for ensuring no contamination occurred during the setup process.
14. In some cases, nested PCR is needed to amplify a working concentration of product for sequencing. Primers should be located internally from the original primer set. Similar cycling parameters should be used without the 50 °C first strand synthesis step. Be sure to use a high-fidelity polymerase and maintain sterile technique. Mock infection controls and no template controls should be used since it is extremely easy to contaminate nested PCRs.
15. Minimize the amount of time the agarose gel is illuminated by UV light to avoid UV-light induced mutagenesis.
16. Be sure that the gel extraction kit used binds PCR products of the appropriate size. Our lab routinely uses the QIAquick Gel Extraction kit (Qiagen), which purifies DNA fragments ranging from 70 bp to 10 kbp.
17. Using nuclease-free water prewarmed to 50 °C can increase the DNA concentration eluted from a purification column.
18. Amplicons can be subcloned into a DNA vector for analysis of individual clones. As long as the polymerase used adds A overhangs onto the PCR amplicon, standard TA cloning techniques can be used. Taq polymerases are recommended for this.
19. Off-target tissue analysis is most beneficial when virus is administered intravenously (IV) or intraperitoneally (IP). Direct injection into a tissue where amplification is necessary to seed other organs will result in lower titers in off-target tissues because of decreased replication in the primary tissue versus inhibition within the off-target tissue itself.
20. Some picornaviruses do not replicate in mouse cells and therefore in vivo models that recapitulate human toxicities are not always available. Generation of transgenic mice expressing human receptors in the appropriate tissues is advised for toxicology studies on these types of viruses. Additionally, the use of immunocompromised animal models (IFNAR mice, SCID mice, etc.) is also highly recommended to determine the full potential of the detargeting configuration to inhibit toxicity.
21. GraphPad Prism software is routinely used in our lab to conduct log-rank statistical analysis to compare the survival distributions of mice treated with unmodified virus versus microRNA-targeted virus.

Acknowledgments

Al and Mary Agnes McQuinn, the Richard M. Schulze Foundation, an NIH Relief Grant from the Mayo Clinic, and Mayo Clinic funded representative work described here.

References

1. Bradley S, Jakes AD, Harrington K, Pandha H, Melcher A, Errington-Mais F (2014) Applications of coxsackievirus A21 in oncology. Oncolytic Virother 3:47–55. https://doi.org/10.2147/OV.S56322
2. Elsedawy NB, Russell SJ, Schulze AJ (2018) MicroRNA-targeted CVA21 infectious RNA is an alternative formulation for oncolytic virotherapy. 2018 International oncolytic virus conference (IOVC) – 11th meeting. Oxford
3. Elsedawy N, Ruiz A, Russell S (2016) Rational design of microRNA-retargeted coxsackievirus A21 infectious RNA for cancer therapy. Mol Ther 24:S29–S29. https://doi.org/10.1016/S1525-0016(16)32876-3
4. Ruiz AJ, Hadac EM, Nace RA, Russell SJ (2016) MicroRNA-detargeted mengovirus for oncolytic virotherapy. J Virol 90(8):4078–4092. https://doi.org/10.1128/JVI.02810-15
5. Gromeier M, Alexander L, Wimmer E (1996) Internal ribosomal entry site substitution eliminates neurovirulence in intergeneric poliovirus recombinants. Proc Natl Acad Sci U S A 93(6):2370–2375
6. Gromeier M, Nair SK (2018) Recombinant poliovirus for cancer immunotherapy. Annu Rev Med 69:289–299. https://doi.org/10.1146/annurev-med-050715-104655
7. Desjardins A, Gromeier M, Herndon JE II, Beaubier N, Bolognesi DP, Friedman AH, Friedman HS, McSherry F, Muscat AM, Nair S, Peters KB, Randazzo D, Sampson JH, Vlahovic G, Harrison WT, McLendon RE, Ashley D, Bigner DD (2018) Recurrent glioblastoma treated with recombinant poliovirus. N Engl J Med 379(2):150–161. https://doi.org/10.1056/NEJMoa1716435
8. Burke MJ (2016) Oncolytic Seneca valley virus: past perspectives and future directions. Oncolytic Virother 5:81–89. https://doi.org/10.2147/OV.S96915
9. Alberts P, Tilgase A, Rasa A, Bandere K, Venskus D (2018) The advent of oncolytic virotherapy in oncology: the Rigvir(R) story. Eur J Pharmacol 837:117–126. https://doi.org/10.1016/j.ejphar.2018.08.042
10. Ruiz AJ, Russell SJ (2015) MicroRNAs and oncolytic viruses. Curr Opin Virol 13:40–48. https://doi.org/10.1016/j.coviro.2015.03.007
11. Kelly EJ, Russell SJ (2009) MicroRNAs and the regulation of vector tropism. Mol Ther 17(3):409–416. https://doi.org/10.1038/mt.2008.288
12. Brown BD, Naldini L (2009) Exploiting and antagonizing microRNA regulation for therapeutic and experimental applications. Nat Rev Genet 10(8):578–585. https://doi.org/10.1038/nrg2628
13. tenOever BR (2013) RNA viruses and the host microRNA machinery. Nat Rev Microbiol 11(3):169–180. https://doi.org/10.1038/nrmicro2971
14. Wightman B, Ha I, Ruvkun G (1993) Posttranscriptional regulation of the heterochronic gene lin-14 by lin-4 mediates temporal pattern formation in C. elegans. Cell 75(5):855–862
15. Lee RC, Feinbaum RL, Ambros V (1993) The C. elegans heterochronic gene lin-4 encodes small RNAs with antisense complementarity to lin-14. Cell 75(5):843–854
16. Ambros V (2004) The functions of animal microRNAs. Nature 431(7006):350–355. https://doi.org/10.1038/nature02871
17. Bartel DP (2004) MicroRNAs: genomics, biogenesis, mechanism, and function. Cell 116(2):281–297
18. Bartel DP (2009) MicroRNAs: target recognition and regulatory functions. Cell 136(2):215–233. https://doi.org/10.1016/j.cell.2009.01.002
19. Brown BD, Gentner B, Cantore A, Colleoni S, Amendola M, Zingale A, Baccarini A, Lazzari G, Galli C, Naldini L (2007) Endogenous microRNA can be broadly exploited to regulate transgene expression according to tissue, lineage and differentiation state. Nat Biotechnol 25(12):1457–1467. https://doi.org/10.1038/nbt1372
20. Kelly EJ, Hadac EM, Greiner S, Russell SJ (2008) Engineering microRNA responsiveness to decrease virus pathogenicity. Nat Med 14

(11):1278–1283. https://doi.org/10.1038/nm.1776

21. Barnes D, Kunitomi M, Vignuzzi M, Saksela K, Andino R (2008) Harnessing endogenous miRNAs to control virus tissue tropism as a strategy for developing attenuated virus vaccines. Cell Host Microbe 4(3):239–248. https://doi.org/10.1016/j.chom.2008.08.003
22. Ruiz AJ, Russell SJ (2017) MicroRNA-based regulation of picornavirus tropism. J Vis Exp (120). https://doi.org/10.3791/55033
23. Kueberuwa G, Cawood R, Tedcastle A, Seymour LW (2014) Tissue-specific attenuation of oncolytic sindbis virus without compromised genetic stability. Hum Gene Ther Methods 25(2):154–165. https://doi.org/10.1089/hgtb.2013.202
24. Dyer A, Baugh R, Chia SL, Frost S, Iris JEJ, Khalique H, Pokrovska TD, Scott EM, Taverner WK, Seymour LW, Lei J (2018) Turning cold tumours hot: oncolytic virotherapy gets up close and personal with other therapeutics at the 11th oncolytic virus conference. Cancer Gene Ther. https://doi.org/10.1038/s41417-018-0042-1
25. Tang S, van Rij R, Silvera D, Andino R (1997) Toward a poliovirus-based simian immunodeficiency virus vaccine: correlation between genetic stability and immunogenicity. J Virol 71(10):7841–7850
26. Association AVM (2013) AVMA guidelines for the euthanasia of animals: 2013 edition, vol 8. American Veterinary Medical Association, Schaumburg, IL
27. Close B, Banister K, Baumans V, Bernoth EM, Bromage N, Bunyan J, Erhardt W, Flecknell P, Gregory N, Hackbarth H, Morton D, Warwick C (1996) Recommendations for euthanasia of experimental animals: part 1. DGXI of the European Commission. Lab Anim 30 (4):293–316. https://doi.org/10.1258/002367796780739871
28. Close B, Banister K, Baumans V, Bernoth EM, Bromage N, Bunyan J, Erhardt W, Flecknell P, Gregory N, Hackbarth H, Morton D, Warwick C (1997) Recommendations for euthanasia of experimental animals: part 2. DGXT of the European Commission. Lab Anim 31 (1):1–32. https://doi.org/10.1258/002367797780600297
29. JoVE Science Education Database (2019) Lab animal research. Blood withdrawal II. JoVE, Cambridge, MA
30. Lukasik A, Wojcikowski M, Zielenkiewicz P (2016) Tools4miRs – one place to gather all the tools for miRNA analysis. Bioinformatics 32(17):2722–2724. https://doi.org/10.1093/bioinformatics/btw189

Chapter 6

Ex Vivo Virotherapy with Myxoma Virus to Treat Cancer

Nancy Y. Villa, Lina S. Franco, and Grant McFadden

Abstract

Myxoma virus (MYXV) is a member of the *Poxviridae* family and the genus *Leporipoxvirus*. In nature MYXV tropism is restricted to lagomorphs, and is specifically pathogenic only for European rabbits (*Oryctolagus cuniculus*), in which this virus causes the lethal systemic disease called myxomatosis. Importantly, although MYXV cannot cause any disease pathology in humans, mice, or any other domestic animals other than rabbit, this virus can productively infect and kill a variety of human and murine cancer cells, from either primary sources or cultured cancer cell lines. Therefore, in the last decade, MYXV has emerged as a novel oncolytic virus against hematologic malignancies and various solid cancers. One novel aspect of MYXV virotherapy is a new systemic virus delivery strategy to cancer sites in the recipient, by which adsorption of the virus to isolated leukocytes is conducted prior to reinfusion of the virus-infected cells back into the recipient, via a procedure called ex vivo virotherapy (EVV, or simply EV2). The EV2 delivery strategy thus exploits the inherent migratory properties of leukocytes to ferry MYXV to tissue sites bearing cancer cells that are accessible to leukocyte chemotaxis. Here, we describe EV2 procedures with MYXV to systemically deliver the virus to sites of disseminated and/or metastatic cancer in situ via infected leukocytes derived from either bone marrow or peripheral blood.

Key words Myxoma virus, Cancer cells, Bone marrow, Peripheral mononuclear cells, Flow cytometry

1 Introduction

Cancer treatments are classically chemotherapy, radiotherapy, surgery, and immune rescue stem cell transplants, but recent new therapies have emerged such as immunotherapy, virotherapy, and adoptive cell therapy (ACT) [1, 2]. For many blood cancers, conventional treatments can include autologous stem cell transplantation (auto-SCT), or allogeneic stem cell transplantation (allo-SCT) for those patients eligible to receive autotransplants or for patients with HLA-matched eligible donors [3]. Among these new treatments, oncolytic virotherapy has started to gain importance and is based on the ability of some viruses to infect and kill preferentially cancer cells versus their nontransformed somatic cell counterparts [4]. Myxoma virus (MYXV) is a normally rabbit-specific poxvirus

Christine E. Engeland (ed.), *Oncolytic Viruses*, Methods in Molecular Biology, vol. 2058,
https://doi.org/10.1007/978-1-4939-9794-7_6,

that has emerged as a novel oncolytic agent to treat a variety of human and murine primary cancers [4].

Oncolytic viruses can be delivered topically (i.e., intratumoral injection) or else delivered systemically, for example by administering the virus intravenously. However, this type of systemic delivery of free virus is not efficient, mainly due to rapid clearance of the virus by the immune system and the circulatory filtration organs, and can even be dangerous for the recipient because the virus can become rapidly accumulated in spleen or liver [5]. The ex vivo virotherapy (EV2) method described here instead utilizes MYXV that has been adsorbed to leukocytes isolated from either the bone marrow (BM) or from peripheral blood (PB) for the cell-assisted delivery of the virus to disseminated cancer sites that are accessible to the migratory leukocytes bearing virus. This EV2 delivery system for the virus is mediated first by ex vivo infection of carrier cells (i.e., the full complement of immune cells derived from BM or PB), and then secondly the reinfusion of the virus-adsorbed leukocytes back into the cancer patient to deliver the virus by cellular chemotaxis into the various tissue sites of the tumor beds. For example, it was recently shown that ex vivo treatment of a donor BM allograft with MYXV that was subsequently infused into a mismatched mouse recipient bearing pre-seeded syngeneic multiple myeloma (MM) dramatically reduced tumor burden in recipient tissue sites as diverse as the bone marrow and the spleen [6].

Although the EV2 method described has demonstrated efficient delivery of MYXV to cancer cells in the bone marrow and spleen via virus-adsorbed BM cells, there are still some important questions to be addressed. First of all, it is important to establish any differences between EV2 using PBMCs or activated immune cells (i.e., T cells, or B cells) as the virus carriers. In a published in vitro study, for example, activated human T cells become productively infected with MYXV and can carry and deliver this virus to cultured human U266 MM cells, thereby inducing infection and killing of these target cancer cells [7]. Likewise, it remains unknown which leukocyte cell types are the most efficient cell carriers of MYXV, and whether specific cell types might exhibit optimal chemotaxis into tumor sites residing in specific tissues. In the future, the delivery of MYXV by carrier cells into specific tumor beds can be studied by tracking the migration of input BM cells or PBMCs (i.e., virus-carrying "donors") using radioisotopes, contrast agents, or fluorophores allowing the cells to be visualized in vivo through fluorescence imaging, magnetic resonance imaging, or positron emission tomography [8, 9]. These questions will help to better understand the scope and limitations of EV2 to better to improve the efficacy of delivery of the virus into specific classes of tumor beds.

2 Materials

As for all biological reagents, diligently follow all waste and biohazard disposal regulations when disposing virus, cell or tissue materials.

2.1 General Reagents

1. Well plates: 6- and 96-well plates.
2. Nanopure and autoclaved water.
3. 70% ethanol.
4. 1× PBS without magnesium and calcium.
5. 1× Hank's balanced salt solution (1× HBSS).
6. 1× fetal bovine serum (FBS).
7. Washing solution: 1× PBS + 2% FBS.
8. 2.5%–1.5% isoflurane.
9. D-Luciferin.
10. Monoclonal antibodies fluorescently tagged for flow cytometry.
11. Mouse FcR blocking reagent.
12. 1× RPMI-1640 medium.
13. 1× MEM medium.
14. 1× Trypsin +0.25% EDTA.
15. 1× GlutaMAX.
16. 1× L-glutamine.
17. 1× penicillin and streptomycin solution.
18. 0.8 mg/mL Geneticin (G418) solution.
19. 0.005% thioglycerol.
20. 1× nonessential amino acids (NEAAs).
21. Cell staining Buffer or fluorescence activated cell sorter (FACS) buffer [1× PBS + 2% FBS + 0.1% sodium azide (NaN_3)] (*see* **Note 1**).
22. 6.4% sodium citrate solution.
23. BD Cytofix fixation solution (*see* **Note 2**).
24. Ficoll-Paque Plus density gradient medium.
25. QuickGel Touch serum protein electrophoresis (SPE) kit containing gels, acid blue stain, QuickGel blotter C, citric acid destain, blade application kit, and electrophoresis chamber.
26. Prophylactic antibiotic for immunocompromised mice: 100 mg/mL Baytril 100 (enrofloxacin) solution.
27. Cancer cells: In our teaching example, we utilized murine multiple myeloma (MM) cells such as mineral-oil induced

plasmocytomas (MPOC) BM expressing the DsRed fluorescence tagged MM cells (MOPC315.BM.DsRed-MM cells) or MOPC315.BM expressing the firefly-luciferase tagged MM cells (MOPC315.BM.FLuc-MM cells), which were derived from a BALB/c mouse strain [10], or C57BL/6-derived VK*MYC MM cells (i.e., the bortezomib resistant VK12598, and the multidrug-resistant VK12563 cells). Melanoma murine B16F10 cell line expressing firefly luciferase [11] originally derived from C57BL/6 mouse strain.

28. Murine bone marrow (BM) and peripheral blood mononuclear cells (PBMCs) isolated from BALB/c, or C57BL/6 mouse strains.
29. Complete RPMI-1640 for growing and maintaining MOPC315.BM.DsRed MM cells: 1× RPMI-1640 supplemented with 20% FBS, 1% nonessential amino acids (NEAAs), 1× GlutaMAX, 0.8 mg/mL geneticin (G418), 0.005% thioglycerol.
30. Complete RPMI-1640 for growing and maintaining MOPC315.BM.FLuc MM cells: 1× RPMI-1640 supplemented with 20% FBS, 1% streptomycin, 1% L-glutamine and 0.005% thioglycerol.
31. Melanoma B16F10-FLuc cells are maintained in Dulbecco's MEM supplemented with 10% FBS, 1× L-glutamine, and 1× penicillin and streptomycin.
32. MYXV reporter constructs expressed either green fluorescence protein (GFP), or TdTomato fluorescent protein (TdTomato).
33. Mouse strains: BALB/c mice 6–8 weeks old, and/or C57BL/6 mice 8–10 weeks old.

2.2 Instruments and Materials

1. Water bath at 37 °C.
2. IVIS Spectrum in vivo imaging system.
3. Flow cytometer equipped with software.
4. Cell counter or hemacytometer.
5. Sterile hood with laminar flow.
6. Dissection forceps and scissors.
7. Incubator, humidified, preset to 37 °C and pre-equilibrated with 5% CO_2.
8. Microcentrifuge or table centrifuge.
9. 27 G½ and 22 G3/4 needles.
10. 1 mL or 3 mL syringes.
11. Vortexer.
12. 0.5 mL, 1.5 mL microcentrifuge tubes, 15-mL, 50 mL centrifuge tubes.

13. 40-μm 70-μm pore size nylon cell strainers.
14. Sonicator.
15. Standard microscope slides.
16. 25 cm^2, 75 cm^2, and 175 cm^2 cell culture flasks.
17. 60 mm tissue culture dish.
18. Gauze pads.
19. 5 mm and/or 4 mm sterile animal lancets.
20. Isoflurane induction chamber.
21. Microtainer serum separator tubes.

3 Methods

Unless indicated otherwise, all the in vivo and in vitro procedures are performed under sterile conditions. In the following sections, we describe an in vivo autologous cell transplantation procedure in combination with virotherapy with MYXV. In addition to this, we described direct in vitro virotherapy with MYXV against primary human samples with different stages of MM.

3.1 Cancer Implantation

BALB/c 6–8 -week-old or C57Bl/6 8–10-week-old mice are used for the experiments described in this book chapter. Before any procedure, mice are first acclimated in the animal facility for at least 1 week according to the Institutional Animal Care and Use Committee (IACUC) protocol. Depending on the cancer model, mice can be sublethally irradiated using a source of X-rays (i.e., 122 cGy for 1 min and 30 s), 24 h before cancer implantation. This sublethal irradiation allows better cancer engraftment for certain models (i.e., BALB/c derived MOPC315.BM.DsRed.MM, or MOPC315.BM.FLuc MM) (*see* **Note 3**). When irradiated, mice are treated with prophylactic antibiotics for 2 weeks (*see* **Note 4**) to avoid opportunistic infections in the immunocompromised mice. The antibiotics are administrated in the drinking water. For other models where immunocompetent mice are implanted with cancer cells, antibiotics are not required (i.e., melanoma or osteosarcoma mouse models).

The number of cancer cells pre-seeded in the animal depends on the cancer model. For example for establishing minimal residual disease (MRD) of MM, 1×10^5 cancer cells are seeded per mouse. For metastatic tumor models 5×10^5 to 2×10^6 cells are usually seeded in each recipient mouse. Briefly, before the cancer implantation, fresh or cryopreserved cancer cells in suspension such as MM and fresh melanoma cells are grown in appropriate medium.

Mice are implanted intravenously (i.v.) via tail vein with the appropriate concentration of cancer cells. In the case of MM, 7 days

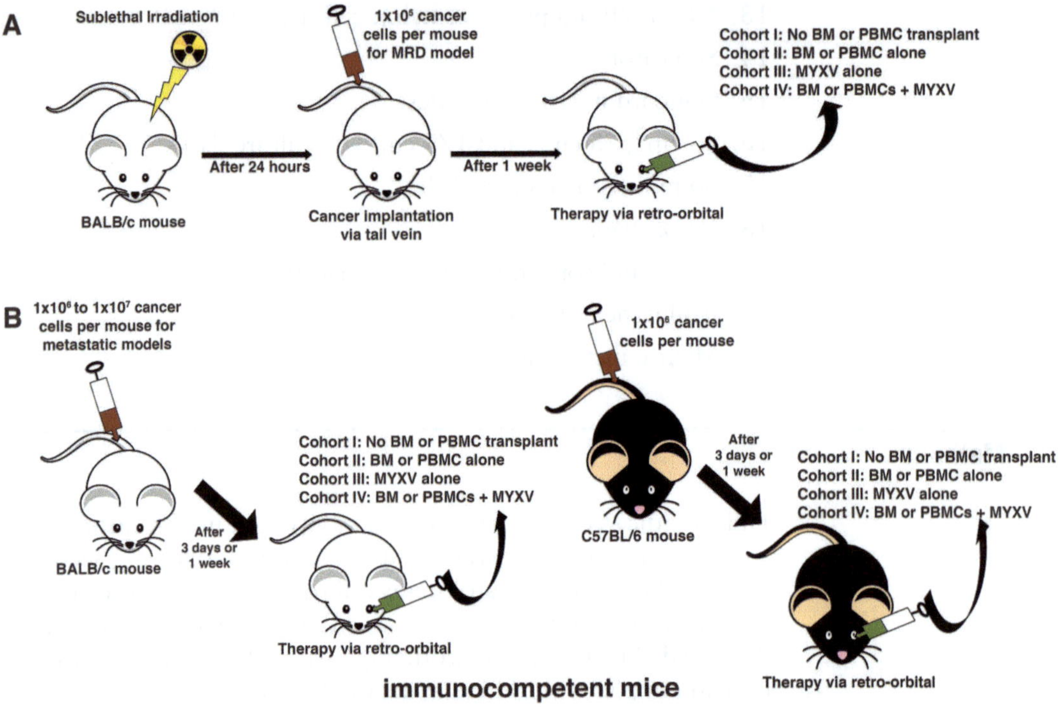

Fig. 1 In vivo cancer implantation and therapy regimen. (**a**) shows the procedure followed for cancer cell implantation in a sublethally irradiated BALB/c mouse. After 24 h, mice are implanted with the respective dose of cancer cells intravenously (i.v.) via tail vein. Depending of the cancer model, the therapy treatment can start 3–7 days after the cancer has been seeded. The route of delivery for the virus and/or leukocytes is via retro-orbital. (**b**) shows the experimental procedure used with immunocompetent mice. In brief, mice (i.e., BALB/c or C57BL/6 mouse strains) are implanted with the respective syngeneic cancer cells via tail vein. At 3–7 days after cancer cell seeding, mice are separated in different cohorts according to the treatment

after cancer implantation mice are transplanted via retro-orbital injection with 1× PBS (Cohort I, no BM cells, or PBMCs transplant but 1× PBS) bone marrow (BM) cells or PBMCs alone (Cohort II), naked MYXV virus systemically delivered (Cohort III), ex vivo mouse BM cells, or PBMCs preloaded with MYXV (Cohort IV). In the case of a solid tumor like melanoma, the mice are treated 3 days post-cancer implantation because this cancer model is very aggressive. Figure 1 summarizes the cancer implantation procedure and the therapy regimen for recipient mice preimplanted with murine cancer cells.

3.2 Isolation of Primary Mouse Bone Marrow (BM)

1. First euthanize healthy mice by CO_2 asphyxiation followed by cervical dislocation. Conduct the subsequent steps under a sterile hood. From 4 mice about 1–3 × 10^8 BM cells are obtained.
2. Place each mouse in a clean blotting sheet and then spray the mouse with 70% ethanol for disinfection.

3. Using sterile dissecting scissors make an incision in each hind leg. Then firmly grasp the skin and pull it downward in order to expose the muscles.
4. Cut the hind leg above the pelvic joint using the dissecting scissors.
5. Making an incision above the claws, remove the lower portion of the hind leg.
6. Transfer the hind leg to a petri dish with sterile 1× PBS for 5–10 min to loosen the muscle tissue.
7. Dissect tibia and femur from surrounding muscles and, using sterile scissors, remove the excess of tissue.
8. Clean the bones with 1× PBS. Then cut the edges of each bone in a way that the BM is exposed.
9. To harvest the BM, first puncture the bottom of a 0.5 mL microcentrifuge tube using an insulin syringe with a 27 G½″ needle, then place the bones into the tube (maximum 2 femurs and tibias per tube). Place the 0.5 mL microcentrifuge tube in a 1.5 mL microcentrifuge tube and centrifuge the tube at 9659 × *g* for 1–3 min to completely flush down the BM pellet. After centrifugation, the bones should appear white. The BM pellet should be visible in the larger tube. Dispose the bones, and resuspend the BM in appropriate medium (i.e., 1× PBS + 2% FBS) (Fig. 2). Use a 40 μm strainer to filter the BM suspension in order to obtain a homogeneous suspension of BM and to eliminate extra-solid tissue. These BM cells are now ready for virus infection (*see* **Note 5**).

3.3 Isolation of Murine Peripheral Blood Mononuclear Cells (PBMCs)

The isolation of murine PBMCs can be performed using submandibular cheek bleeding and via intracardiac puncture. From 7–10 BALB/c or C57BL/6 mice about 1 × 10^8 PBMCs are obtained.

1. For the submandibular cheek bleeding procedure, use a 5 mm lancet for adult mice, or a 4 mm lancet for small or young mice. In brief, restrain the mouse with one hand and apply some pressure to the maxillary vein, then using the point of the lancet apply pressure holding the lancet in a perpendicular angle to the bleeding site, then release until the blood flows. Placing the collection tube containing anticoagulant (see the step below) just below the puncture site, quickly collect the blood. After this apply gentle pressure with a gauze pad until the bleeding is stopped. After this proceed with the intracardiac puncture procedure. In brief, euthanize the mouse by asphyxiation with CO_2. Once the mouse stops breathing insert a 22 G3/4″ needle attached to a 1 or 3 mL syringe left and under the sternum, and then slowly insert it into the heart. Hold the needle and syringe until all blood has been collected.

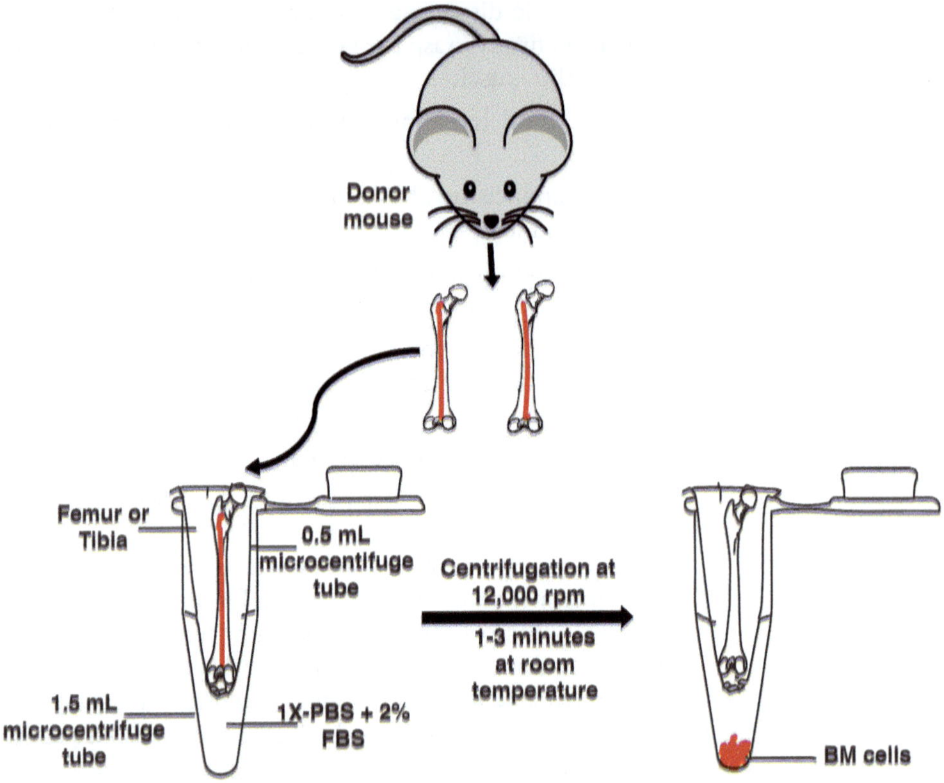

Fig. 2 Harvesting murine bone marrow (BM) cells. To isolate murine BM leukocytes, first euthanize the donor mice. Isolate femurs and tibias and clean them. Cut the edges of the bones to expose the BM. Place the cut bones in a 0.5 mL eppendorf tube as shown in the figure, Spin the cells down at 14,800 × *g* for 1–3 min. Suspend the collected BM pellet in appropriate media

2. Mix the whole blood isolated in the step described above with the 6.4% anticoagulant sodium citrate (*see* **Note 6**) to a cell to anticoagulant ratio of 1:1. Then dilute this mixture to 7 mL using sterile 1× PBS + 2% FBS at room temperature (i.e., 20 °C).
3. Add 3 mL of Ficoll-Paque plus density medium to a 15 mL centrifuge tube (*see* **Note 7**).
4. Carefully layer the diluted blood sample onto the Ficoll-Paque plus density medium (Fig. 3). It is important not to mix the Ficoll-Paque plus density medium with the diluted blood sample.
5. Centrifuge at 400 × *g* for 30 min at 20 °C. Turn off the centrifuge break.
6. Using a sterile serological pipette or a pipette with a 1 mL pipette tip draw off the upper layer, which contains plasma. Then carefully remove the layer of mononuclear cells at the interface (Fig. 3).

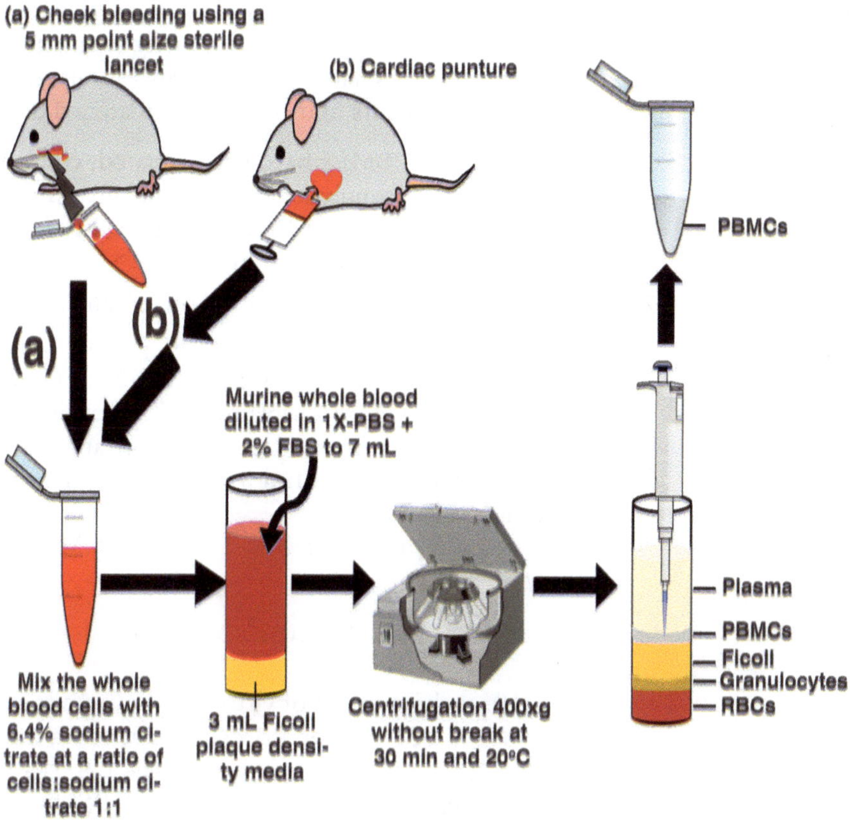

Fig. 3 Isolation of murine peripheral mononuclear cells (PBMCs) from peripheral blood. A summary of the all steps required to harvest murine PBMCs are described in the figure. First, isolate whole blood (WB) via submandibular cheek bleeding and following by intracardiac puncture. Mix the blood with 6.4% sodium citrate to avoid coagulation. Dilute the sample with 1× PBS + 2% FBS and layer the diluted sample onto the Ficoll-plaque density media. Centrifuge at 400 × *g* for 30 min, at 20 °C, turning off the centrifuge brake. Using a sterile serological pipette, or a pipette with a 1 mL sterile pipette tip, carefully remove the PBMC layer, which corresponds to the mononuclear cells. Wash the PBMCs with 1× PBS + 2% FBS. After this, resuspend the cells in a small volume of media (i.e., 1–2 mL depending on the cell density) and count the number of viable cells using the cell counter

7. Transfer the mononuclear cells into a sterile 15 mL centrifuge tube.
8. Estimate the volume of the mononuclear cells and then add at least 3–4 volumes of 1× PBS + 2% FBS and centrifuge at 212 × *g* for 5 min at 20 °C using a table centrifuge.
9. Carefully aspirate the supernatant and resuspend the pellet in 1 mL of complete RPMI-1640 medium (*see* Subheading 2).
10. Count the number of viable cells using a cell counter.

3.4 Ex Vivo Preloading BM or PBMCs with MYXV Versus Systemic Delivery of Free Virus

For hematopoietic cancers, like MM, standard therapies include autologous stem cell transplant (ASCT) along with chemotherapy but allogeneic cell transplantation (allo-SCT) can also be performed when the recipients find compatible matched donors [3].

1. Once BM cells or PBMCs have been isolated, count the number of cells and determine their viability using an automatic cell counter, or a manual cellometer. About 2×10^6 BM cells or PBMCs are transplanted per mouse.
2. Thaw the specific MYXV construct on ice and sonicate it for 1 min also on ice.
3. The following applies per mouse: about $1–2 \times 10^6$ BM cells or PBMCs are required for transplantation. Accordingly, incubate murine BM cells, or PBMCs with the MYXV at a multiplicity of infection (MOI) of 10 for 1 h at 37 °C to allow virus adsorption to the cells.

   For example, to infect 2×10^6 cells at MOI = 10, we multiply.

   $$\left[2 \times 10^6 \cdot (10) = 2 \times 10^7\right].$$

   2×10^7 is the number of virus infectious units required to infect 2×10^6 BM cells or PBMCs. Depending on the virus titer, the volume of the added virus varies. Let us assume that the MYXV preparation has a titer of 2×10^{10} focus forming units (ffu) per mL (ffu/mL). Therefore,

   2×10^7 ffu = 2×10^{10} ffu/mL $\cdot$ (V_x),

   V_x = 0.002 mL or 2 μL of virus stock. This is the amount of virus required to infect 2×10^6 murine BM cells or PBMCs.
4. After 1 h of virus adsorption, depending on whether the experiment is limited to analyzing only cell-adsorbed virus, wash the unbound virus using 1× PBS + 2% FBS followed by centrifugation in a microcentrifuge at 268 × *g* for 2 min (*see* **Note 8**). Aspirate the supernatant and resuspend the pellet in 100 μL 1× PBS. For more than one mouse simply determine the volumes accordingly (*see* **Note 9**). Keep the samples on ice all the time.
5. Before treatment, anesthetize the mice using 1.5–2.5% of isoflurane. Place the mice in the isoflurane induction chamber and wait until the mice are completed sedated.
6. Once the mice are completely sedated, inject 100 μL of the suspended BM or PBMCs ex vivo preloaded with MYXV to each recipient mouse via retro-orbital injection using a 1 mL insulin syringe with a 27 G×½ needle (Cohort IV). For systemic delivery of MYXV, simply dilute the virus stock in 1× PBS to a concentration of 2×10^7 ffu/mL (*see* the above description). Deliver 100 μL of this virus suspension to sedated mice via retro-orbital injection as described above (Cohort III).

As controls, transplant 2×10^6 BM cells or PBMCs that have not been exposed to the virus (Cohort II). Deliver 100 μL of the cell suspension to sedated mice via retro-orbital injection as described above. Finally, Cohort I corresponds to not BM or PBMC transplantation but only 1× PBS (i.e., deliver 100 μL of 1× PBS per mouse as described before) (*see* **Note 10**).

7. Mice are monitored on a daily basis by different symptom system scores, depending on the model, and by in vivo imaging system (IVIS). Once the mice display symptoms such as hunched position, weight loss, lethargy, paralysis (the latter is a typical symptom for multiple myeloma), abdominal distension, labored breathing etc., immediate euthanization of the mice is required.
8. Once the mice are sacrificed, perform necropsy and isolate tumors in order to determine tumor burden.

3.5 Tracking Tumor Burden in Real Time Using IVIS Spectrum Imaging System

In order to monitor tumor burden in real time, the IVIS spectrum system can be used. One advantage of using this instrument is the possibility of imaging both bioluminescent and fluorescent reporters in real time. If the test cancer cells used express firefly luciferase reporter protein, the procedures described here focus on monitoring the proliferation (or regression) of firefly luciferase tagged tumor cells. Figure 4 briefly describes the in vivo imaging to track tumor progression. After 9 to 10 days post-cancer implantation it is possible to monitor cancer cell engraftment in the mouse body as follows:

1. Inject 150 mg/Kg body weight D-luciferin intraperitoneally (i.p.) to each mouse.
2. Anesthetize each mouse with isoflurane [10]. Use isoflurane at a final concentration of 2.5% to 1.5%.
3. Once the mouse is unconscious, place it into the IVIS imaging machine to quantify cancer cell load based on the intensity of the luciferase signal. Images should be acquired from 10–15 min post-substrate injection. In brief, position each mouse ventral, dorsal, left, and right flanks. In each set of images include mice from all cohorts, if possible. In this case, we included mice that have not received any treatment (Cohort I), mice transplanted with BM or PBMCs (Cohort II), mice that systemically received MYXV only (Cohort III), and transplants with ex vivo-preloaded BM cells or PBMCs with MYXV) (Cohort IV). Luminescence of the tumor-bearing mice is quantified using average photons/s/cm^2 [10]. The progression or regression of the cancer can be monitored every week in vivo.

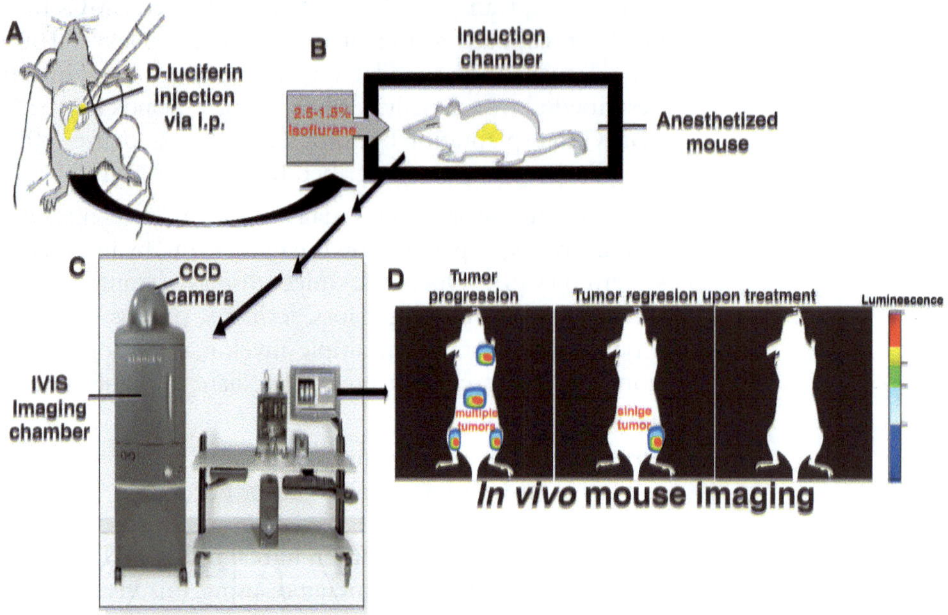

Fig. 4 Imaging in vivo tumor cells in live mice. In order to track the cell load of tumors expressing the luciferase reporter protein: (**a**) mice are first injected intraperitoneally (i.p.) with 100 μL of D-luciferin (i.e., 150 mg/Kg body weight). (**b**) Mice are then anesthetized in using isoflurane. (**c**) Once the mice are completely sedated, three mice are placed in the IVIS imaging chamber and imaged. (**d**), Progression or regression of tumors can be monitored in real time by quantifying the tumor luciferase signal over time

3.6 Flow Cytometry to Quantify Tumor Burden Postmortem in Mice Implanted with Cancer Cells

For hematological cancers like MM, BM and spleen are the major tumor microenvironment niches. However, under metastatic conditions, solid tumors can also be observed outside of these tissues. For MM when the BALB/c-derived MOPC315.BM.FLuc.MM is implanted, cancer progression is monitored using the IVIS imaging. Additionally the percentages of MM in BM, spleen or solid tumors (if any) can be quantified using flow cytometry. In this model, cancer cells are $CD138^+CD4^+$ double positive.

On the other hand, for the BALB/c derived MOPC315.BM. DsRed.MM model the percentages of $CD138^+DsRed^+$ cells can be quantified from BM, spleen or solid tumors (if any) postmortem using flow cytometry with the red fluorescence channel.

In the case of the transplantable C57BL/6 VK∗MYC multiple myeloma model, cancer progression is monitored by measuring the monoclonal spike (M-spike, or paraprotein) in mouse serum, using QuickGel Touch serum protein electrophoresis (SPE) kit and the electrophoresis chamber, according to manufacturer's recommendations and published methodologies [12]. The M-Spike can be measured 4 weeks after the cancer implantation, and then every 1–2 weeks. To collect serum, mice are cheek bled and 100–200 μL per mouse of whole blood is collected in a microtainer serum separator tube. After centrifugation in a microcentrifuge at

6708 × *g* for 5 min, serum (top layer in the microtainer tube) is collected and frozen at −80 °C for further quantification of M-Spike using SPEP according to published procedures [12]. The levels of multiple myeloma (i.e., $CD138^{+}B220^{-}$) in BM and spleen are quantified postmortem using flow cytometry.

1. Once the mice have reached the end-point, this is when symptoms such as lethargy, paralysis, weight loss, labored breathing, visible solid tumors (i.e., 2 cm in diameter) are reached, sacrifice the mice by CO_2 asphyxiation followed by cervical dislocation.
2. Isolate and process BM, spleen, and solid tumor(s) (if any). Process BM in identical fashion as described in Subheading 3.2.
3. To isolate the spleen, make an incision in the abdominal cavity and remove the spleen, which is located to the left side of the abdomen, below the stomach. Place the spleen in a 60 mm tissue culture dish with 1–2 mL 1× PBS and then slice the excised spleen into small pieces using the whole rough side of 2-standard microscope slides (*see* **Note 11**).
4. For solid tumors proceed in the same fashion as described for processing of the spleen.
5. Place the fragments of the spleen onto a 40 μL or 70 μL strainer attached to a 50 mL centrifuge tube. Using the plunger end of a 3 mL syringe, press the excised spleen through the strainer. Wash the cells through the strainer using 1× PBS.
6. Centrifuge the splenocytes at 300 × *g* for 5 min at 20 °C using a table centrifuge. Discard the supernatant and resuspend the pellet in appropriate cell staining buffer. Continue with staining for cancer antigens or any other immune cell population.
7. Aliquots of 50–100 μL of BM, spleen or solid tumor cells are placed in 96-well plate round bottom. Incubate the cells in 95 μL cell staining buffer with 5 μL mouse FcR blocking reagent. Incubate for 10 min at 4 °C. Then proceed with the staining against surface antigens of interest according to the manufacturer's recommendations (*see* **Note 12**). Immediately after this, resuspend the cells in 300 μL of cell staining buffer and run the samples using a flow cytometer instrument (*see* **Note 13**).

3.7 In Vitro Virotherapy with MYXV Against Primary Human Samples Contaminated with MM

Samples described in this section are provided only when the patient has formally consented to donate an aliquot of his/her PBMCs or BM for research purposes. All the samples are nonidentified to maintain the privacy of the patient. Accordingly, immediately process fresh primary human samples from patients with MM (i.e., BM or peripheral blood) as follows:

1. Count about 0.5 mL of fresh primary human BM cells using the cell counter to determine concentration and viability.
2. Immediately infect BM cells with MYXV at different MOIs (i.e., MOI = 10, 1, or 0.1). Before infection thaw the MYXV virus construct and sonicate on ice. Resuspend a minimum of 1×10^6 BM cells in a small volume (i.e., 100–200 μL) of warmed complete RPMI and then incubate with MYXV at different MOIs. As mentioned elsewhere in this chapter, mock-treat cells (i.e., without adding the virus, control), or inoculate with the fluorescent MYXV at each MOI at 37 °C for 1 h to allow virus adsorption. Using a 6× well plate, incubate mock- or MYXV-treated BM cells in 2–3 mL complete medium for 24 h at 37 °C to allow virus infection.
3. Process about 10 mL of peripheral blood (PB) in identical fashion as described in Subheading 3.3, using Ficoll-Paque plus density gradient to isolate the PBMCs. Count isolated PBMCs as described before to determine concentration and viability, and use at minimum 1×10^6 cells for virus infection following the same conditions as those described above.
4. After 24 h of virus infection stain BM cells or PBMCs with fluorescent antibodies for surface proteins. For example, the surface protein CD138 is the classic marker for MM. Therefore, fluorescent antibodies against CD138 are used in this protocol. Flow cytometry is used to quantify the percentages of $CD138^+$ cells in BM or PBMCs. Other readouts are: the percentages of $CD138^+$ cells that are infected with MYXV (i.e., $CD138^+GFP^+$ or $CD138^+TdTomato^+$) and the percentages of $CD138^+GFP^+$ or $CD138^+GFP^-$, $CD138^+TdTomato^+$, $CD138^+TdTomato^-$ and the percentage of $CD138^+$ cell death that is induced by the virus infection. The near-IR LIVE/DEAD cell staining conjugated with APC/Cy7 fluorochrome can be used to assess cell death. Concentrations of antibodies and dyes as well as the time and conditions of incubation are performed according to the manufacturer's recommendation.

4 Notes

1. As an alternative FACS buffer can contain 1× PBS + 0.5–1% bovine serum albumin (BSA) + 0.1% sodium azide (NaN_3).
2. As an alternative, cells can be fixated using 1–4% final concentration of paraformaldehyde.
3. For some models, like the transplantable C57BL/6 VK*MYC MM or solid tumor models such as melanoma and

osteosarcoma, sublethal irradiation of the recipient mice it is not required.

4. As an alternative method to collect murine BM, use a 1 mL insulin syringe with a 27 G½″ needle and flush the content of the marrow with 2 mL of 1× Hank's balanced salt solution (1× HBSS). Use a 40 μL strainer and filter the BM to eliminate any leftover tissue. Transfer the content of the BM into a 50 mL centrifuge tube, and spin the cells down at 300 × *g* for 5 min, 20 °C and discard the supernatant. Resuspend the BM pellet in appropriate medium for further applications.
5. Add 2.25 mL of Baytril into 400 mL sterile water (this is 0.56% v/v). After 1 week replace the medicated water for a fresh preparation.
6. Besides sodium citrate it is possible to use heparan sulfate. In this latter case dilute the whole blood cells with the heparan sulfate at a ratio of 1:1.
7. Alternatively, Histopaque-1077 can be used as a density gradient. After mixing the blood sample in a 1:1 proportion with 1× PBS + 2% FBS, a septated tube with the density gradient previously added, can be used (i.e., SeptMate) for separation of PBMCs from red blood cells. The addition of the blood sample to the tube needs to be done through the tube walls.

   The tube is centrifuged at 305 × *g* for 10 min with the brake on, using a table centrifuge. After centrifugation the top layer corresponds to PBMCs and can be removed and washed once with 1× PBS + 2% FBS.
8. For MYXV infection of BM or PBMCs, it is important to use appropriate medium (i.e., 1× RPMI-1640 supplemented with nutrients and antibiotics for BM or PBMCs that eventually are going to be used to transplant recipient mice with multiple myeloma).
9. In cases where it is not critical to distinguish between free virus and leukocyte-bound virus, after 1 h of virus adsorption to the leukocytes, the unbound virus does not need to be washed.
10. A minimum of 8–10 mice are required per cohort per experiment. Thus the amounts of cells and virus particles need to determined accordingly. Importantly, if 10 mice are going to be transplanted always prepare transplant samples as if 15–20 mice were going to be transplanted.
11. As an alternative method to process spleen and isolated splenocytes, place the spleen in a 60-mm tissue culture dish containing 1–2 mL of 1× PBS, or 1× HBSS. Then using a 1 mL insulin syringe with a 27 G½″ needle flush the content of the spleen several times. Then transfer the spleen to a 40 μm or 70 μm pore size strainer attached to a 50 mL centrifuge tube

and continue processing the spleen as described elsewhere in this book chapter.

12. For antigen staining, resuspend the cells in 50 μL to 100 μL cell staining buffer and add the appropriate amount of fluorochrome antibody against the antigen of interest. Incubate the cells at 4 °C for 15 min to 1 h. After this wash the excess of antibody and discard the supernatant.
13. It is possible stop the experiment at this point. However, cells need to be fixed using 100–200 μL of fixation solution at 4 °C for 15–30 min. After this, wash the fixation buffer and remove the supernatant. Resuspend the cells in 300 μL cell staining buffer. Keep the suspended cells at 4 °C for 24 h and then run the samples in the flow cytometer.

References

1. Ruella MKM (2014) Adoptive immunotherapy for cancer. Immunol Rev 257(1):14–38
2. Rosenberg SARN (2015) Adoptive cell transfer as personalized immunotherapy for human cancer. Science 348(6230):62–68
3. Sahebi FGL, Kanate AS, Eikema DJ, Knelange NS, Alvelo OFD, Koc Y, Blaise D, Bashir Q, Moraleda JM, Dreger P, Sanchez JF, Ciurea S, Schouten H, Shah NN, Verbeek M, Rösler W, Diez-Martin JL, Schoenland S, D'Souza A, Kröger N, Hari P (2018) The outcome of haplo-identical transplantation in patients with relapsed multiple myeloma: an EBMT/CIBMTR report. Biol Blood Marrow Transplant S1083-8791(18):30575–30575
4. Stanford MM, Barrett JW, Nazarian SH, Werden S, McFadden G (2007) Oncolytic virotherapy synergism with signaling inhibitors: rapamycin increases myxoma virus tropism for human tumor cells. J Virol 81(3):1251–1260
5. Ferguson MSLN, Wang Y (2012) Systemic delivery of oncolytic viruses: hopes and hurdles. Adv Virol 2012:805629
6. Lilly CL, Villa NY, Lemos de Matos A, Ali HM, Dhillon JS, Hofland T, Rahman MM, Chan W, Bogen B, Cogle C, McFadden G (2017) Ex vivo oncolytic virotherapy with myxoma virus arms multiple allogeneic bone marrow transplant leukocytes to enhance graft versus tumor. Mol Ther Oncolytics 4:31–40. https://doi.org/10.1016/j.omto.2016.12.002
7. Villa NY, Wasserfall CH, Meacham AM, Wise E, Chan W, Wingard JR, McFadden G, Cogle CR (2015) Myxoma virus suppresses proliferation of activated T lymphocytes yet permits oncolytic virus transfer to cancer cells. Blood 125(24):3778–3788
8. Lee HW, Gangadaran P, Kalimuthu S, Ahn BC (2016) Advances in molecular imaging strategies for in vivo tracking of immune cells. Biomed Res Int 2016:1946485
9. Seth APH, Hong KS (2017) Current perspective on in vivo molecular imaging of immune cells. Molecules 22(6):E881
10. Hofgaard PO, Jodal HC, Bommert K, Huard B, Caers J, Carlsen H, Schwarzer R, Schünemann N, Jundt F, Lindeberg MM, Bogen B (2012) A novel mouse model for multiple myeloma (MOPC315.BM) that allows noninvasive spatiotemporal detection of osteolytic disease. PLoS One 7(12):e51892
11. Uluçkan O, Eagleton MC, Floyd DH, Morgan EA, Hirbe AC, Kramer M, Dowland N, Prior JL, Piwnica-Worms D, Jeong SS, Chen R, Weilbaecher K (2008) APT102, a novel adpase, cooperates with aspirin to disrupt bone metastasis in mice. J Cell Biochem 104 (4):1311–1223
12. Chesi M, Robbiani DF, Sebag M, Chng WJ, Affer M, Tiedemann R, Valdez R, Palmer SE, Haas SS, Stewart AK, Fonseca R, Kremer R, Cattoretti G, Bergsagel PL (2008) AID-dependent activation of a MYC transgene induces multiple myeloma in a conditional mouse model of post-germinal center malignancies. Cancer Cell 13(2):167–180. https://doi.org/10.1016/j.ccr.2008.01.007

Chapter 7

Immunomodulation in Oncolytic Measles Virotherapy

Laura Dietz and Christine E. Engeland

Abstract

With the recognition of oncolytic virotherapy as an immunotherapy, the distinct interactions between oncolytic agents and the immune system have come into focus. The role of the immune system in oncolytic virotherapy is somewhat ambiguous: While preexisting or arising immunity directed against viral antigens may preclude efficient viral replication and spread, immunity directed against tumor antigens is considered essential for long-term treatment success. Aside from the antiviral and antitumor immune status of the patient, the specific immunological microenvironment in a given tumor adds an additional layer of complexity.

In this review we focus on the case of measles virus, which has long been known for its multifaceted interplay with the immune system. The high prevalence of measles-neutralizing antibodies in the general population may pose additional challenges. The oncolytic measles virus vector platform offers manifold opportunities for tumor-targeted immunomodulation. This review provides a survey of immunomodulation in the context of measles virotherapy including strategies to suppress or circumvent antiviral immunity as well as enhance antitumor immunity that have been pursued in preclinical and clinical studies. Understanding and selective manipulation of the intricate balance between antiviral and antitumor immunity will be crucial to develop the full potential of oncolytic virotherapy.

Key words Oncolytic virus, Cancer immunotherapy, Measles virus, Immunomodulation, Antiviral immunity

1 Introduction

The regression of tumors in association with naturally occurring measles infections was already reported in the 1970s [1, 2]. However, it took several decades along with advances in virology and molecular genetics before cancer treatment with nonpathogenic, live-attenuated measles vaccine strains could be developed clinically [3]. Meanwhile, oncolytic measles viruses have been investigated in phase I–II clinical trials, demonstrating safety and showing first promising results in terms of efficacy [4].

Measles vaccine strains have an excellent safety record, having been applied for standard immunization of infants for over 50 years. As a result of prior vaccination or natural infection, the vast

Christine E. Engeland (ed.), *Oncolytic Viruses*, Methods in Molecular Biology, vol. 2058,
https://doi.org/10.1007/978-1-4939-9794-7_7,

majority of adults are immune to measles virus (MV). While this adds to the safety profile of oncolytic MV, antiviral immunity may hamper treatment efficacy by premature clearance of the therapeutic vector [3].

Antiviral immunity is triggered after sensing of measles vaccine viruses by pattern recognition receptors including RIG-I, MDA5, and toll-like receptors (TLRs) upon infection, resulting in induction of type-I interferon (IFN), which has been reported to be mainly driven by defective interfering RNAs [5, 6]. While the IFN response contains viral replication, it also orchestrates induction of MV-specific humoral and cellular immune responses.

In contrast to MV vaccine strains, wild-type, pathogenic MV strains counteract IFN responses via their C and V proteins, but induce nF-κB signaling and inflammasome formation. After a 10–14 day period of viral replication, MV-specific CD4+, CD8+ and antibody responses arise [7]. Concomitant with the generation of MV-specific immune responses, pathogenic MV infection leads to generalized immunosuppression associated with lymphopenia, suppression of lymphocyte proliferation and a T_H2-skewed cytokine production by CD4+ T-cells, which is not observed after vaccination [8, 9].

The oncoselectivity of MV vaccine strains seems to be mainly determined by defects in type-I IFN signaling in malignant cells and IFN-competent tumors may show primary resistance to MV [10–13]. Upon infection, MV-permissive tumor cells secrete inflammatory cytokines and ultimately undergo immunogenic cell death [14]. MV oncolysis has been shown to mediate activation and maturation of dendritic cells, cross-presentation of tumor antigens and priming of tumor-specific cytotoxic CD8+ T-cells [6, 14, 15].

With the recent resurgence of cancer immunotherapy, these immunomodulatory properties of MV oncolysis have come into focus. In order to fully exploit the potential of MV for cancer immunotherapy, a detailed understanding and selective manipulation of the interactions between oncolytic MV, the specific tumor microenvironment and the immune system are crucial. The immune system can affect oncolytic MV therapy in two opposing ways: On the one hand, antiviral immunity may hamper therapeutic efficacy by premature clearance of the oncolytic agent. On the other hand, antitumor immunity may be vital for long-term therapeutic success. Therefore, counteracting antiviral immunity and enhancing antitumor immunity may be necessary to optimize oncolytic treatment regimens. This review aims to classify and discuss current strategies to selectively modulate immune responses to improve the efficacy of oncolytic MV therapy.

2 Counteracting Antiviral Immunity

Systemically administered MV can be subjected to clearance by the humoral immune response, which may reduce or completely abrogate therapeutic efficacy. Systemic clearance can be circumvented by intratumoral administration or alleviated by intraperitoneal injection for tumors located in the abdominal cavity. However, systemic administration is inevitable when targeting distant metastases or hematological malignancies. The majority of adults exhibit protective MV antibody titers due to vaccination or natural infection. Since oncolytic MVs are derived from vaccine strains, they comprise epitopes that are recognized by antibodies generated after vaccination. Besides baseline anti-MV antibodies in preimmunized patients, multiple administration rounds can result in a boost of anti-MV antibody production [16]. Further humoral immune response mechanisms comprise complement activation and antibody-mediated phagocytosis. In addition, systemic clearance can be facilitated by the mononuclear phagocytic system (MPS) found in liver and spleen [17]. Furthermore, MV exposure induces cellular antiviral immunity and CD8+ T-cells also contribute to clearance [7].

In principle, several approaches are conceivable to overcome immune-mediated virus clearance during oncolytic virotherapy, either (1) suppressing the patient's antiviral immune response or (2) shielding viral particles to avoid their detection or (3) enhancing viral immune evasion mechanisms.

2.1 Suppressing Antiviral Immunity

When considering suppression of antiviral immunity during oncolytic therapy, one must take into account that some cancer patients are already immunocompromised, (e.g., those suffering from myeloma or chemotherapy-induced myelosuppression). Immunosuppression may increase the risk for infections and promote cancer progression by hampering antitumor immune responses. Nevertheless, since immune-mediated viral clearance may impede successful oncolytic therapy, suppression of antiviral immunity may be warranted in specific situations.

One strategy to suppress antibody-mediated clearance is the coadministration of cyclophosphamide (CPA) during oncolytic therapy. This alkylating compound causes immunosuppression by selective depletion of hematopoietic cells. CPA is routinely used for chemotherapy in several tumor entities and for immunosuppression in patients with autoimmune diseases or organ transplants. Therefore, experience with dosage exists and can be adapted for suppression of the humoral immune response in the context of systemic oncolytic virus application. Experiments in mice showed that antibody-mediated anti-MV immunity can be suppressed by CPA [18]. As demonstrated by studies in mice, squirrels, monkeys, and humans a single-dose treatment regimen is not sufficient to suppress humoral immune responses [16, 18, 19]. A preclinical study by Peng et al. proposes a multiple-dose CPA regimen with doses routinely used in oncology [16]. The completion of a phase I clinical trial treating multiple myeloma patients with oncolytic MV and CPA was reported in 2017. In this trial, a single dose of CPA was applied prior to intravenously administered oncolytic MV. No severe toxicities were observed and first encouraging signs of efficacy were reported [20]. Currently, two clinical phase II studies are active which investigate oncolytic MV in combination with a CPA regimen (clinicaltrials.gov NCT00450814, NCT02192775). CPA could not only suppress vaccination-derived antibody titers, but also enable multiple rounds of administration by preventing antibody boosts [16].

Experiments with polyinosinic acid (poly(I)), which is known to competitively inhibit scavenger receptors, revealed that systemic clearance of oncolytic MV is also facilitated through the uptake of viruses by macrophages, primarily in the MPS. Liu et al. demonstrated that coadministration of poly(I) prevents uptake of oncolytic MV by macrophages and thereby enhances delivery [17]. However, the substance poly(I) is not clinically approved and has not yet been tested in humans. There are alternatives such as fucoidan and PolyI:C, which have already been applied clinically. The latter is commonly used as an adjuvant in vaccination trials and induces IFN-α production. Poly(I) and PolyI:C belong to a class of RNA compounds which block scavenger receptors and exhibit different immunomodulating properties [21]. Therefore, additional immunomodulation can be achieved using these compounds.

Another alternative to avoid macrophage-mediated clearance of MV could be the coadministration of chlodronate-loaded liposomes which leads to macrophage depletion [22]. However, this approach depletes all phagocytes, including dendritic cells, which could be detrimental to oncolytic vaccination effects.

The relative contribution of antibody neutralization and MV sequestration by macrophages to systemic clearance of oncolytic MV has not yet been investigated extensively, but may need to be addressed to optimize oncolytic MV treatment regimens.

2.2 Shielding Viral Particles

In contrast to an immunosuppressive strategy, shielding MV particles from the immune system may be less critical as this does not alter the patient's physiology. One strategy to disguise MV particles from the patient's immune system is to use cells infected with oncolytic MV for delivery. Presumably such virus-infected carrier cells are less prone to antibody neutralization during blood stream circulation than naked virus particles. After the carrier cells have reached the tumor site, tumor cells and other cells in the tumor microenvironment are infected by cell-to-cell transmission. This transmission mode plays a predominant role for viral spread in natural MV infections and is therefore likely to constitute an efficient and robust delivery strategy [23].

Many cell types including myeloma cells, T-cells, monocytes, mesenchymal stem cells and endothelial cells have been used as carriers. For the choice of an appropriate carrier cell, several factors must be taken into account: Different cell types differ in their susceptibility to MV and the amount of viral progeny they release [24–27]. Further, certain cell types might mediate a directed transport of oncolytic MV to specific tissues or tumor sites. For example, myeloma cells home to the bone marrow and mesenchymal stem cells migrate to tumor sites as they are part of the local stroma [26, 28].

Liu et al. showed that intravenously administered lethally irradiated MV-infected myeloma cell carriers improved survival of severe combined immunodeficient (SCID) mice bearing disseminated myeloma xenografts in presence of anti-MV antibodies at titers similar to those found in myeloma patients compared to naked MV [24]. Ong et al. showed that intraperitoneally administered infected mesenchymal stem cells were able to double survival rates of passively immunized mice bearing intraperitoneal ovarian tumor xenografts compared to the control group that received naked virus particles [27]. Based on these preclinical results achieved with MV-infected mesenchymal stem cells, a clinical phase I/II trial is currently recruiting to test infected mesenchymal stem cells in patients with recurrent ovarian cancer (NCT02068794).

Remarkably, many cell carrier systems were able to protect MV from high concentrations of neutralizing antibodies in vitro, but did not protect MV from lower concentrations of neutralizing antibodies in vivo [27]. This finding, in conjunction with an observed enrichment of cell carriers in the liver, suggests that infected cell carriers may be protected from neutralizing antibodies but may also be subjected to endocytosis by the MPS.

Aside from biological shielding, viral particles can be coated with polymers known to lower immunogenic potential. This strategy has also been adapted to MV. Nosaki et al. reported to have successfully coated oncolytic MV with an ionic polymer. These coated oncolytic MV showed enhanced efficacy compared to naked viruses when injected into tumor xenografts of immunized mice [29]. Whether these polymer-coated viruses are superior to naked viruses in terms of antibody resistance when administered systemically still needs to be tested.

For MV, only one serotype is known [23]. Thus, serotype switching in order to overcome an antibody-mediated immune response is not possible. Alternatively, the MV envelope glycoproteins H and F which are the targets of neutralizing antibodies can be exchanged by the respective proteins of closely related viruses such as canine distemper virus (CDV) or Tupaia paramyxovirus (TPMV). Both CDV and TPMV belong to the family *Paramyxoviridae* and are not pathogenic to humans. Importantly, MV antibodies do not cross-react with these viruses and humans are generally seronegative for CDV- and TPMV-neutralizing antibodies. However, only the exchange of MV H and F by homologous proteins of CDV efficiently generated viral particles that were comparable to unmodified MV regarding infectivity and replication [30]. When exchanging MV H and F by TPMV homologs, particle assembly was impaired and despite extensive protein engineering, no variant was found that allowed efficient particle assembly and viral spread [31]. Single-chain antibody fragments fused to the CDV H protein could target chimeric viral particles to CEA-positive cells. Finally, CEA-targeted chimeric viruses were able to reduce tumor growth in MV-immunized mice in an immunocompetent colon adenocarcinoma model, MC38cea [30].

Neutralization epitopes of the MV H protein have been characterized previously [32]. Although the MV envelope glycoproteins seem to be intolerant toward mutations [33], recently engineered MV variants with H protein epitope mutations and an F protein homolog were reported that showed resistance to antibody neutralization with human serum [34]. Another study, aiming to develop an improved MV vaccine for infants, showed that enhanced H protein expression and surface display can decrease maternal anti-MV antibody-mediated neutralization [35]. However, MV variants resistant to antibody neutralization could cause considerable safety concerns. A key safety feature of oncolytic MV is the high prevalence of immunity in the general population, which precludes uncontrolled replication and spread. Thus, any engineered MV variants which are not susceptible to neutralizing antibodies must undergo a scrutinous risk assessment.

2.3 Enhancing Viral Immune Evasion

One of the primary defense mechanisms against RNA viruses, including MV, is the induction of a type-I IFN response in infected cells. IFN-α and -β function in an autocrine and paracrine manner, inducing interferon-stimulated genes (ISGs), which trigger an antiviral state. Therefore, many RNA viruses have evolved evasion mechanisms that counteract the host IFN response. In the case of MV, gene products encoded by the *P* gene, the V and C proteins, act as inhibitors of IFN. V and C from wild-type MV isolates are strong antagonists of IFN, whereas attenuated vaccine strains are more sensitive to IFN due to mutations in the *P* gene [36].

In order to create an MV strain that is less sensitive to IFN, Haralambieva et al. exchanged the *P* gene of an Edmonston-tag strain by a wild-type *P* gene. This chimeric MV caused complete tumor regression in SCID mice bearing subcutaneous myeloma tumor xenografts. Interestingly, the parental vaccine strain also led to complete tumor regression, but showed slower tumor regression kinetics [37].

An even more aggressive oncolytic MV was created by not only exchanging the *P* gene of the Edmonston-tag strain by its wild-type homolog, but the *N* and *L* genes in addition. It was hypothesized that this novel chimeric strain, featuring a complete wild-type replication machinery and a reduced sensitivity to IFN, would spread more efficiently due to faster replication and decreased IFN sensitivity. This oncolytic MV was tested in nude mice bearing subcutaneous human renal cell carcinoma xenografts. In contrast to the parental Edmonston-tag strain and its derivative with a wild-type *P* gene only, the novel strain led to tumor remission, whereas the other two strains only delayed tumor growth. Moreover, the novel strain was able to significantly increase the survival rate compared to Edmonston-tag containing a wild-type *P* gene in this model [38].

Although MV derivatives harboring wild-type genes exhibited enhanced oncolytic potential in preclinical models, their reduced sensitivity to IFN poses a considerable safety risk. IFN resistance is a main determinant of MV pathogenicity and IFN incompetence of tumor cells is one determinant of MV tumor selectivity. Several studies have demonstrated that cotreatment with antagonists of the IFN response including JAK inhibitors such as ruxolitinib can overcome IFN-mediated MV resistance of tumor cells [10, 13, 39]. Whether such combination approaches to counteract antiviral immunity are safe and effective in vivo remains to be determined.

Moreover, antiviral immune responses may benefit antitumor immunity by recruiting immune cells and reversing the immunosuppressive tumor microenvironment, thereby facilitating the induction of adaptive antitumor immune responses [40]. Thus, rather than counteracting antiviral immunity, enhancing antitumor immunity may be a more viable approach.

3 Enhancing Antitumor Immunity

MV oncolysis releases tumor antigens in the context of an acute viral infection which triggers a multifaceted immune response involving type-I IFN, danger-associated and pathogen-associated molecular patterns (DAMPs and PAMPs), immunogenic cell death, inflammatory cytokines, and immune cell influx, creating an immune milieu that can provide signals 2 and 3, that is, costimulatory molecules and cytokines, for priming of T-cells. Thus, the natural response to oncolytic MV bears the potential to induce antitumor immunity via in situ tumor vaccination effects. Beyond local tumor control, these effects can extend also to metastatic sites and include a memory response that can protect from tumor recurrence.

To enhance these effects, MV has been functionalized as a vector for tumor-targeted expression of immunomodulators. A challenge for study of MV is that murine tissues lack expression of the MV receptor CD46 and show limited permissiveness for this primate-adapted virus. Thus, viruses retargeted to murine tumor surface proteins, tumor cell lines engineered to express CD46 or CD46-transgenic mice have been used in models of MV oncolysis [18, 41, 42]. While analyses in immunocompetent models are mandatory to evaluate the immunomodulatory properties of a specific MV variant, these can be complemented by experiments with xenograft models in immunodeficient mice to test direct oncolytic effects. Despite these challenges with preclinical models, proof-of-concept for the immunotherapeutic effects of several MV variants have been obtained, which are discussed in the following sections.

3.1 IFN-β

IFN-β secretion is one of the first reactions upon MV infection and initiates further antiviral immune responses [7]. Moreover, IFN-β is known to exhibit antiproliferative and antiangiogenic effects and has therefore long been considered for antitumor treatment including cancer immunotherapy.

Thus, expression of additional IFN-β by MV was hypothesized to potentiate antitumor efficacy. Li et al. tested this strategy in nude mice bearing human mesothelioma xenografts. Mice treated with IFN-β-encoding MV showed a significant increase in CD68+ tumor-infiltrating leukocytes, delayed tumor growth and significantly prolonged median survival compared to mice treated with control MV. However, iodide uptake experiments in mice infected with MV derivatives encoding the sodium-iodide symporter (NIS) suggested that the additional expression of IFN-β decreased viral replication rates at later time points after treatment [43]. This finding is in line with the major role of type-I IFN in containing MV vaccine strain replication and highlights the ambivalent role of this cytokine for oncolytic MV therapy. On the one hand, IFN restricts MV spread. On the other hand, IFN-β potently induces acute inflammation that can evolve into an antitumor immune response engaging innate and adaptive cellular immune responses [40]. Along these lines, there is some debate as to which extent of viral replication is actually necessary to achieve tumor control and whether direct virus-mediated tumor cell lysis or immune-mediated tumor cell killing prevail.

3.2 GM-CSF

Granulocyte macrophage-colony stimulating factor (GM-CSF) is a cytokine mainly known for regulation of monocytes, macrophages and dendritic cells. GM-CSF directly acts on the innate immune system by stimulating the production of granulocytes and monocytes in the bone marrow as well as their recruitment to sites of inflammation. By promoting maturation of antigen-presenting cells, it bridges innate and adaptive immunity [44].

GM-CSF has been encoded in many oncolytic vectors with the aim to enhance tumor vaccination effects. Of note, talimogene laherparepvec, the first FDA- and EMA-approved oncolytic therapy, also harbors this transgene. GM-CSF-encoding Edmonston B-derived MV was first investigated in Raji xenograft-bearing SCID mice. This immunocompromised mouse model only allows evaluation of effects on the innate immune response, as SCID mice lack mature B- and T-cells. Nevertheless, an enhanced therapeutic effect of MV-GM-CSF compared to MV was observed, which was associated with an increased neutrophil infiltration in treated tumors [45]. Grossardt et al. tested MV-GM-CSF in a fully immunocompetent murine colon adenocarcinoma model, MC38cea. In this setting, MV-GM-CSF increased survival rates compared to control MV, achieving complete tumor regression and long-term tumor-free survival in 40% of individuals. Furthermore, upon treatment with MV-GM-CSF tumor-specific cytotoxic lymphocytes could be isolated. Tumor rechallenge experiments demonstrated induction of durable, protective antitumor immunity in animals that exhibited complete tumor remission [46]. These data suggest that the expression of GM-CSF enhances the tumor vaccination effects of MV-mediated oncolysis.

3.3 IL-12

Interleukin-12 (IL-12) is mainly expressed by macrophages and dendritic cells upon sensing of pathogens. IL-12 activates cell-mediated immune responses driven by CD8+ T-cells and NK-cells and functions as a major determinant for T_H1 polarization including induction of IFN-γ and tumor necrosis factor (TNF)-α [47].

In a study by Veinalde et al., a panel of MV encoding different immunomodulators targeting distinct phases of the "cancer immunity cycle" [48] were tested in the MC38cea murine colon adenocarcinoma model, which had previously been used for investigation of MV-GM-CSF. MV-IL-12 was identified as the most effective MV variant, achieving complete tumor regression in 90% of treated animals [49]. Compared to the complete regression rate after MV-GM-CSF treatment, this documents exceptional efficacy. As observed in experiments with MV-GM-CSF, animals with complete primary tumor regression were immune against secondary tumor engraftment. An increase of the transcription factor *Tbet* suggested T_H1-polarized T-cell responses in mice treated with MV-IL-12. However, depletion experiments revealed that CD4+ T helper cells were not the main mediators of treatment efficacy, but rather CD8+ cytotoxic T-cells. Analysis of tumor-infiltrating lymphocytes revealed an increase in activated NK-cells and CD8+ T-cells, while cytokine profiling showed significant increases in intratumoral IFN-γ and TNF-α [49].

Of note, this study used Schwarz vaccine-based MV, which differs from Edmonston B-derived vaccine strains in its immunomodulatory properties [50, 51]. To date, a systematic comparison of different MV vaccine strains for oncolytic immunotherapy has not been conducted, but may be worthwhile, given the differential immunomodulatory properties of different MV vaccine strains [51].

Further, encoding cytokines such as IL-12, which can be highly toxic when administered systemically, in an oncotropic vector allows for tumor-restricted expression, thereby increasing the therapeutic window.

3.4 NAP

The neutrophil-activating protein (NAP) originates from the bacterium *Helicobacter pylori*. It is one of the main virulence factors of *H. pylori* as it attracts and activates neutrophils, which in consequence establish chronic infection without pathogen clearance. Mechanistically, NAP induces inflammatory cytokine secretion in neutrophils and monocytes via activation of TLR2. The induced cytokine profile including IL-12 and IL-23 can promote T_H1 polarization [52].

Iankov et al. hypothesized that encoding secretable NAP in an MV vector can enhance therapeutic efficacy by inducing inflammation and a T_H1 response [53]. Transthoracically injected MV-NAP demonstrated therapeutic benefits in a pleural effusion model of metastatic breast cancer in athymic nude mice, doubling the median survival of mice compared to treatment with parental MV. Immunological effects were monitored by analyzing pleural fluid of mice, which contained significantly increased neutrophils and increased levels of IL-6, IL-12, IL-23 and TNF-α after treatment with MV-NAP. As athymic mice lack T-cells, T_H1 activation could not be assessed directly, but IL-12 as a major determinant of T_H1 polarization was detected.

3.5 Checkpoint Inhibitors

The recent resurgence of cancer immunotherapy has been driven by inhibitors of the immune checkpoints cytotoxic T lymphocyte antigen 4 (CTLA-4), programmed death-1 (PD-1) and its ligand PD-L1. Under physiological conditions, these coinhibitory molecules support self-tolerance by limiting T-cell activation. However, they also contribute to tumor immune evasion. CTLA-4 is expressed on T-cells. Although its predominant mechanism of action remains controversial, it is known that CTLA-4 can compete with costimulatory CD28 for the ligands CD80 and CD86, which are expressed on antigen-presenting cells. This interaction is mainly relevant in the context of initial activation of both naïve and memory T-cells in lymphoid organs. In contrast, PD-1 and its ligand PD-L1 are more relevant in regulating T-cell activation in the periphery. PD-1 is often expressed on tumor-infiltrating T-cells and associated with T-cell exhaustion or anergy, while upregulation of PD-L1 in tumor tissue can contribute to tumor immune evasion [54, 55].

Oncolytic MVs of the Edmonston B vaccine strain encoding antibodies against CTLA-4 (MV-aCTLA-4), or PD-L1 (MV-aPD-L1) have been developed [56]. In a B16-derived fully immunocompetent murine melanoma model which expresses PD-L1, both viruses led to an increase in intratumoral CD3+ leukocytes and a decrease in intratumoral regulatory T-cells. Thus, immunostimulatory activities were demonstrated. However, only MV-aPD-L1 prolonged survival significantly. Coculture experiments with tumor cells and splenocytes collected at different time points suggested that in MV-aCTLA-4-treated mice tumor-specific IFN-γ secretion arose at early time points, whereas in MV-aPD-L1-treated mice tumor-specific IFN-γ secretion arose at later time points after treatment. This probably reflects the different roles of CTLA-4 and PD-1 in T-cell coinhibition. Testing the novel MVs in a human melanoma xenograft model demonstrated that expression of checkpoint inhibitors did not impair the oncolytic activity of MV [56].

MV Schwarz encoding an anti-PD-L1 antibody has also been tested in the MC38cea model. In this setting, complete tumor regression was achieved in approximately half of the treated animals, which were subsequently immune to tumor re-engraftment, indicating induction of long-term antitumor immunity [49].

3.6 BiTEs

Bispecific T-cell engagers (BiTEs) are synthetic molecules which consist of two covalently linked single-chain variable fragments (scFv). With one scFv binding specifically to T-cells and the other scFv binding to a tumor surface antigen of choice, BiTEs link tumor cells and T-cells independent of T-cell receptor specificity. This spatial proximity allows formation of an artificial immunological synapse and induction of tumor cell lysis. Although CD19-targeted BiTEs have achieved clinical success against B-cell malignancies, efficacy against solid tumors remains limited. Further, the challenge to identify suitable tumor surface antigens as well as the need for continuous infusion hamper broader clinical application [57].

To address these challenges, Speck and Heidbuechel et al. generated MV encoding BiTEs and tested their efficacy in B16, MC38cea and patient-derived colorectal cancer spheroid models [41]. MV-encoded BiTEs elicited significantly increased secretion of IFN-γ, TNF-α, IL-2, and IL-6 levels and cytotoxicity in cocultures of tumor cells and T-cells. In mice bearing patient-derived colorectal cancer spheroid xenografts MV-BiTE treatment in presence of human peripheral blood mononuclear cells (PBMCs) significantly prolonged survival compared to MV-BiTE treatment or PBMC transfer alone. In the immunocompetent MC38cea model, which exhibits high baseline T-cell infiltration, MV-BiTE treatment did not provide a significant benefit compared to control MV. In contrast, in the B16 model, MV-BiTE treatment led to an increase in tumor-infiltrating T-cells, an increased ratio of effector to regulatory T-cells (T_{eff}/T_{reg}) and prolonged survival. However, this was also associated with upregulation of T-cell exhaustion markers, indicating potential for further improvement.

The two distinct phenotypes observed in the MC38cea and B16 models upon MV-BiTE treatment most likely reflect different tumor immune microenvironments (TIMEs) [58]. The "immune desert" B16 benefits from MV-BiTE-mediated T-cell recruitment, while the T-cell-inflamed MC38cea requires a different treatment approach to stimulate tumor-specific T-cells (e.g., MV-IL-12) [49]. These findings point at the potential of oncolytic MV as a platform for tumor-targeted delivery of immunomodulators selected according to the individual TIME.

4 Conclusion and Perspectives

As outlined in this review, the immune system can affect oncolytic MV therapy by multiple mechanisms. Humoral immunity including preexisting neutralizing antibodies may preclude effectiveness, especially in case of systemic MV treatment. While the role of cellular antiviral immunity in the context of oncolytic MV therapy remains to be determined, cellular antitumor immunity is desirable to enable long-term systemic tumor control. Virus-mediated tumor lysis has been shown to induce tumor-specific T-cell responses as a result of in situ vaccination upon exposure of tumor antigens after virus-mediated lysis, with the DAMPs and PAMPs accompanying viral infection acting as adjuvants. However, these innate immune activators also trigger IFN as a key component of the antiviral immune response that limits MV replication and spread. These considerations highlight the complexity of the interplay between the immune system and oncolytic MV. Conclusively, a major determinant for successful oncolytic MV therapy is selective immune modulation (Fig. 1). On one side, deliberate immunosuppression may be justified to achieve substantial delivery and oncolysis. On the other side, purposeful immune activation may add to fully realize the immunotherapeutic potential of oncolytic MV. Toward this end, oncolytic MV can be utilized for expression of immunomodulatory genes. Due to its versatility in hosting a variety of transgenes, oncolytic MV can be conceived as a platform for tumor-targeted immunotherapy which can be adapted to the specific TIME, enabling personalized oncolytic cancer therapy.

The intricate relationships between the oncolytic virus, the immune system and the individual TIME result in complex dynamics that ultimately determine therapeutic outcome. These dynamics can hardly be accessed experimentally in a highly time-resolved

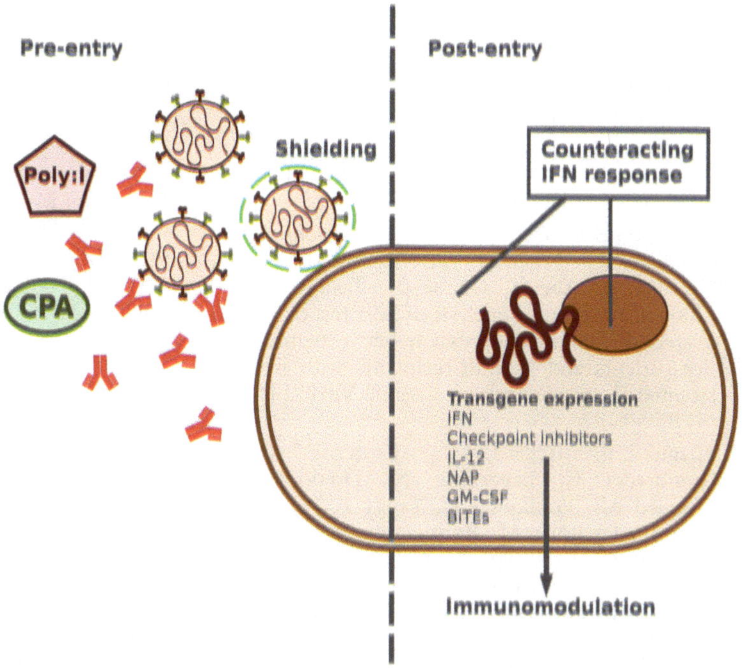

Fig. 1 Immunomodulation in oncolytic measles virotherapy. The schematic depicts strategies for selective immunomodulation in the context of oncolytic measles virotherapy summarized in this review

manner, but experimental data can be used to create mathematical models [59]. Such models can help to understand these dynamics, which can support further oncolytic MV preclinical study design and development of rational combination regimens for preclinical testing [60]. Finally, translational studies in appropriate preclinical models as well as correlative research in clinical trials are necessary to further decipher the underlying immunology and identify the most effective strategies for targeted immunomodulation in oncolytic virotherapy.

References

1. Bluming AZ, Ziegler JL (1971) Regression of Burkitt's lymphoma in association with measles infection. Lancet 2(7715):105–106
2. Mota HC (1973) Infantile Hodgkin's disease: remission after measles. Br Med J 2(5863):421
3. Russell SJ, Peng KW (2009) Measles virus for cancer therapy. Curr Top Microbiol Immunol 330:213–241
4. Robinson S, Galanis E (2017) Potential and clinical translation of oncolytic measles viruses. Expert Opin Biol Ther 17(3):353–363. https://doi.org/10.1080/14712598.2017.1288713
5. Shingai M, Ebihara T, Begum NA, Kato A, Honma T, Matsumoto K, Saito H, Ogura H, Matsumoto M, Seya T (2007) Differential type I IFN-inducing abilities of wild-type versus vaccine strains of measles virus. J Immunol 179(9):6123–6133
6. Guillerme JB, Boisgerault N, Roulois D, Menager J, Combredet C, Tangy F, Fonteneau JF, Gregoire M (2013) Measles virus vaccine-infected tumor cells induce tumor antigen cross-presentation by human plasmacytoid dendritic cells. Clin Cancer Res 19(5):1147–1158. https://doi.org/10.1158/1078-0432.ccr-12-2733

7. Griffin DE (2016) The immune response in measles: virus control, clearance and protective immunity. Viruses 8(10). https://doi.org/10.3390/v8100282
8. Griffin DE (2010) Measles virus-induced suppression of immune responses. Immunol Rev 236:176–189. https://doi.org/10.1111/j.1600-065X.2010.00925.x
9. Okada H, Sato TA, Katayama A, Higuchi K, Shichijo K, Tsuchiya T, Takayama N, Takeuchi Y, Abe T, Okabe N, Tashiro M (2001) Comparative analysis of host responses related to immunosuppression between measles patients and vaccine recipients with live attenuated measles vaccines. Arch Virol 146 (5):859–874
10. Allagui F, Achard C, Panterne C, Combredet C, Labarriere N, Dreno B, Elgaaied AB, Pouliquen D, Tangy F, Fonteneau JF, Gregoire M, Boisgerault N (2017) Modulation of the type I interferon response defines the sensitivity of human melanoma cells to Oncolytic measles virus. Curr Gene Ther 16 (6):419–428. https://doi.org/10.2174/1566523217666170102110502
11. Achard C, Boisgerault N, Delaunay T, Roulois D, Nedellec S, Royer PJ, Pain M, Combredet C, Mesel-Lemoine M, Cellerin L, Magnan A, Tangy F, Gregoire M, Fonteneau JF (2015) Sensitivity of human pleural mesothelioma to oncolytic measles virus depends on defects of the type I interferon response. Oncotarget 6(42):44892–44904. https://doi.org/10.18632/oncotarget.6285
12. Noll M, Berchtold S, Lampe J, Malek NP, Bitzer M, Lauer UM (2013) Primary resistance phenomena to oncolytic measles vaccine viruses. Int J Oncol 43(1):103–112. https://doi.org/10.3892/ijo.2013.1914
13. Kurokawa C, Iankov ID, Anderson SK, Aderca I, Leontovich AA, Maurer MJ, Oberg AL, Schroeder MA, Giannini C, Greiner SM, Becker MA, Thompson EA, Haluska P, Jentoft ME, Parney IF, Weroha SJ, Jen J, Sarkaria JN, Galanis E (2018) Constitutive interferon pathway activation in tumors as an efficacy determinant following Oncolytic virotherapy. J Natl Cancer Inst 110(10):1123–1132. https://doi.org/10.1093/jnci/djy033
14. Donnelly OG, Errington-Mais F, Steele L, Hadac E, Jennings V, Scott K, Peach H, Phillips RM, Bond J, Pandha H, Harrington K, Vile R, Russell S, Selby P, Melcher AA (2013) Measles virus causes immunogenic cell death in human melanoma. Gene Ther 20(1):7–15. https://doi.org/10.1038/gt.2011.205
15. Achard C, Guillerme JB, Bruni D, Boisgerault N, Combredet C, Tangy F, Jouvenet N, Gregoire M, Fonteneau JF (2017) Oncolytic measles virus induces tumor necrosis factor-related apoptosis-inducing ligand (TRAIL)-mediated cytotoxicity by human myeloid and plasmacytoid dendritic cells. Oncoimmunology 6(1):e1261240. https://doi.org/10.1080/2162402x.2016.1261240
16. Peng KW, Myers R, Greenslade A, Mader E, Greiner S, Federspiel MJ, Dispenzieri A, Russell SJ (2013) Using clinically approved cyclophosphamide regimens to control the humoral immune response to oncolytic viruses. Gene Ther 20(3):255–261. https://doi.org/10.1038/gt.2012.31
17. Liu YP, Tong C, Dispenzieri A, Federspiel MJ, Russell SJ, Peng KW (2012) Polyinosinic acid decreases sequestration and improves systemic therapy of measles virus. Cancer Gene Ther 19 (3):202–211. https://doi.org/10.1038/cgt.2011.82
18. Ungerechts G, Springfeld C, Frenzke ME, Lampe J, Parker WB, Sorscher EJ, Cattaneo R (2007) An immunocompetent murine model for oncolysis with an armed and targeted measles virus. Mol Ther 15(11):1991–1997. https://doi.org/10.1038/sj.mt.6300291
19. Myers RM, Greiner SM, Harvey ME, Griesmann G, Kuffel MJ, Buhrow SA, Reid JM, Federspiel M, Ames MM, Dingli D, Schweikart K, Welch A, Dispenzieri A, Peng KW, Russell SJ (2007) Preclinical pharmacology and toxicology of intravenous MV-NIS, an oncolytic measles virus administered with or without cyclophosphamide. Clin Pharmacol Ther 82(6):700–710. https://doi.org/10.1038/sj.clpt.6100409
20. Dispenzieri A, Tong C, LaPlant B, Lacy MQ, Laumann K, Dingli D, Zhou Y, Federspiel MJ, Gertz MA, Hayman S, Buadi F, O'Connor M, Lowe VJ, Peng KW, Russell SJ (2017) Phase I trial of systemic administration of Edmonston strain of measles virus genetically engineered to express the sodium iodide symporter in patients with recurrent or refractory multiple myeloma. Leukemia 31(12):2791–2798. https://doi.org/10.1038/leu.2017.120
21. Zare F, Magnusson M, Mollers LN, Jin T, Tarkowski A, Bokarewa M (2008) Single-stranded polyinosinic acid oligonucleotides trigger leukocyte production of proteins belonging to fibrinolytic and coagulation cascades. J Leukoc Biol 84(3):741–747. https://doi.org/10.1189/jlb.0506345
22. van Rooijen N, Sanders A, van den Berg TK (1996) Apoptosis of macrophages induced by liposome-mediated intracellular delivery of clodronate and propamidine. J Immunol Methods 193(1):93–99

23. Fields BN, Knipe DM, Howley PM (2013) Fields virology, 6th edn. Wolters Kluwer Health/Lippincott Williams & Wilkins, Philadelphia
24. Liu C, Russell SJ, Peng KW (2010) Systemic therapy of disseminated myeloma in passively immunized mice using measles virus-infected cell carriers. Mol Ther 18(6):1155–1164. https://doi.org/10.1038/mt.2010.43
25. Iankov ID, Blechacz B, Liu C, Schmeckpeper JD, Tarara JE, Federspiel MJ, Caplice N, Russell SJ (2007) Infected cell carriers: a new strategy for systemic delivery of oncolytic measles viruses in cancer virotherapy. Mol Ther 15 (1):114–122. https://doi.org/10.1038/sj.mt.6300020
26. Mader EK, Maeyama Y, Lin Y, Butler GW, Russell HM, Galanis E, Russell SJ, Dietz AB, Peng KW (2009) Mesenchymal stem cell carriers protect oncolytic measles viruses from antibody neutralization in an orthotopic ovarian cancer therapy model. Clin Cancer Res 15 (23):7246–7255. https://doi.org/10.1158/1078-0432.ccr-09-1292
27. Ong HT, Hasegawa K, Dietz AB, Russell SJ, Peng KW (2007) Evaluation of T cells as carriers for systemic measles virotherapy in the presence of antiviral antibodies. Gene Ther 14 (4):324–333. https://doi.org/10.1038/sj.gt.3302880
28. Komarova S, Kawakami Y, Stoff-Khalili MA, Curiel DT, Pereboeva L (2006) Mesenchymal progenitor cells as cellular vehicles for delivery of oncolytic adenoviruses. Mol Cancer Ther 5 (3):755–766. https://doi.org/10.1158/1535-7163.mct-05-0334
29. Nosaki K, Hamada K, Takashima Y, Sagara M, Matsumura Y, Miyamoto S, Hijikata Y, Okazaki T, Nakanishi Y, Tani K (2016) A novel, polymer-coated oncolytic measles virus overcomes immune suppression and induces robust antitumor activity. Mol Ther Oncolytics 3:16022. https://doi.org/10.1038/mto.2016.22
30. Miest TS, Yaiw KC, Frenzke M, Lampe J, Hudacek AW, Springfeld C, von Messling V, Ungerechts G, Cattaneo R (2011) Envelope-chimeric entry-targeted measles virus escapes neutralization and achieves oncolysis. Mol Ther 19(10):1813–1820. https://doi.org/10.1038/mt.2011.92
31. Hudacek AW, Navaratnarajah CK, Cattaneo R (2013) Development of measles virus-based shielded oncolytic vectors: suitability of other paramyxovirus glycoproteins. Cancer Gene Ther 20(2):109–116. https://doi.org/10.1038/cgt.2012.92
32. Munoz-Alia MA, Casasnovas JM, Celma ML, Carabana J, Liton PB, Fernandez-Munoz R (2017) Measles virus Hemagglutinin epitopes immunogenic in natural infection and vaccination are targeted by broad or genotype-specific neutralizing monoclonal antibodies. Virus Res 236:30–43. https://doi.org/10.1016/j.virusres.2017.04.018
33. Fulton BO, Sachs D, Beaty SM, Won ST, Lee B, Palese P, Heaton NS (2015) Mutational analysis of measles virus suggests constraints on antigenic variation of the glycoproteins. Cell Rep 11(9):1331–1338. https://doi.org/10.1016/j.celrep.2015.04.054
34. Dyer A, Baugh R, Chia SL (2019) Turning cold tumours hot: oncolytic virotherapy gets up close and personal with other therapeutics at the 11th Oncolytic virus conference. Cancer Gene Ther 26(3–4):59–73. https://doi.org/10.1038/s41417-018-0042-1
35. Julik E, Reyes-Del Valle J (2017) A recombinant measles vaccine with enhanced resistance to passive immunity. Viruses 9(10):E265. https://doi.org/10.3390/v9100265
36. Ohno S, Ono N, Takeda M, Takeuchi K, Yanagi Y (2004) Dissection of measles virus V protein in relation to its ability to block alpha/beta interferon signal transduction. J Gen Virol 85(Pt 10):2991–2999. https://doi.org/10.1099/vir.0.80308-0
37. Haralambieva I, Iankov I, Hasegawa K, Harvey M, Russell SJ, Peng KW (2007) Engineering oncolytic measles virus to circumvent the intracellular innate immune response. Mol Ther 15(3):588–597. https://doi.org/10.1038/sj.mt.6300076
38. Meng X, Nakamura T, Okazaki T, Inoue H, Takahashi A, Miyamoto S, Sakaguchi G, Eto M, Naito S, Takeda M, Yanagi Y, Tani K (2010) Enhanced antitumor effects of an engineered measles virus Edmonston strain expressing the wild-type N, P, L genes on human renal cell carcinoma. Mol Ther 18 (3):544–551. https://doi.org/10.1038/mt.2009.296
39. Stewart CE, Randall RE, Adamson CS (2014) Inhibitors of the interferon response enhance virus replication in vitro. PLoS One 9(11): e112014. https://doi.org/10.1371/journal.pone.0112014
40. Gujar S, Pol JG, Kim Y, Lee PW, Kroemer G (2018) Antitumor benefits of antiviral immunity: an underappreciated aspect of Oncolytic virotherapies. Trends Immunol 39 (3):209–221. https://doi.org/10.1016/j.it.2017.11.006
41. Speck T, Heidbuechel JPW, Veinalde R, Jaeger D, von Kalle C, Ball CR, Ungerechts G, Engeland CE (2018) Targeted BiTE expression by an oncolytic vector augments therapeutic efficacy against solid tumors.

Clin Cancer Res 24(9):2128–2137. https://doi.org/10.1158/1078-0432.CCR-17-2651

42. Hutzler S, Erbar S, Jabulowsky RA, Hanauer JRH, Schnotz JH, Beissert T, Bodmer BS, Eberle R, Boller K, Klamp T, Sahin U, Muhlebach MD (2017) Antigen-specific oncolytic MV-based tumor vaccines through presentation of selected tumor-associated antigens on infected cells or virus-like particles. Sci Rep 7 (1):16892. https://doi.org/10.1038/s41598-017-16928-8
43. Li H, Peng KW, Dingli D, Kratzke RA, Russell SJ (2010) Oncolytic measles viruses encoding interferon beta and the thyroidal sodium iodide symporter gene for mesothelioma virotherapy. Cancer Gene Ther 17(8):550–558. https://doi.org/10.1038/cgt.2010.10
44. Ushach I, Zlotnik A (2016) Biological role of granulocyte macrophage colony-stimulating factor (GM-CSF) and macrophage colony-stimulating factor (M-CSF) on cells of the myeloid lineage. J Leukoc Biol 100(3):481–489. https://doi.org/10.1189/jlb.3RU0316-144R
45. Grote D, Cattaneo R, Fielding AK (2003) Neutrophils contribute to the measles virus-induced antitumor effect: enhancement by granulocyte macrophage colony-stimulating factor expression. Cancer Res 63 (19):6463–6468
46. Grossardt C, Engeland CE, Bossow S, Halama N, Zaoui K, Leber MF, Springfeld C, Jaeger D, von Kalle C, Ungerechts G (2013) Granulocyte-macrophage colony-stimulating factor-armed oncolytic measles virus is an effective therapeutic cancer vaccine. Hum Gene Ther 24(7):644–654. https://doi.org/10.1089/hum.2012.205
47. Trinchieri G (2003) Interleukin-12 and the regulation of innate resistance and adaptive immunity. Nat Rev Immunol 3(2):133–146. https://doi.org/10.1038/nri1001
48. Chen DS, Mellman I (2013) Oncology meets immunology: the cancer-immunity cycle. Immunity 39(1):1–10. https://doi.org/10.1016/j.immuni.2013.07.012
49. Veinalde R, Grossardt C, Hartmann L, Bourgeois-Daigneault MC, Bell JC, Jager D, von Kalle C, Ungerechts G, Engeland CE (2017) Oncolytic measles virus encoding interleukin-12 mediates potent antitumor effects through T cell activation. Oncoimmunology 6(4):e1285992. https://doi.org/10.1080/2162402x.2017.1285992
50. Combredet C, Labrousse V, Mollet L, Lorin C, Delebecque F, Hurtrel B, McClure H, Feinberg MB, Brahic M, Tangy F (2003) A molecularly cloned Schwarz strain of measles virus vaccine induces strong immune responses in macaques and transgenic mice. J Virol 77 (21):11546–11554
51. Bankamp B, Takeda M, Zhang Y, Xu W, Rota PA (2011) Genetic characterization of measles vaccine strains. J Infect Dis 204(Suppl 1): S533–S548. https://doi.org/10.1093/infdis/jir097
52. Amedei A, Cappon A, Codolo G, Cabrelle A, Polenghi A, Benagiano M, Tasca E, Azzurri A, D'Elios MM, Del Prete G, de Bernard M (2006) The neutrophil-activating protein of helicobacter pylori promotes Th1 immune responses. J Clin Investig 116(4):1092–1101. https://doi.org/10.1172/jci27177
53. Iankov ID, Allen C, Federspiel MJ, Myers RM, Peng KW, Ingle JN, Russell SJ, Galanis E (2012) Expression of immunomodulatory neutrophil-activating protein of helicobacter pylori enhances the antitumor activity of oncolytic measles virus. Mol Ther 20 (6):1139–1147. https://doi.org/10.1038/mt.2012.4
54. Quezada SA, Peggs KS (2013) Exploiting CTLA-4, PD-1 and PD-L1 to reactivate the host immune response against cancer. Br J Cancer 108(8):1560–1565. https://doi.org/10.1038/bjc.2013.117
55. Rowshanravan B, Halliday N, Sansom DM (2018) CTLA-4: a moving target in immunotherapy. Blood 131(1):58–67
56. Engeland CE, Grossardt C, Veinaide R, Bossow S, Lutz D, Kaufmann JK, Shevchenko I, Umansky V, Nettelbeck DM, Weichert W, Jager D, von Katie C, Ungerechts G (2014) CTLA-4 and PD-L1 checkpoint blockade enhances Oncolytic measles virus therapy. Mol Ther 22(11):1949–1959. https://doi.org/10.1038/mt.2014.160
57. Klinger M, Benjamin J, Kischel R, Stienen S, Zugmaier G (2016) Harnessing T cells to fight cancer with BiTE(R) antibody constructs--past developments and future directions. Immunol Rev 270(1):193–208. https://doi.org/10.1111/imr.12393
58. Binnewies M, Roberts EW, Kersten K, Chan V (2018) Understanding the tumor immune microenvironment (TIME) for effective therapy. Nat Med 24(5):541–550. https://doi.org/10.1038/s41591-018-0014-x
59. Bajzer Z, Carr T, Josic K, Russell SJ, Dingli D (2008) Modeling of cancer virotherapy with recombinant measles viruses. J Theor Biol 252(1):109–122. https://doi.org/10.1016/j.jtbi.2008.01.016
60. Santiago DN, Heidbuechel JPW, Kandell WM, Walker R, Djeu J, Engeland CE, Abate-Daga D, Enderling H (2017) Fighting cancer with mathematics and viruses. Viruses 9(9): E239. https://doi.org/10.3390/v9090239

Chapter 8

A Functional Assay to Determine the Capacity of Oncolytic Viruses to Induce Immunogenic Tumor Cell Death

Tiphaine Delaunay, Carole Achard, Marc Grégoire, Frédéric Tangy, Nicolas Boisgerault, and Jean-François Fonteneau

Abstract

Oncolytic immunotherapy efficacy relies partially on the induction of immunogenic tumor cell death following infection with oncolytic viruses (OV) to induce an antitumor immune response. Here, we describe a method to determine if an OV is able to induce such an immunogenic tumor cell death. This method consists in testing whether tumor cells lysed by an OV are able to induce the maturation of human monocyte-derived immature dendritic cells (Mo-iDC).

Key words Antigen-presenting cells, Dendritic cells, Immunogenic cell death, Oncolytic viruses, Flow cytometry

1 Introduction

Oncolytic immunotherapy is a powerful approach to stimulate patient antitumor immune responses [1]. Following infection, oncolytic viruses (OV) are able to induce immunogenic death of tumor cells, which is characterized by the release of tumor-associated antigens (TAA) and danger signals of cellular origin coined as damage-associated molecular patterns (DAMPs) and of viral origin that are part of the pathogen-associated molecular patterns (PAMPs). Antigen-presenting cells (APC), such as myeloid dendritic cells (DC), in contact with OV-infected cells, capture tumor antigens and undergo maturation after the recognition of DAMPs and PAMPs by pattern recognition receptors. Mature DC are then able to present TAA to T lymphocytes to induce an antitumor immune response.

Our team was one of the first to report in vitro this phenomenon using the Schwarz strain of Measles Virus (MV), which is spontaneously oncolytic against many cancers including mesothelioma [2]. We showed that MV-infected tumor cells were able to

Christine E. Engeland (ed.), *Oncolytic Viruses*, Methods in Molecular Biology, vol. 2058,
https://doi.org/10.1007/978-1-4939-9794-7_8,

induce the maturation of human monocyte-derived immature dendritic cells (Mo-iDC), whereas UV-B irradiated tumor cells were not. This maturation was characterized by an increased expression of MHC and costimulatory molecules (CD80, CD83, CD86, CD40, and HLA-DR) along with the production of pro-inflammatory cytokines (IL-6, IFN-α, TNF-α, IL-1β, and IL-12p70). Furthermore, DC exposed to MV-infected mesothelioma cells were used to stimulate T lymphocytes of healthy donors and were able to prime the amplification of $CD8^{+}$ T lymphocytes specific for mesothelin, a tumor differentiation antigen. The induction of immunogenic cell death by attenuated MV was also observed by the Melcher team [3]. They reported that MV stimulates the release from dying tumor cells of an inflammatory set of cytokines and chemokines (IL-6, IL-8, CCL5, IFN-α, IFN-β, and IFN-λ) and DAMPs (HMGB1). These inflammatory proteins induce the maturation of Mo-iDC characterized by the increased expression of CD80 and CD86. They also showed that DC exposed to MV-infected tumor cells are able to induce a tumor-specific cytotoxic T cell response. We then later showed that MV and MV-infected tumor cells are potent activators of plasmacytoid DC (pDC), a subset of DC involved in antiviral immune responses that is able to produce large amounts of IFN-α in response to viruses [4–6].

Thus, Mo-iDC that are easily obtainable from peripheral blood lymphocytes (PBMC) can be used as a tool to determine if an OV is able to trigger immunogenic death of tumor cells. Here, we propose a functional assay based on the maturation of Mo-iDC applicable to other OV in order to study their capacity to induce immunogenic cell death. Maturation of Mo-iDC is measured by expression analysis of the maturation marker CD83 on the surface of DC by flow cytometry. Negative controls are conditions where Mo-iDC are cultured alone, with living tumor cells or with non-immunogenic apoptotic tumor cells obtained after UV-B irradiation. In these conditions DC should not express CD83. The positive control consists in a condition in which Mo-iDC are cultured with bacterial lipopolysaccharide (LPS). LPS induces maturation of DC characterized by the expression of CD83 on their surface [7]. Test conditions are Mo-iDC cultured with the OV alone or with tumor cells lyzed by the OV. We hope this method can be useful to document the induction of immunogenic tumor cell death by an OV.

2 Materials

1. Healthy donor peripheral blood mononuclear cells.
2. Magnetic beads for negative selection of monocytes (Stemcell Technologies or Miltenyi Biotec).

3. Complete RPMI for DC culture: RPMI1640 medium supplemented with L-glutamine (2 mM), penicillin (100 U/mL), streptomycin (100 μg/mL), 20 mg/mL human serum albumin.
4. RPMI for tumor cell culture: RPMI1640 medium supplemented with L-glutamine (2 mM), penicillin (100 U/mL), streptomycin (100 μg/mL), 10% fetal calf serum.
5. Granulocyte macrophage colony-stimulating factor (GM-CSF).
6. Interleukin-4 (IL-4).
7. Oncolytic virus and tumor cell culture of interest.
8. UV-B source (Bio-Link from Vilber-Lourmat or Stratalinker from Stratagene).
9. Lipopolysaccharide (LPS).

    Phosphate buffer saline (PBS) containing 0.1% bovine serum albumin (BSA).
10. Fluorescence-conjugated monoclonal antibodies against CD83 and HLA-DR (BD Biosciences or Biolegend) and corresponding fluorescence-conjugated isotype control antibodies.
11. PBS containing 4% (v/v) paraformaldehyde.
12. Flow cytometer.

3 Methods

3.1 Monocyte-Derived iDC Production

1. Purify monocytes from healthy donor peripheral blood mononuclear cells by magnetic bead negative selection as recommended in the manufacturer's protocol or by counterflow centrifugation [8].
2. Incubate in a 6-well plate with 3 mL per well, 2×10^6 monocytes per mL in complete RPMI supplemented with 1000 U/mL of granulocyte macrophage colony-stimulating factor (GM-CSF) and 200 U/mL of interleukin-4 (IL-4). Then culture the cells during 6 days at 37 °C with 5% CO_2.
3. Three days after the start of the culture, replace half of the medium with fresh complete RPMI containing GM-CSF (1000 U/mL) and IL-4 (200 U/mL).
4. Harvest and count living Mo-iDC after 6 days.

3.2 Oncolytic Virus Infection and UV-B Irradiation

1. Plate 2 wells of a 6-well plate with 5×10^5 tumor cells to be infected or not by the OV. On a separate 6-well plate, plate 1 well with 5×10^5 tumor cells for UV-B irradiation.

2. On the following day, infect the tumor cells with the studied OV at appropriate multiplicities of infection (MOI). For MV we used an MOI of 1 [2]. Culture infected tumor cells at 37 °C until at least 50% are lyzed by the OV (1–3 days depending on the OV and the MOI). Harvest OV-infected tumor cells by flushing gently the plate with their culture supernatant and store them on ice before the coculture with Mo-iDC.
3. In parallel, irradiate the tumor cells with UV-B (312 nm, 10 J/cm^2, without the plate lid and then culture them during 24–48 h. We recommend using a dose of UV-B adapted to each individual tumor cell line to get at least 50% of apoptotic cells. UV-B irradiation doses may range from 0.2 to 10 J/cm^2 depending on the tumor cell line. It is recommended that the culture plate is set on ice for extended irradiation times so to avoid heat shock.

3.3 DC/Tumor Cell Coculture

1. Incubate 1×10^6 immature Mo-iDC in 1 mL/well of complete RPMI in 24-well plates. Add 1 mL of either:
   (a) Medium alone (negative control).
   (b) 1×10^6 living tumor cells in their culture medium (negative control).
   (c) 1×10^6 UV-irradiated tumor cells (negative control) in their culture medium.
   (d) Medium containing 2 μg/mL of LPS (1 μg/mL final) (positive control).
   (e) Medium containing the OV. For MV we used an MOI of 1 (test condition) [2].
   (f) 1×10^6 MV-infected tumor cells (test condition) in their culture medium.
2. Culture the iDC for 20–24 h at 37 °C with 5% CO_2.

3.4 DC Maturation Analysis

1. After 20–24 h, harvest and count the DC. For each condition, resuspend 1×10^5 DC in PBS containing 0.1% BSA and 1 μg/mL of fluorescence-conjugated monoclonal antibodies against both CD83 and HLA-DR, or corresponding fluorescence-conjugated isotype control antibodies. Incubate in the dark at 4 °C. After 30 min, wash the cells three times with PBS 0.1% BSA before fixing them for 10 min in PBS containing 4% (v/v) paraformaldehyde to inactivate the virus. Transfer the cells into suitable tubes for flow cytometry analysis.
2. Analyze DC maturation by flow cytometry (Fig. 1). First, draw a gate to select DC based on their FSC/SSC profile. Within this population, draw a second gate on $HLA\text{-}DR^+$ cells to differentiate DC from tumor cells. Analyze CD83 expression only on $HLA\text{-}DR^+$ positive cells. In the negative control conditions,

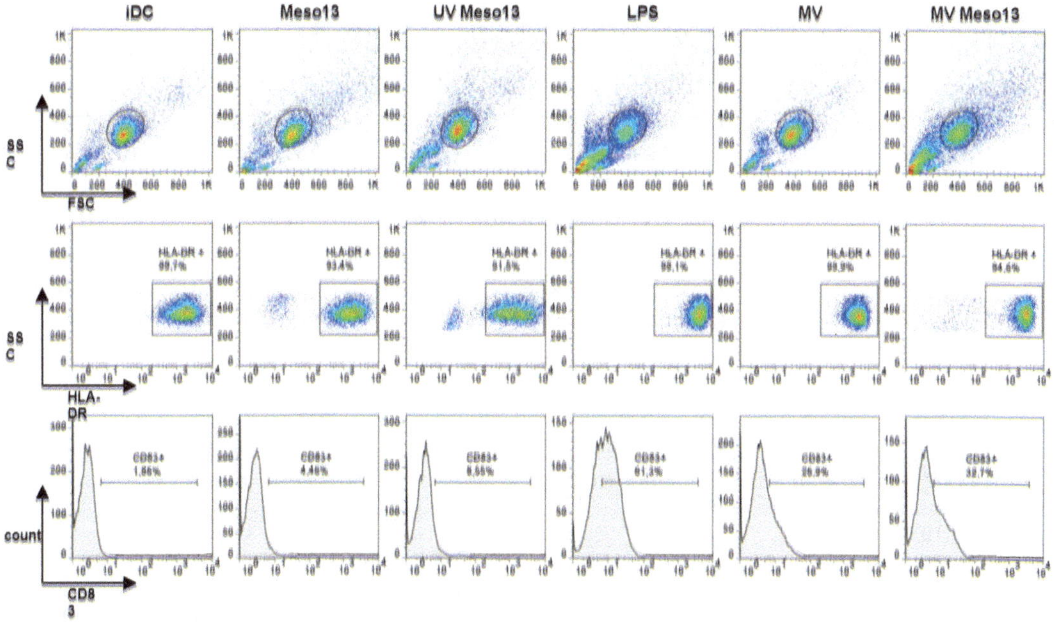

Fig. 1 MV-induced immunogenic mesothelioma cell death. Monocyte-derived immature DC were cultured alone or with living Meso13 mesothelioma cell line (Meso13), UV-B irradiated Meso13 (UV Meso13), LPS, MV, or MV-infected Meso13 (MV Meso13). After 24h, cells were harvested and stained with fluorescence-conjugated monoclonal antibodies against HLA-DR and CD83. CD83 fluorescence was analyzed on HLA-DR + cells that correspond to dendritic cells

that is, Mo-iDC cultured alone, or with living or UV-irradiated tumor cells, the maturation marker CD83 should not be expressed. In the positive control condition, a majority of Mo-iDC cultured with LPS should express the maturation marker CD83. In the test conditions, CD83 expression allows to determine whether the OV alone or OV-infected tumor cells are able to induce DC maturation.

4 Notes

1. Other maturation markers can be studied such as CD80, CD86 or CD40 costimulatory molecules. These molecules are already expressed by immature DC and their expression increases on mature DC.
2. You may collect the coculture supernatants to perform ELISA assays against IL-10 and IL-12p70. IL-10 is an immunosuppressive cytokine, whereas IL-12p70 is necessary for the induction of Th1 and cytotoxic T cell response. The IL-10/IL-12p70 ratio will be inversely correlated with Mo-iDC maturation.

3. Tumor cells should be free of *Mycoplasma* infection, otherwise UV-irradiated tumor cells used as a negative control will be able to trigger Mo-iDC maturation.
4. Immunogenic cell death can also be checked on both myeloid and plasmacytoid DC from the blood. These DC can be sorted from PBMCs by magnetic sorting using the EasySep™ Human pan-DC pre-enrichment kit (Stem cells) and flow cytometry sorting [5, 7]. When using the pDC, we recommend to add IL-3 as it is necessary for pDC survival when they are cultured alone or with the OV only.

References

1. Lichty BD, Breitbach CJ, Stojdl DF, Bell JC (2014) Going viral with cancer immunotherapy. Nat Rev Cancer 14(8):559–567. https://doi.org/10.1038/nrc3770
2. Gauvrit A, Brandler S, Sapede-Peroz C, Boisgerault N, Tangy F, Gregoire M (2008) Measles virus induces oncolysis of mesothelioma cells and allows dendritic cells to cross-prime tumor-specific CD8 response. Cancer Res 68 (12):4882–4892. https://doi.org/10.1158/0008-5472.can-07-6265
3. Donnelly OG, Errington-Mais F, Steele L, Hadac E, Jennings V, Scott K, Peach H, Phillips RM, Bond J, Pandha H, Harrington K, Vile R, Russell S, Selby P, Melcher AA (2013) Measles virus causes immunogenic cell death in human melanoma. Gene Ther 20(1):7–15. https://doi.org/10.1038/gt.2011.205
4. Achard C, Guillerme JB, Bruni D, Boisgerault N, Combredet C, Tangy F, Jouvenet N, Gregoire M, Fonteneau JF (2017) Oncolytic measles virus induces tumor necrosis factor-related apoptosis-inducing ligand (TRAIL)-mediated cytotoxicity by human myeloid and plasmacytoid dendritic cells. Oncoimmunology 6(1):e1261240. https://doi.org/10.1080/2162402x.2016.1261240
5. Fonteneau JF, Guillerme JB, Tangy F, Gregoire M (2013) Attenuated measles virus used as an oncolytic virus activates myeloid and plasmacytoid dendritic cells. Oncoimmunology 2(5):e24212. https://doi.org/10.4161/onci.24212
6. Guillerme JB, Boisgerault N, Roulois D, Menager J, Combredet C, Tangy F, Fonteneau JF, Gregoire M (2013) Measles virus vaccine-infected tumor cells induce tumor antigen cross-presentation by human plasmacytoid dendritic cells. Clin Cancer Res 19(5):1147–1158. https://doi.org/10.1158/1078-0432.ccr-12-2733
7. Royer PJ, Tanguy-Royer S, Ebstein F, Sapede C, Simon T, Barbieux I, Oger R, Gregoire M (2006) Culture medium and protein supplementation in the generation and maturation of dendritic cells. Scand J Immunol 63 (6):401–409. https://doi.org/10.1111/j.1365-3083.2006.001757.x
8. Coulais D, Panterne C, Fonteneau JF, Gregoire M (2012) Purification of circulating plasmacytoid dendritic cells using counterflow centrifugal elutriation and immunomagnetic beads. Cytotherapy 14(7):887–896. https://doi.org/10.3109/14653249.2012.689129

Chapter 9

Design and Production of Newcastle Disease Virus for Intratumoral Immunomodulation

Gayathri Vijayakumar and Dmitriy Zamarin

Abstract

Newcastle disease virus (NDV) is an avian paramyxovirus that has been extensively studied as an oncolytic agent, in addition to being an economically important pathogen in the poultry industry. The establishment of a reverse genetics system for this virus has enabled the development of genetically modified recombinant NDV viruses with improved oncolytic and immunotherapeutic properties. In this chapter, we describe the materials and methods involved in the in vitro cloning and rescue of NDV expressing murine 4-1BBL as well as the in vivo evaluation of NDV expressing 4-1BBL in a B16-F10 murine melanoma model.

Key words Newcastle disease virus, Reverse genetics, Virus rescue, Oncolytic vector, Immunotherapy

1 Introduction

Newcastle Disease Virus (NDV) is an avian paramyxovirus and an economically important pathogen for the poultry industry [1]. NDV, or avian paramyxovirus type 1 (APMV-1), is classified as a member of the genus *Avulavirus* belonging to the family *Paramyxoviridae* [2]. It consists of a negative-sense, single-stranded RNA genome that is nonsegmented and is approximately 15,186 nucleotides (nt) in length [3]. Additionally, it encodes for six proteins including the nucleoprotein (NP), the phosphoprotein (P), the matrix protein (M), the fusion protein (F), the hemagglutinin-neuraminidase (HN), and the large polymerase protein (L) [4]. NDV isolates can be grouped into velogenic (highly virulent), mesogenic (intermediate), or lentogenic (nonvirulent) strains, depending on the virulence and pathogenicity in avian species and this is correlated with the cleavage site in the F protein [5]. NDV LaSota is a naturally occurring lentogenic (nonvirulent) strain commonly used as a live attenuated vaccine in the poultry industry, and has been demonstrated to be an effective and safe vaccine vector in multiple studies [6, 7].

Christine E. Engeland (ed.), *Oncolytic Viruses*, Methods in Molecular Biology, vol. 2058,
https://doi.org/10.1007/978-1-4939-9794-7_9,

Besides being an important avian pathogen, NDV has been explored as a candidate for oncolytic immunotherapy since 1955 [8]. Pioneering preclinical work by several labs including Sinkovics and Cassell from the 1960s sought to establish NDV as an anti-neoplastic agent [9, 10], with Cassell and Garrett observing in 1965 the development of anti-tumor immunity after viral oncolytic therapy with NDV [11, 12]. This precipitated the development of NDV oncolysates as an adjunctive immunotherapeutic vaccine for post-operative management of malignant melanoma patients, with two Phase II clinical studies involving 32 and 51 patients with Stage II malignant melanoma [13]. A 10 year and a 15 year follow-up with these patients showed that over 60% and 55% were alive and free of recurrent disease, respectively, which was superior to historical controls [14, 15]. Another Phase II study involving 208 patients with locally advanced renal cell carcinoma was carried out with NDV oncolysates using the NDV 73T strain combined with low-dose recombinant interleukin-2 (IL-2) and interferon-α (IFNα), demonstrating improved disease-free survival [16]. Later on, an attenuated veterinary vaccine strain NDV-MTH-68/N was tested out in a variety of previously chemorefractory tumors by the Hungarian scientist Dr. Laszlo Csatary in a placebo-controlled Phase II clinical trial demonstrating an overall improvement in survival rates as compared to the control group [17, 18]. Furthermore, other strains such as the lentogenic NDV-HUJ strain [19], the lytic NDV PV701 [20, 21], and ATV-NDV [22] have been studied in patients with glioblastoma multiforme (GBM) and a variety of other advanced cancers.

As evidenced, decades of research have demonstrated the natural and selective oncolytic capabilities of NDV in different mammalian cancer cell lines, animal tumor models, and clinical trials [23–25]. The establishment of a reverse genetics systems for this virus has been particularly important as it has enabled the development of genetically modified recombinant NDV viruses with improved oncolytic and immunostimulatory properties [26, 27]. Several strategies have been attempted so far, including but not limited to the introduction of a polybasic cleavage site in the F protein of a lentogenic Hitchner B1 NDV strain, engineering of viruses to express IFNγ, GM-CSF, IL-2, or TNFα, as well as the generation of viruses which possess both modifications [28–33]. Recent preclinical studies in a bilateral flank syngeneic melanoma model as well as in a bladder cancer model have shown that intratumoral administration of NDV in conjunction with systemic delivery of immune checkpoint blockade antibodies that target CTLA-4 and PD-1, could effectively activate both the innate and adaptive immune pathways to induce an abscopal effect. This abscopal effect was characterized by an enhanced immune infiltration into distant nontreated tumors [34, 35]. Additionally, T cell effector function could be further enhanced within the tumor microenvironment with the stimulation of a relevant co-stimulatory pathway by the intratumoral delivery of a recombinant NDV expressing the ICOS ligand (ICOSL) [36]. These studies demonstrate that stimulation of an adaptive immune response within the injected tumor has a potential to induce a stronger systemic anti-tumor immune response, which could result in superior anti-tumor efficacy not just against the injected, but also distant tumors.

Such strategies provide for a possibility of intratumoral delivery of immunostimulatory ligands that otherwise carry an increased potential for systemic toxicity. In this book chapter, we describe detailed methods for the generation of recombinant NDV with enhanced immunostimulatory effects. As a model, we use NDV expressing 4-1BB ligand (4-1BBL), a ligand that binds to the 4-1BB (CD137) receptor, signaling through which provides co-stimulatory function for cytotoxic T cells and NK cells, which has been explored for anti-cancer therapy using agonistic antibodies [37]. An agonistic antibody to 4-1BB has been evaluated in a phase I study of 83 patients with advanced malignancies [38]. Evidence of clinical activity was seen across a wide dose range, but the trial was suspended due to hepatotoxicity seen at higher doses. Intratumoral delivery of NDV expressing 4-1BBL has a potential to provide the initial costimulatory signal needed for T cell activation, while avoiding systemic toxicity. The chapter is divided into five sections with Subheadings 1 and 2 describing the cloning and generation of NDV-LaSota full-length cDNA plasmid encoding 4-1BBL as well as the generation of the helper plasmids. Subheading 3 details the procedures for setting up rescue transfections, RT-PCR confirmation of positive rescues and tittering of NDV-LaSota viruses expressing 4-1BBL. Finally, Subheadings 3.6 and 4 demonstrate confirmation of surface expression of 4-1BBL in NDV-infected cells by flow cytometry and will demonstrate the ability of the virus to drive immune cell infiltration into both treated and nontreated tumors in vivo.

2 Materials

2.1 Construction of Recombinant Full-Length NDV (rNDV) Encoding 4-1BBL and Helper Plasmids

1. Codon-optimized synthesized gene (sequence available upon request).
2. *Taq* DNA polymerase, High Fidelity.
3. Primer set (forward and reverse, 10 μM stock concentration).
4. dNTP mix.
5. 10× Tris–acetate–EDTA (TAE) electrophoresis buffer: Mix 11.4 mL of glacial acetic acid, 3.72 g of EDTA, 48.4 g Tris base, 900 mL deionized water. Adjust volume to 1 L with

deionized water. Dilute 100 mL of 10× stock to 1 L with deionized water to make 1× TAE buffer.

6. Ethidium Bromide.
7. 1.5% Agarose-TAE gel: Measure 1.5 g agarose in 100 mL TAE. Microwave for 2–3 min until the agarose is completely dissolved. Add 2–3 μL of ethidium bromide once the agarose solution has cooled down to about 60 °C.
8. Gel loading dye, without SDS.
9. DNA Ladder (0.1–10.0 kb).
10. Gel and PCR cleanup kit.
11. SacII restriction enzyme.
12. Antarctic phosphatase.
13. In-Fusion® HD Cloning Plus.
14. Stable competent *E. coli*—high efficiency.
15. Luria–Bertani (LB)—Ampicillin agar plates: Mix one pouch of premixed, presterilized LB growth medium including agar and ampicillin in 200 mL deionized water in an autoclaved flask. Gently heat the mixture in a microwave, without bringing it to a full boil so that it completely dissolves without degrading the ampicillin. Pour out the media into petri dishes and let it cool at room temperature.
16. LB broth supplemented with ampicillin (100 μg/mL).
17. DNA Miniprep Kit.
18. Plasmid Midiprep Kit for purification of high purity, endotoxin-free DNA.
19. 1.5 mL microcentrifuge tubes.
20. Micropipette tips, sterile.
21. PCR thermal cyclers.
22. Dry block heaters.
23. UV transilluminator.
24. NanoDrop spectrophotometer.
25. Water bath.
26. Super optimal broth with catabolic repressor (SOC) media.
27. Microbiological incubators.
28. Sterile 1 L baffled flasks.
29. Shaking incubators.
30. Refrigerated tabletop centrifuge.

2.2 Rescue and Amplification of Recombinant NDV Expressing the Transgene

1. BSRT-7 cells.
2. Vero cells (ATCC, CCL-81).
3. Complete DMEM (cDMEM): Supplement Dulbecco's Modified Eagle Medium (DMEM), high glucose, pyruvate. With 10% fetal bovine serum (FBS) and 1% penicillin–streptomycin (P/S).
4. Trypsin–EDTA (0.05%), phenol red.
5. 1× PBS, pH 7.4.
6. Opti-MEM™ Reduced Serum Medium, GlutaMAX™ Supplement.
7. High efficiency transfection reagent.
8. 9–11-day-old specific pathogen-free (SPF) embryonated chicken eggs.
9. One Step RT-PCR with hot-start, thermostable *Taq* polymerase with enhanced specificity.
10. RNA Mini Kit.
11. Turkey red blood cells (RBC).
12. 1% PBS-BSA: Mix 1 mL of 10× BSA and 9 mL of 1× PBS.
13. 4% formaldehyde, methanol-free fixative solution.
14. Goat anti-rabbit IgG (H+L) highly cross-adsorbed secondary antibody, Alexa Fluor 488.
15. 6-well tissue culture plates.
16. 1.5 mL microcentrifuge tubes.
17. Micropipette tips, sterile.
18. Class II biological safety cabinet.
19. 37 °C, 5% CO_2 humidified incubator.
20. Cell scrapers.
21. Forceps.
22. Disposable spatula.
23. 15 and 50 mL conical centrifuge tubes.
24. Refrigerated tabletop centrifuge.
25. V-bottom 96-well plates.
26. 96-well tissue culture plates.
27. PCR thermal cyclers.
28. Water bath.
29. UV transilluminator.
30. Cryogenic vials.

2.3 Confirmation of Ligand Expression

1. Nonenzymatic EDTA-based cell dissociation solution.
2. FACS buffer: 1× PBS with 1% FBS.
3. 6-well tissue culture plates.
4. Refrigerated tabletop centrifuge.
5. 1.5 mL microcentrifuge tube.
6. 96-well round bottom plate.
7. Flow cytometer.

2.4 B16-F10 Tumor Implantation and Intratumoral Treatment

1. B16-F10 melanoma cells (ATCC, CRL-6475).
2. Complete DMEM, Nutrient Mixture F-12 (cDMEM/F12). Supplement DMEM/F12 with 10% FBS and 1% P/S.
3. 0.4% Trypan Blue Solution.
4. T175 flasks.
5. 15 and 50 mL conical centrifuge tubes.
6. Micropipette tips, sterile.
7. Refrigerated tabletop centrifuge.
8. Automated cell counter or hemacytometer.
9. 70 μm strainers.
10. 1 mL Sub-Q Syringe.
11. 30G × 1/2 in. PrecisionGlide™ specialty use sterile hypodermic needle.
12. 70% ethanol swabs.
13. Small animal trimmers.
14. Insulin syringes with 6 mm × 31G needles.

2.5 Isolation of Tumor-Infiltrating Lymphocytes

1. RPMI-0: RPMI-1640 Medium.
2. Complete RPMI (cRPMI): Supplement RPMI-0 with 10% FBS and 1% P/S.
3. 2× Liberase/DNAse solution: Final concentration of Liberase in the sample will be 1.67 Wunsch U/mL. Prepare fresh. Add 7.5 mL of RPMI-0 to a 5 mg (26 U) vial of Liberase™ TL Research Grade and leave it on ice for 30 min. For each 7.5 mL of Liberase solution, add 150 μL of 100× DNAse (from a 20 mg/mL stock).
4. Preweighed 1.5 mL microcentrifuge tubes for tumors.
5. 6-Well plates for grinding tumors.
6. 70 μm strainers.
7. 15 mL centrifuge tubes.
8. Refrigerated tabletop centrifuge.
9. Flow cytometer.

3 Methods

3.1 Construction of Recombinant Full-Length NDV (rNDV) Encoding an Immunomodulatory Transgene (4-1BBL)

pNDV-LaSota containing the full-length antigenomic cDNA of the NDV LaSota strain was constructed by reverse transcription-polymerase chain reaction (RT-PCR) of virion-derived genomic RNA with the addition of a unique restriction enzyme site SacII (nt 1747 to 1752), into a low-copy transcription vector as shown in Fig. 1 [39]. The antigenomic cDNA is under the control of the T7 promoter followed by a noncoding self-cleaving hepatitis delta virus (HDV) ribozyme site and a T7 terminator.

As the viral RNA polymerase synthesizes and transcribes the genome in a polar and sequential manner by a start and stop mechanism, starting from the 3′ entry site, there is a gradient of expression with respect to the position of the transgene relative to the 3′ start end [40]. Furthermore, studies have shown that the optimal insertion site for a transgene for efficient expression is in between the P and M genes [41]. Foreign genes are inserted into the SacII cloning site as a transcriptional unit in between the P and M gene and are flanked by short conserved sequence motifs that guide transcription, namely the gene start and gene end sequences [42].

1. Digest and linearize pNDV-LaSota plasmid (2–5 μg) with 2–5 μL of SacII restriction enzyme for 1 h at 37 °C in a total reaction volume of 50 μL. Heat-inactivate the enzyme by incubating at 65 °C for 20 min (*see* **Note 1**).
2. To prevent recircularization of the linearized pNDV-LaSota plasmid, incubate the above reaction mixture with Antarctic phosphatase for 1 h at 37 °C. Purify with the gel and PCR cleanup kit following the manufacturer's instruction.

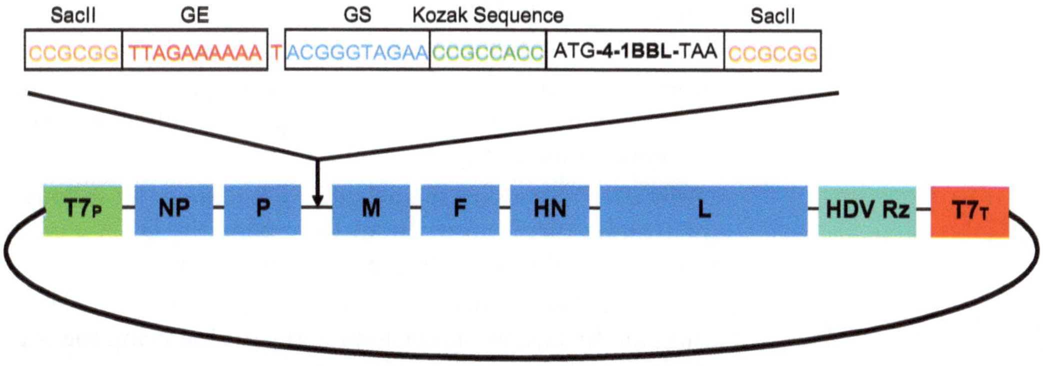

Fig. 1 Construction of a full-length antigenomic NDV plasmid containing the transgene for murine 4-1BBL (m4-1BBL). The T7 RNA Polymerase transcribes the antigenome starting at the T7 promoter (T7p). The HDV ribozyme site ensures the cleavage of the transcript. The ORF for 4-1BBL is flanked at the beginning with the gene end (GE) and gene start (GS) signals followed by the Kozak sequence and with the SacII site at the end after the stop codon

Table 1
List of primers

Primer	Sequence
4-1BBL cloning for	GCACCGAGTTCCCCCCCGCGGTTAGAAAAAATACGGG TAGAACCGCCACCATGGACCGCGCCGTTAGC
4-1BBL cloning Rev	AGTTGGACCTTGGGTCCGCGGATTATCATTCCCATGGGTTGTC
Sequencing/RT-PCR primer for	GATATCTAGATTAGAAAAAATACGGGTAGAACCGCCACC
Sequencing/RT-PCR primer Rev	CAAAGTACAGCCCAATTGTCC

Determine the DNA concentration using a NanoDrop Spectrophotometer (*see* **Note 1**).

3. Generate a codon-optimized insert using gene synthesis. Make sure to add a Kozak consensus sequence before the start codon to enhance translation efficiency of the inserted transgene. In addition, take care to ensure that the cloned transgene follows the "rule of six" to ensure optimal viral replication. This can be achieved with careful primer design, with adding additional stop codons to the transgene (*see* **Notes 2–4**).
4. Amplify the insert by PCR with primers specifically designed to include 15 nucleotides of 5′ overhangs that are complementary to the 5′ and 3′ cloning ends of the linearized pNDV-LaSota plasmid (Table 1). These overhangs are necessary for the In-Fusion® HD system-based cloning.
5. Mix the PCR reactions with the gel loading dye and load into individual wells on a 1.5% Agarose-TAE gel in a gel box filled with 1× TAE buffer, including one well with the DNA ladder. Run the gel at 80–150 V for 45–60 min or until the dye front is 80% of the way down the gel.
6. After visualization on a UV transilluminator, cut the gene band of the appropriate size from the gel and place it into a microcentrifuge tube.
7. Purify the DNA using the gel and PCR cleanup kit following the manufacturer's instruction. Determine the DNA concentration using the NanoDrop Spectrophotometer.
8. Clone the insert into the linearized pNDV-LaSota plasmid using the In-Fusion® HD Cloning Plus kit following the manufacturer's instructions.
9. Transform 50 μL of Stable Competent *E. coli* cells with 2.5 μL of the In-Fusion reaction mixture for 30 min on ice (*see* **Note 5**).
10. Heat-shock the transformation reaction for 30 s at 42 °C and incubate it for 2 min on ice.

11. Add 1000 μL of SOC medium to the transformation reaction and incubate for 1 h at 37 °C in a shaker.
12. Centrifuge the transformation reaction at 3800 × *g* for 5 min. Discard the supernatant and resuspend the pellet in 100 μL of SOC medium. Plate all of the transformation onto a 10 cm LB-Ampicillin agar plate and incubate overnight at 30 °C.
13. The next day, pick individual colonies and grow them overnight in a shaker with LB broth supplemented with ampicillin at 30 °C.
14. Isolate plasmid DNA using the Miniprep Kit following the manufacturer's instructions and sequence the plasmids with the screening primers listed in Table 1 to confirm presence of the transgene and absence of any mutations introduced during the PCR or cloning procedure.
15. The correct clone is then grown overnight in 200 mL of LB broth supplemented with ampicillin at 30 °C.
16. Isolate high purity, endotoxin-free plasmid DNA using the plasmid Midiprep Kit following the manufacturer's instructions. Measure DNA concentration using the NanoDrop Spectrophotometer.

3.2 Construction of the NDV NP, P, and L and Expression Plasmids

The rescue of rNDV viruses requires the cotransfection of cDNA encoding the ribonucleoprotein complex or RNP, which consists of the viral polymerase (P and L proteins), the nucleoprotein (NP) and the full-length antigenomic RNA of the virus.

Genes encoding each of the L, P, and NP proteins are cloned into a pTM1 expression vector under the transcriptional control of a T7 promoter. The expression vector also includes T7 terminator, a pUC19 origin of replication and an ampicillin resistance gene. The full-length DNAs of each gene for construction of the plasmids are obtained by RT-PCR amplification from viral RNA. Additionally a codon-optimized T7 RNA polymerase is cloned into a pCAGGs expression vector (pCAGGs-T7opt) [43].

3.3 Rescue and Amplification of Recombinant NDV Expressing the Transgene

The plasmid-based rescue of recombinant NDV viruses as shown in Fig. 2 entails the transfection of cells that constitutively express T7 polymerase such as BSRT7 cells, with (1) plasmid DNA encoding the L, NP, and P proteins of NDV under the transcriptional control of a T7 promoter (pTM1-L, pTM1-NP, and pTM1-P, respectively), (2) pCAGGs expression vector encoding T7 RNA polymerase, and (3) a plasmid DNA encoding the NDV antigenome with the transgene under the transcriptional control of a T7 promoter. Subsequently, the rescue of infectious NDV viruses is facilitated by the amplification and transcription of the NDV antigenome in embryonated chicken eggs.

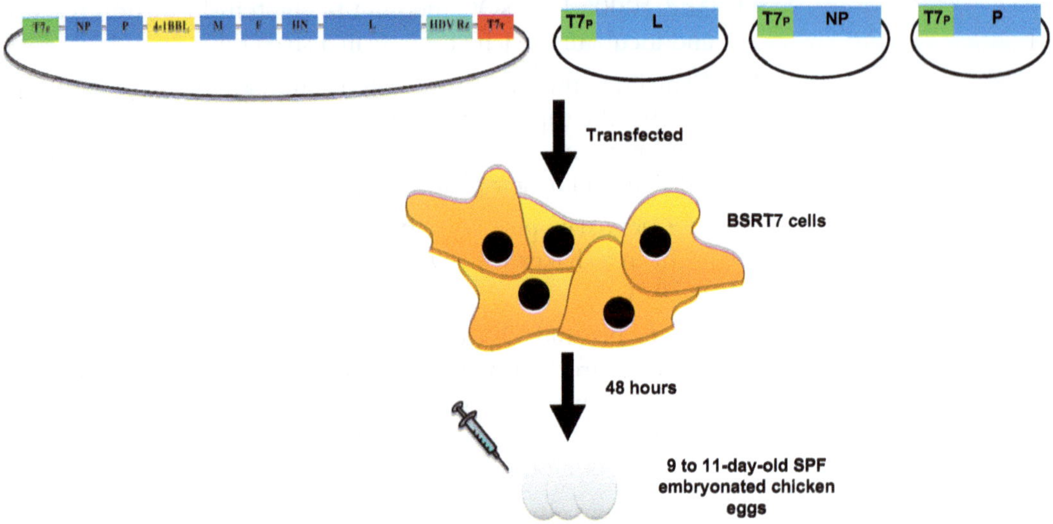

Fig. 2 Rescue of recombinant NDVs expressing m4-1BBL. BSRT7 cells are cotransfected with plasmids encoding L, NP, and P proteins under the transcriptional control of a T7 promoter and the plasmid encoding the NDV antigenome with the cloned transgene (4-1BBL) under the transcriptional control of a T7 promoter. The supernatants of transfected cells are injected into the allantoic cavities of 9–11-day-old embryonated chicken eggs

3.3.1 Setting Up Rescue Transfections

1. Maintain BSRT7 cells in cDMEM. One day before rescue transfections, split BSRT7 cells into 6-well tissue culture plates at a concentration of 3×10^5 cells per well to achieve 75–80% confluency on the day of transfection (*see* **Note 6**).
2. The next day, perform transfections in the tissue culture hood. Warm the transfection reagent TransIT-LT1 to room temperature and gently vortex before using (*see* **Note 7**).
3. Prepare the transfection plasmid cocktail in 250 μL of Opti-MEM Reduced-Serum Medium in a sterile 1.5 mL microcentrifuge tube. For a single transfection, add 1 μg of pTM1-NP, 0.5 μg of pTM1-P, 0.5 μg of pTM1-L, 1 μg of pCAGGs-T7opt, and 2 μg of pNDV-LaSota into the microcentrifuge tube with 250 μL of Opti-MEM and mix gently by pipetting (*see* **Note 7**).
4. Add dropwise 15 μL of the TransIT-LT1 reagent to the plasmid–Opti-MEM mixture and mix gently by pipetting. Incubate the transfection plasmid cocktail mix at room temperature for 30 min (*see* **Note 8**).
5. Add the transfection plasmid cocktail mix dropwise to the wells and gently rock the plate back and forth to evenly distribute the transfection plasmid cocktail mix. Incubate at 37 °C, 5% CO_2 for 48 h.
6. Scrape the cells along with the tissue culture supernatant into a sterile 1.5 mL microcentrifuge tube and inoculate 500 μL

directly into the allantoic cavity of 9–11-day-old SPF embryonated chicken eggs. Incubate the eggs at 37 °C for 2–3 days.

7. After incubation, place the infected eggs at 4 °C overnight.
8. Harvest the allantoic fluid in a biosafety cabinet by carefully taping the apical section of the egg, over the air sac with a spoon after sterilizing the eggs with 70% ethanol.
9. Remove the cracked eggshells with forceps and expose the allantoic fluid by carefully peeling back the allantoic membrane without damaging the blood vessels and the yolk.
10. Press down the embryo with a disposable spatula. Collect the allantoic fluid and transfer into a 15 mL centrifuge tube on ice (*see* **Note 9**).
11. Centrifuge the allantoic fluid at 250 × *g* for 5 min at 4 °C to remove any debris and transfer the supernatant into a new sterile 15 mL centrifuge tube.

3.3.2 Hemagglutination (HA) Assay Confirming Positive rNDV Rescue

1. A positive rescue is confirmed by a hemagglutination (HA) assay using 0.5% turkey RBC in 1× PBS, with approximately 10^6 plaque forming units (pfu) per mL giving a positive signal.
2. HA assays are carried out in V-bottom 96-well plates with negative controls (either 1× PBS or uninfected allantoic fluid) and/or positive controls (NDV virus grown in allantoic fluid) included to validate the assay.
3. Pipet 50 μL of 1× PBS per well into all wells, except for the first row.
4. Pipet 100 μL of the allantoic fluid from individual eggs into the wells in the first row of the plate and perform twofold serial dilutions across the plate.
5. Pipet 50 μL of 0.5% turkey RBC into all the wells and gently tap the plate to mix.
6. Incubate the plate at 4 °C (or on ice) for 30–45 min or until a clear pellet is observed in the negative control wells. Absence of a red pellet in rows 2, 3 and 4 as shown in Fig. 3 demonstrates positive viral rescue of rNDV-41BBL.

3.3.3 RT-PCR Confirmation of Viral Transgene

1. Confirm presence of the transgene by extracting the viral RNA using the RNA Mini Kit following the manufacturer's instructions.
2. Set up an RT-PCR for the transgene with the screening primers listed in Table 1 using wild-type NDV (NDV-WT) as control. Run the products of the RT-PCR reaction on a 1.5% Agarose-TAE gel in a gel box filled with 1× TAE buffer, with one well containing the DNA ladder.

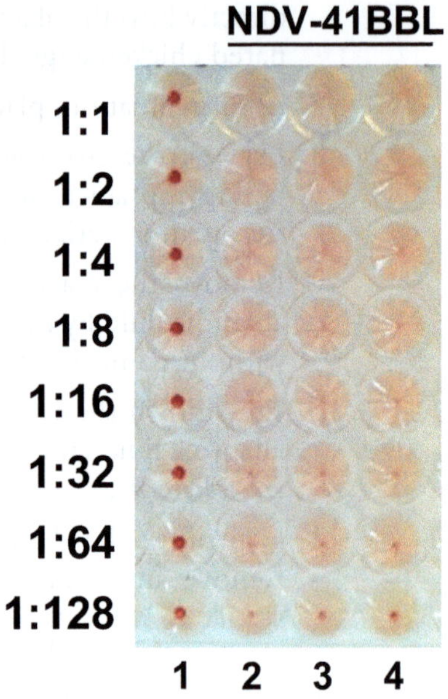

Fig. 3 Hemagglutination assay (HA) confirming positive rNDV rescue. Positive rescues with the presence of viral particles induce hemagglutination, whereas negative rescues or the absence of viruses induce a red pellet in the bottom of the well. The first row is the negative control (1× PBS) and rows 2, 3, and 4 show positive rescue of rNDVs expressing 4-1BBL

3. Once the bands have been confirmed to be of the right size as shown in Fig. 4, gel extract the appropriate bands using the gel and PCR cleanup kit following the manufacturer's instructions.
4. Confirm sequence fidelity of the extracted DNA by Sanger sequencing using the primers listed in Table 1.
5. Once the sequence has been confirmed, grow stocks of the newly rescued virus by dilution purification. Inoculate 9–10-day-old SPF embryonated chicken eggs with 100 μL each of 10^{-7} serially diluted virus in 1× PBS.
6. Repeat **steps 7–11** in Subheading 3.3.1 and steps in Subheading 3.3.2 to collect the allantoic fluid and confirm positive viral growth by an HA assay.
7. Aliquot 0.5 mL of the confirmed virus in cryogenic tubes and store up to 10 years at −80 °C.

3.3.4 Titration of rNDV Stocks by Immunofluorescence

1. One day before infection set up a 96-well plate of Vero cells at a concentration of 2×10^4 cells/mL (*see* **Note 10**).
2. On the day of the infection, thaw out one vial each of the rescued virus and wild-type NDV (NDV-WT) as control. Perform a fivefold serial dilution of the viruses in a separate sterile 96-well plate with 1× PBS.

Fig. 4 Confirmation of presence of transgene by RT-PCR. Viral RNA was extracted from positive rNDV rescues and using primers mentioned in Table 1, RT-PCR was set up to confirm transgene presence. Viral RNA from NDV-WT was used as a positive control

3. Aspirate the medium from the Vero cells in the 96-well plate.
4. Pipet 50 μL of the virus dilutions from the dilution plate into the 96-well plate with the Vero cells using a multichannel pipette.
5. Incubate at 37 °C, 5% CO_2 for 1 h. Gently rock the plates every 15 min to ensure the cells are evenly infected with the virus dilutions.
6. Aspirate the virus dilutions off the Vero cells and add 100 μL of cDMEM.
7. Aspirate the medium from the infected Vero cells and wash the cells once with 100 μL of 1× PBS.
8. Fix cells with 100 μL of Fixative Solution at 4 °C for 10 min protected from light.
9. Aspirate the 4% formaldehyde and wash once with 100 μL of 1× PBS.
10. Block cells with 1% PBS-BSA for 2 h at 4 °C.
11. Incubate cells with the primary antibody either overnight at 4 °C or at room temperature for 2 h (Primary antibody: anti-NDV polyclonal rabbit serum diluted in 1× PBS at 1:500).
12. Aspirate the primary antibody and wash the cells four times with 100 μL of 1× PBS.
13. Incubate the cells with the secondary antibody for 1 h at room temperature protected from light (Secondary antibody: anti-rabbit-Alexa Fluor 488 diluted in 1× PBS at 1:1000).

14. Aspirate the secondary antibody and wash the cells four times with 100 μL of 1× PBS.
15. Count the infected green cells, starting from the undiluted wells until you find a well down the dilution series with a countable number of cells. Count the cells in the entire well.
16. The titer of the virus can be determined by the following formula:

$$\text{Titer} = \text{Number of green cells} \times \text{Dilution factor} \times (1000\,\mu L/\text{Volume of infection})$$

3.4 Confirmation of Ligand Expression

For confirmation of expression of the transgene in the infected cells, one can utilize any human or murine cancer cell line known to be susceptible to NDV infection. For the purposes of subsequent testing in syngeneic tumor models, B16-F10 melanoma is commonly used. For confirmation of surface expression, flow cytometry is a commonly used method.

1. Split B16-F10 cells into 6-well tissue culture plates to achieve approximately 80% confluency the following day. Ensure that a sufficient number of wells are available to allow for treatment with 1× PBS, wild-type NDV (NDV-WT), and rNDV-4-1BBL.
2. On the day of infection, dilute NDV to 8×10^6 pfu/mL in cDMEM.
3. Aspirate the medium, rinse with 2 mL PBS, and infect the cells at an MOI of 2 with 200 μL of diluted control or rNDV-4-1BBL. Incubate the plates at room temperature for 1 h. Gently rock the plates every 15 min to ensure even distribution of the virus over the cells and to prevent the cells from drying.
4. After 1 h, add 2 mL of cDMEM to each plate. Incubate for 24 h at 37 °C.
5. Aspirate the medium and wash the cells with 2 mL of 1× PBS twice. Add 1 mL of Cellstripper™ solution to each well and incubate at 37 °C for up to 5 min. Cells may require pipetting or a mechanical stripper to detach from the plate.
6. Collect the cells in microcentrifuge tubes and spin down at 4 °C at 2700 × *g* for 5 min.
7. Meanwhile, prepare the antibody staining solution in FACS buffer according to the recommended staining concentration.
8. Resuspend the cells in 100 μL of antibody staining solution and transfer to a 96-well round bottom plate. Cover and incubate on ice for 30 min.

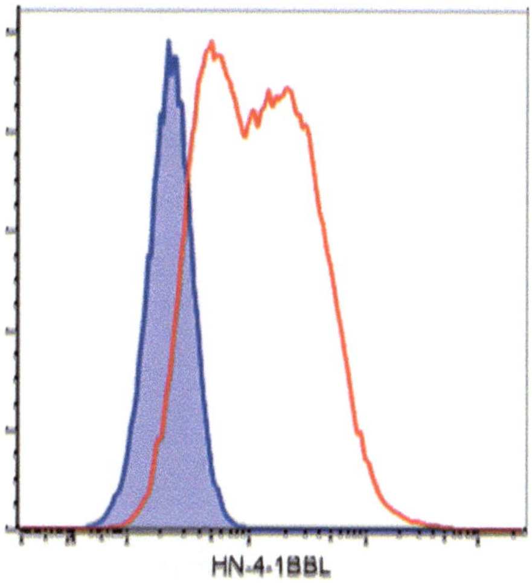

Fig. 5 Confirmation of surface expression of the transgene (4-1BBL) in B16-F10 infected cells by flow cytometry. B16-F10 cells were infected with NDV-WT or rNDV-4-1BBL at an MOI of 2. Twenty-four hours later, the cells were collected using Cell Stripper, processed for staining of surface 4-1BBL and analyzed by flow cytometry. Blue: NDV-WT, red: rNDV-m4-1BBL

9. Spin down the plate at 650 × *g* in a refrigerated tabletop centrifuge for 2 min. Discard the staining solution by inverting the plate.
10. Resuspend the cells in 200 μL of FACS buffer, spin down, and discard. Repeat the wash **step 2** more times.
11. Resuspend the cells in 200 μL FACS buffer and analyze on a flow cytometer (Fig. 5).

3.5 Model Assay of Efficacy Using Intratumoral Administration

As a model system, the murine B16-F10 melanoma tumor derived from C57BL/6 mice is the most widely used [44]. To assess the therapeutic efficacy of the virus both in the treated tumor and the nontreated tumor, a bilateral flank B16-F10 melanoma model is established to characterize the local and abscopal effect (Fig. 6). Once tumors have been established and treated with intratumoral injections of rNDV-41BBL, both the treated and the nontreated tumors will be analyzed for immune infiltration, as a readout for antitumor efficacy.

3.5.1 B16-F10 Tumor Implantation

1. B16-F10 cells are maintained in cDMEM/F12 (*see* **Note 11**).
2. Two days prior to implantation, split B16-F10 cells and expand into several T175 cm^2 flasks to achieve 60–75% confluency on the day of implantation. The number of T175 cm^2 flasks depends on the number of mice that need to be implanted.
3. On the day of implantation, rinse B16-F10 cells with 1× PBS and detach using 3 mL of trypsin–EDTA (0.05%).

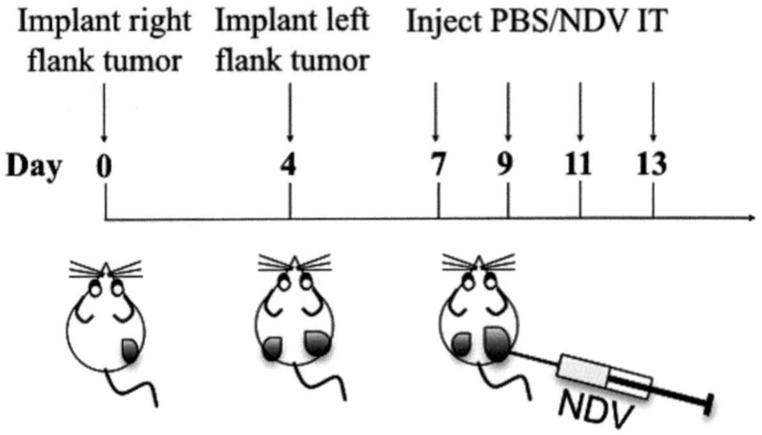

Fig. 6 Treatment schema for intratumoral administration of rNDV-4-1BBL in a B16-F10 murine melanoma tumor model. Bilateral flank tumor models were established by injecting C57BL/6 mice with 2×10^5 B16-F10 cells in the right flank intradermally (i.d.) on day 0 and 1×10^5 cells in the left flank on day 4. On days 7, 9, 11, and 13, mice were treated with four 100 μL intratumoral injections of rNDV-4-1BBL in 1× PBS (1×10^7 pfu). On day 15 the animals were euthanized and the tumors were processed for analysis of tumor-infiltrating lymphocytes

4. Once the cells have lifted off, add 7 mL of cDMEM/F12 medium and pipet up and down to obtain a single-cell suspension.
5. Transfer the cells from all of the flasks into a 50 mL conical centrifuge tube and pellet the cells by spinning it at 250 × *g* for 5 min at 4 °C.
6. Decant the supernatant and resuspend the pellet in 10 mL of ice-cold 1× PBS and pipet it through a 70 μm cell strainer into a new sterile 50 mL conical centrifuge tube.
7. Count the viable cells on a hemocytometer or an automated cell counter using Trypan Blue (*see* **Note 12**).
8. Resuspend the tumor cells at a concentration of 2×10^6 cells/mL in ice-cold 1× PBS. This final cell suspension is kept on ice throughout the implantation. It is critical to implant the mice as quickly as possible, as viability of the cells decreases over time.
9. 4–6 week old female C57BL/6 mice are used for implantation. After the mice have been anesthetized appropriately, shave the fur off the right flank and wipe the area with a 70% ethanol swab.
10. Resuspend the cells by inverting the tube kept on ice 2–3 times and using a 1 mL Sub-Q syringe without the needle attached to

it to draw up the cells. Attach a new sterile 30G × 1/2 needle to the syringe for implantation.

11. Gently insert the needle very superficially under the skin and inject 100 μL of cell suspension intradermally. Avoid injecting the underlying muscle and gently withdraw the needle. Place the mice back in the cage and allow them to recover.
12. For tumor implantation on the left flank on Day 4, repeat the above steps and dilute the cells to 1×10^6 cells/mL in ice-cold 1× PBS before implantation. Repeat **steps 9–11** for implantation.
13. On Day 6 randomize the mice among the different treatment cohorts prior to the start of the experiment.

3.5.2 Intratumoral Injection of Tumors with rNDV-4-1BBL

1. On Day 7, thaw out the virus on ice and depending on the titer of the virus, dilute it down so that 1×10^7 pfu is delivered into the tumor in a volume of 100 μL.
2. Anesthetize the tumor-implanted mice appropriately.
3. Draw up 100 μL of the virus in an insulin syringe or a syringe with a 31G needle. Slowly glide the needle into the tumor, towards the middle and inject.
4. Place the mice back in the cage and allow them to recover.
5. Repeat **steps 1–4** on days 9, 11 and 13 with freshly thawed out virus.

3.5.3 Isolation of Tumor-Infiltrating Lymphocytes

1. Euthanize animals and remove tumors. Be careful not to remove too much connective tissue with the tumors. Work with one animal cage at a time.
2. Place the tumors into the preweighed microcentrifuge tubes.
3. After each cage, weigh and record the tube weight and add 500 μL of RPMI-0 (without serum) to each tube and place on ice.
4. When all tumors have been collected in tubes, prepare 2× Liberase/DNAse mix and put on ice.
5. Mince the tumors with scissors into small pieces.
6. Add 0.5 mL RPMI-0/DNAse/Liberase solution to tumors and incubate at 37 °C on a shaker for 30 min.
7. At the end of the incubation, transfer the tumors into 70 μm strainers in 6-well plates.
8. Pass the tumors through the strainers using the back end of a 1 or 3 mL syringe plunger.
9. Rinse the strainers with 3 mL of cRPMI.
10. Transfer to prelabeled 15 mL centrifuge tubes.
11. Spin down the cells at 400 × *g* for 5 min.

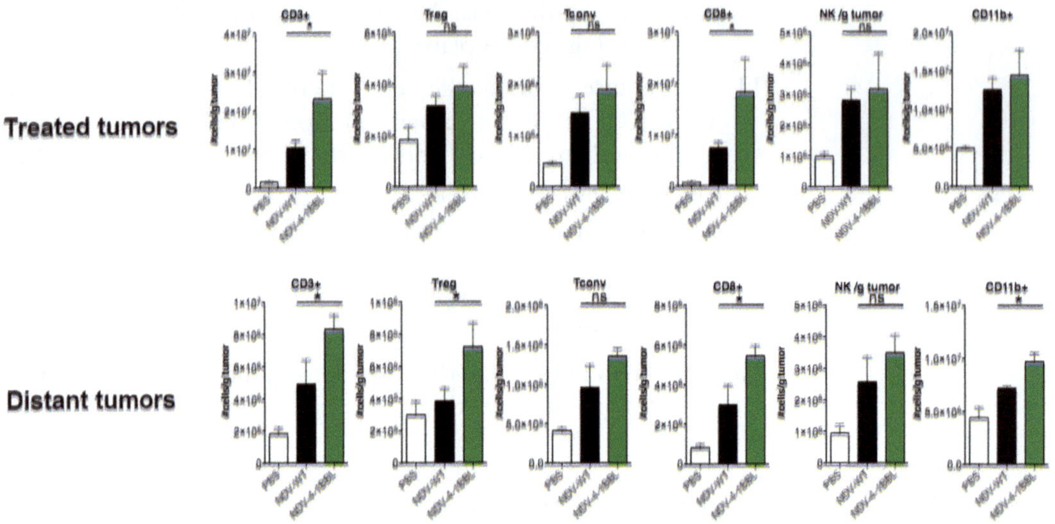

Fig. 7 NDV-HN-4-1BBL induces increased T cell infiltration in the injected and distant tumors. Animals bearing bilateral flank B16 melanoma tumors were treated intratumorally with the indicated virus as previously. After four treatments, animals were euthanized and tumor-infiltrating lymphocytes from bilateral tumors were analyzed by flow cytometry

12. Resuspend the cells in cRPMI.
13. Proceed with labeling for flow cytometry.

Following dissociation, the treated and nontreated tumors were analyzed for tumor-infiltrating lymphocyte (TIL) infiltration by flow cytometry. As shown in Fig. 7, introduction of 4-1BBL into NDV resulted in enhanced T cell infiltration in both virus-injected and distant tumors, when compared to the parental NDV.

3.6 Conclusions

The ability to genetically engineer NDV to express immunostimulatory agents has a potential to enhance its immune-potentiating properties. We demonstrated that introduction of immunomodulatory genes into NDV has the potential to potentiate T cell infiltration into both treated and distant tumors. Such properties have the potential to obviate the need for systemic administration of NDV, as they rely primarily on NDV-induced anti-tumor immune responses. Administration of such agents within the context of systemic immune checkpoint blockade carries a potential to have a strong systemic anti-tumor effect as we demonstrated previously.

The agents for intratumoral modulation may span a range of genes targeting both innate and adaptive arms of the immune system. Optimal targets for intratumoral modulation using NDV are currently unknown and may be influenced by the differences in tumor types, levels of virus replication, or even differences in the composition in tumor microenvironment in the same tumor type. Selection of optimal agents for intratumoral modulation may require evaluation of recombinant NDV in several syngeneic tumor models in order to determine the generalizability of the treatment to different tumor types and to define potential biomarkers that may guide tumor selection.

4 Notes

1. When digesting pNDV-LaSota plasmid with SacII restriction enzyme, overnight incubation at 37 °C prior to Antarctic Phosphatase treatment can enhance linearization efficiency with reduction in background. Run the linearized vector on a 1.5% Agarose-TAE gel to ensure optimal digestion. After visualization on a UV transilluminator, excise the appropriate band using a razor blade or a scalpel and place it into a 1.5 mL microcentrifuge tube. Purify the linearized vector using the gel and PCR cleanup kit following the manufacturer's instructions.
2. When designing primers to PCR-amplify the insert for InFusion cloning, the optimal length of homologous overlap between the terminus of the linearized plasmid and insert can vary between 15 nts for a single insert and 20 nts for multiple inserts. Desalted, high-quality PCR primers are preferred.
3. Codon optimization of the transgene can enhance gene expression.
4. Insertion of large transgenes (>3 kb) can attenuate the virus .
5. When selecting a competent *E. coli* strain for cloning applications, select one that is compatible with cloning large unstable DNA.
6. BSRT7 cells that are used for transfection should be maintained at a low passage number and under mycoplasma-free conditions to enhance transfection efficiency.
7. While any high efficiency transfection reagent could be used, for this particular protocol we use *Trans*IT®-LT1 Transfection Reagent from Mirus. Transfection protocol can be modified based on reagent of choice.
8. When adding the TransIT-LT1 Reagent to the plasmid-Opti-MEM mix, avoid touching the sides of the plastic tube.
9. Avoid damaging the yolk sac membrane and minimize the amount of blood in the allantoic fluid when extracting it from the egg.
10. When plating Vero cells for virus titration, use poly D-lysine coated plates to prevent the cells from sloughing off during the wash steps.

11. Maintain B16-F10 cells at a low passage number to enhance implantation efficiency.
12. Make sure the viability of the B16-F10 cells to be implanted is >90%.

Acknowledgements

D.Z. is funded by the Liz Tilberis Award from the Ovarian Cancer Research Alliance, and the Department of Defense Ovarian Cancer Research Academy (OC150111). D.Z. is a member of the Parker Institute for Cancer Immunotherapy, which supports the MSKCC Cancer Immunotherapy Program. MSKCC is supported by the NCI Core grant P30 CA008748.

Competing financial interests: D.Z. is an inventor on a patent concerning the uses of recombinant Newcastle Disease Virus for cancer therapy.

References

1. Lancaster JE (2007) A history of Newcastle disease with comments on its economic effects. Worlds Poult Sci J 32(2):167–175. https://doi.org/10.1079/WPS19760001
2. Lamb RA, Parks GD (2007) Paramyxoviridae: The Viruses and Their Replication. In: Knipe DM, Howley PM (eds) Fields Virology, 5th edition, vol 1. Lippincott Williams and Wilkins, Philadelphia, PA, pp 1449–1496.
3. Nagai Y, Hamaguchi M, Toyoda T (1989) Molecular biology of Newcastle disease virus. Prog Vet Microbiol Immunol 5:16–64
4. Krishnamurthy S, Samal SK (1998) Nucleotide sequences of the trailer, nucleocapsid protein gene and intergenic regions of Newcastle disease virus strain Beaudette C and completion of the entire genome sequence. J Gen Virol 79 (Pt 10):2419–2424. https://doi.org/10.1099/0022-1317-79-10-2419
5. Panda A, Huang Z, Elankumaran S, Rockemann DD, Samal SK (2004) Role of fusion protein cleavage site in the virulence of Newcastle disease virus. Microb Pathog 36(1):1–10
6. Bukreyev A, Huang Z, Yang L, Elankumaran S, St. Claire M, Murphy BR, Samal SK, Collins PL (2005) Recombinant Newcastle disease virus expressing a foreign viral antigen is attenuated and highly immunogenic in primates. J Virol 79(21):13275
7. Kim SH, Samal SK (2016) Newcastle disease virus as a vaccine vector for development of human and veterinary vaccines. Viruses 8 (7):183. https://doi.org/10.3390/v8070183
8. Flanagan AD, Love R, Tesar W (1955) Propagation of Newcastle disease virus in Ehrlich ascites cells in vitro and in vivo. Proc Soc Exp Biol Med 90(1):82–86
9. Sinkovics J (1957) Studies on the biological characteristics of the Newcastle disease virus (NDV) adapted to the brain of newborne mice. Arch Gesamte Virusforsch 7(4):403–411
10. Eaton MD, Levinthal JD, Scala AR (1967) Contribution of antiviral immunity to oncolysis by Newcastle disease virus in a murine lymphoma. J Natl Cancer Inst 39(6):1089–1097
11. Cassel WA, Garrett RE (1965) Newcastle disease virus as an antineoplastic agent. Cancer 18 (7):863–868. https://doi.org/10.1002/1097-0142(196507)18:7<863::AID-CNCR2820180714>3.0.CO;2-V
12. Cassel WA, Garrett RE (1966) Tumor immunity after viral oncolysis. J Bacteriol 92(3):792
13. Murray DR, Cassel WA, Torbin AH, Olkowski ZL, Moore ME (1977) Viral oncolysate in the management of malignant melanoma. II. Clinical studies. Cancer 40(2):680–686
14. Cassel WA, Murray DR (1992) A ten-year follow-up on stage II malignant melanoma patients treated postsurgically with Newcastle disease virus oncolysate. Med Oncol Tumor Pharmacother 9(4):169–171

15. Batliwalla FM, Bateman BA, Serrano D, Murray D, Macphail S, Maino VC, Ansel JC, Gregersen PK, Armstrong CA (1998) A 15-year follow-up of AJCC stage III malignant melanoma patients treated postsurgically with Newcastle disease virus (NDV) oncolysate and determination of alterations in the CD8 T cell repertoire. Mol Med 4(12):783–794
16. Kirchner HH, Anton P, Atzpodien J (1995) Adjuvant treatment of locally advanced renal cancer with autologous virus-modified tumor vaccines. World J Urol 13(3):171–173
17. Csatary LK, Eckhardt S, Bukosza I, Czegledi F, Fenyvesi C, Gergely P, Bodey B, Csatary CM (1993) Attenuated veterinary virus vaccine for the treatment of cancer. Cancer Detect Prev 17 (6):619–627
18. Csatary LK, Gosztonyi G, Szeberenyi J, Fabian Z, Liszka V, Bodey B, Csatary CM (2004) MTH-68/H oncolytic viral treatment in human high-grade gliomas. J Neuro-Oncol 67(1-2):83–93
19. Freeman AI, Zakay-Rones Z, Gomori JM, Linetsky E, Rasooly L, Greenbaum E, Rozenman-Yair S, Panet A, Libson E, Irving CS, Galun E, Siegal T (2006) Phase I/II trial of intravenous NDV-HUJ oncolytic virus in recurrent glioblastoma multiforme. Mol Ther 13(1):221–228. https://doi.org/10.1016/j.ymthe.2005.08.016
20. Pecora AL, Rizvi N, Cohen GI, Meropol NJ, Sterman D, Marshall JL, Goldberg S, Gross P, O'Neil JD, Groene WS, Roberts MS, Rabin H, Bamat MK, Lorence RM (2002) Phase I trial of intravenous administration of PV701, an oncolytic virus, in patients with advanced solid cancers. J Clin Oncol 20(9):2251–2266. https://doi.org/10.1200/jco.2002.08.042
21. Lorence RM, Pecora AL, Major PP, Hotte SJ, Laurie SA, Roberts MS, Groene WS, Bamat MK (2003) Overview of phase I studies of intravenous administration of PV701, an oncolytic virus. Curr Opin Mol Ther 5(6):618–624
22. Schirrmacher V (2016) Fifty years of clinical application of Newcastle disease virus: time to celebrate! Biomedicine 4(3). https://doi.org/10.3390/biomedicines4030016
23. Reichard KW, Lorence RM, Cascino CJ, Peeples ME, Walter RJ, Fernando MB, Reyes HM, Greager JA (1992) Newcastle disease virus selectively kills human tumor cells. J Surg Res 52(5):448–453
24. Zorn U, Dallmann I, Grosse J, Kirchner H, Poliwoda H, Atzpodien J (1994) Induction of cytokines and cytotoxicity against tumor cells by Newcastle disease virus. Cancer Biother 9 (3):225–235
25. Sinkovics JG, Horvath JC (2000) Newcastle disease virus (NDV): brief history of its oncolytic strains. J Clin Virol 16(1):1–15
26. Peeters BP, de Leeuw OS, Koch G, Gielkens AL (1999) Rescue of Newcastle disease virus from cloned cDNA: evidence that cleavability of the fusion protein is a major determinant for virulence. J Virol 73(6):5001–5009
27. Molouki A, Peeters B (2017) Rescue of recombinant Newcastle disease virus: a short history of how it all started. Arch Virol 162 (7):1845–1854. https://doi.org/10.1007/s00705-017-3308-2
28. Vigil A, Park MS, Martinez O, Chua MA, Xiao S, Cros JF, Martinez-Sobrido L, Woo SL, Garcia-Sastre A (2007) Use of reverse genetics to enhance the oncolytic properties of Newcastle disease virus. Cancer Res 67 (17):8285–8292. https://doi.org/10.1158/0008-5472.can-07-1025
29. Vigil A, Martinez O, Chua MA, Garcia-Sastre A (2008) Recombinant Newcastle disease virus as a vaccine vector for cancer therapy. Mol Ther 16(11):1883–1890. https://doi.org/10.1038/mt.2008.181
30. Zamarin D, Martinez-Sobrido L, Kelly K, Mansour M, Sheng G, Vigil A, Garcia-Sastre A, Palese P, Fong Y (2009) Enhancement of oncolytic properties of recombinant Newcastle disease virus through antagonism of cellular innate immune responses. Mol Ther 17 (4):697–706. https://doi.org/10.1038/mt.2008.286
31. Song KY, Wong J, Gonzalez L, Sheng G, Zamarin D, Fong Y (2010) Antitumor efficacy of viral therapy using genetically engineered Newcastle disease virus [NDV(F3aa)-GFP] for peritoneally disseminated gastric cancer. J Mol Med (Berl) 88(6):589–596. https://doi.org/10.1007/s00109-010-0605-6
32. Li P, Chen CH, Li S, Givi B, Yu Z, Zamarin D, Palese P, Fong Y, Wong RJ (2011) Therapeutic effects of a fusogenic Newcastle disease virus in treating head and neck cancer. Head Neck 33 (10):1394–1399. https://doi.org/10.1002/hed.21609
33. Zamarin D, Palese P (2012) Oncolytic Newcastle disease virus for cancer therapy: old challenges and new directions. Future Microbiol 7 (3):347–367. https://doi.org/10.2217/fmb.12.4
34. Zamarin D, Holmgaard RB, Subudhi SK, Park JS, Mansour M, Palese P, Merghoub T, Wolchok JD, Allison JP (2014) Localized oncolytic virotherapy overcomes systemic tumor resistance to immune checkpoint blockade immunotherapy. Sci Transl Med 6(226):226ra232.

https://doi.org/10.1126/scitranslmed.3008095

35. Oseledchyk A, Ricca JM, Gigoux M, Ko B, Redelman-Sidi G, Walther T, Liu C, Iyer G, Merghoub T, Wolchok JD, Zamarin D (2018) Lysis-independent potentiation of immune checkpoint blockade by oncolytic virus. Oncotarget 9(47):28702–28716. https://doi.org/10.18632/oncotarget.25614
36. Zamarin D, Holmgaard RB, Ricca J, Plitt T, Palese P, Sharma P, Merghoub T, Wolchok JD, Allison JP (2017) Intratumoral modulation of the inducible co-stimulator ICOS by recombinant oncolytic virus promotes systemic antitumour immunity. Nat Commun 8:14340. https://doi.org/10.1038/ncomms14340
37. Watts TH (2005) TNF/TNFR family members in costimulation of T cell responses. Annu Rev Immunol 23:23–68. https://doi.org/10.1146/annurev.immunol.23.021704.115839
38. Sznol M, Hodi FS, Margolin DF, McDermott D, Ernstoff JM, Kirkwood JM, Wojtaszek C, Feltquate D, Logan T (2008) Phase I study of BMS-663513, a fully human anti-CD137 agonist monoclonal antibody, in patients (pts) with advanced cancer (CA). Paper presented at the 2008 ASCO Annual Meeting, Chicago, IL
39. Nakaya T, Cros J, Park MS, Nakaya Y, Zheng H, Sagrera A, Villar E, Garcia-Sastre A, Palese P (2001) Recombinant Newcastle disease virus as a vaccine vector. J Virol 75 (23):11868–11873. https://doi.org/10.1128/jvi.75.23.11868-11873.2001
40. Zhao H, Peeters BP (2003) Recombinant Newcastle disease virus as a viral vector: effect of genomic location of foreign gene on gene expression and virus replication. J Gen Virol 84 (Pt 4):781–788. https://doi.org/10.1099/vir.0.18884-0
41. Zhao W, Zhang Z, Zsak L, Yu Q (2015) P and M gene junction is the optimal insertion site in Newcastle disease virus vaccine vector for foreign gene expression. J Gen Virol 96 (Pt 1):40–45. https://doi.org/10.1099/vir.0.068437-0
42. Marcos F, Ferreira L, Cros J, Park MS, Nakaya T, Garcia-Sastre A, Villar E (2005) Mapping of the RNA promoter of Newcastle disease virus. Virology 331(2):396–406. https://doi.org/10.1016/j.virol.2004.10.040
43. Beaty SM, Park A, Won ST, Hong P, Lyons M, Vigant F, Freiberg AN, tenOever BR, Duprex WP, Lee B (2017) Efficient and robust Paramyxoviridae reverse genetics systems. mSphere 2(2). https://doi.org/10.1128/mSphere.00376-16
44. Overwijk WW, Restifo NP (2001) B16 as a mouse model for human melanoma. Curr Protoc Immunol. Chapter 20:Unit 20.1. https://doi.org/10.1002/0471142735.im2001s39

Chapter 10

Analysis of Immunological Treatment Effects of Virotherapy in Tumor Tissue

Krishna Das, Carles Urbiola, Bart Spiesschaert, Philipp Mueller, and Guido Wollmann

Abstract

In addition of being directly tumoricidal, oncolytic viruses have emerged as potent partners for established and investigational immunotherapies due to their immune-stimulatory effects. The shifting focus on virus-mediated immune modulation calls for a comprehensive analysis of the tumor microenvironment (TME) and the factors orchestrating the antiviral and antitumor immune response. The oncolytic VSV-GP studied in our lab is a safe and potent antitumor agent with a fast replication cycle and killing of a broad range of different cancer types. It induces a robust local inflammatory conversion of the TME and drives a strong adaptive immune response toward the tumor. Here we present our multidisciplinary approach to study VSV-GP treatment effects in tumors by assessing both immune cells (tumor-infiltrating lymphocytes and tumor-associated macrophages) and immune-regulatory factors (cytokines) as well as characterizing immune signatures using an immune-targeted NanoString gene expression system.

Key words Tumor-infiltrating lymphocytes, Tumor-associated macrophages, NanoString, Oncolytic virus, Cytokine multiplex, LEGENDplex, VSV-GP

1 Introduction

Oncolytic viruses had originally been conceived as biological therapeutics to directly target and kill cancer cells. Over the past 25 years, this initial tumor-centric view has expanded to recognize that oncolytic viruses are powerful immune stimulators with potential to induce antitumor immunity and overcome the immunosuppressive tumor microenvironment [1, 2]. Oncolytic virus-induced immunogenic cell death can release tumor-associated antigens, which are cross-presented to T cells by endogenous antigen-presenting cells [3]. Furthermore, some oncolytic viruses with transgene capacity can be "armed" with additional therapeutic elements [4], such as immune-modulating genes, checkpoint

Krishna Das, Carles Urbiola, and Bart Spiesschaert contributed equally to this work.

Christine E. Engeland (ed.), *Oncolytic Viruses*, Methods in Molecular Biology, vol. 2058,
https://doi.org/10.1007/978-1-4939-9794-7_10,

inhibitors, suicide genes, or antiangiogenic factors [5]. To fully capture these immune-modulating effects of virotherapy in all their complexity, we routinely employ a multidisciplinary approach for in-depth analysis of virus-treated tumors. This includes FACS analysis to characterize specific cellular immune components, cytokine multiplex quantifications, and NanoString assays with the PanCancer Mouse Immune Profiling gene panel [6] to dissect immune-related RNA expression levels in the course of virotherapy.

1.1 Cytometric Analysis of Tumor-Infiltrating Lymphocytes and Tumor-Associated Macrophages

T-cell inflamed or "hot" tumors are more likely to respond to immunotherapeutic interventions compared to immune deserts [7]. While this notion generally holds true, the pathogenic nature of viral agents triggers antiviral immune responses that recruit cytotoxic T-cells without specificity for (noninfected) tumor cells. Hence, the therapeutic outcome of oncolytic virotherapy also depends to no small extent on the balance between antiviral and antitumor immunity to increase the response against tumor-associated antigens [2]. Even if in some cases it is postulated that antiviral immunity can augment the antitumor therapeutic effects [8], the identification of tumor-specific immune cells is paramount. While immunohistochemistry provides important information about the spatial distribution of the infiltrating immune cells, traditional methods are quite restricted in the number of markers that can be assessed simultaneously. Flow cytometry allows detailed phenotyping of various cell subsets that are present in the tumor microenvironment. Addressing the dual nature of antiviral and antitumor immunity, T cell specificity and functionality can be measured by flow cytometry as well. Recent developments allow for the detection of a wide variety of T cell specificities [9]. Another pertinent question in OV-mediated immune activation is the phenotype composition and the kinetics of memory and effector T cell induction [10]. Furthermore, immunosuppressive cells like myeloid-derived suppressor cells (MDSCs), regulatory T cells (Tregs), and tumor-associated macrophages (TAMs) present in the tumor microenvironment dampen the antitumor immune response [11]. Phenotyping and quantification of immunosuppressive cells can aid in identifying resistance mechanisms and predicting combination partners for virotherapy to overcome resistance or further improve therapeutic outcome [12].

1.2 Selective Transcription Analysis with Immune-Oncology Focus

To assess treatment responses in the tumor microenvironment in a broad manner, we use the targeted gene expression profiling system NanoString. The nCounter Analysis System allows for direct multiplexed measurement of gene expression from low amounts of mRNA extracted from the treated tumors [13]. This setup allows identification of different tumor-infiltrating immune cell populations via gene expression signatures [14]. Furthermore, it detects changes in immunological function and response to the treatment, such as immune checkpoint regulation. Overall, this assay provides a broad overview of the immunological impact of oncolytic virus treatment in the tumor microenvironment. The gene expression assay method is based on direct detection of RNA using combination of specific fluorescent molecular reporter probes, which carry a unique fluorescent molecular barcode on their 5′ end, and capture probes which carry a biotin on the 3′ end. Each RNA molecule is detected by ligating to a specific probe with a unique fluorescent molecular barcode, which represents the specific bound RNA. The expression level of a gene is measured by counting the number of times the fluorescent barcode for that gene is detected [13, 15]. The reliability of gene expression profiling is influenced by the quality of the extracted RNA. Thus, an optimized extraction protocol that ensures high quality RNA is essential.

1.3 Multiplex Cytokine Analysis

Complementing the analysis of the cellular immune factors via FACS assays, the study of cytokines provides insight into the factors that mediate proper immune function and regulation. In the context of OV treatment, a particular emphasis is on the dynamic of pro-inflammatory Th1 cytokines and suppressive Th2-associated cytokines [8]. For VSV-based virotherapy, the strong induction of pro-inflammatory cytokines was linked to antitumor activity even when single-cycle virus was used [16] highlighting the role of the cytokine milieu orchestrating antitumor immunity. To detect changes in the composition of cytokines within the tumor, we employ the LEGENDplex bead-based assay. Other commonly used cytokine multiplex technologies are for example Luminex Technology multiplex assays or V-PLEX from Meso Scale Discovery (MSD). The use of multiplex assays is not essential, but facilitates the acquisition of quantitative data for many cytokines from very small samples.

Vesicular stomatitis virus (VSV) is a well-studied oncolytic virus that acts as a potent immune-modulator in addition to its rapid lytic action [17]. To abrogate the neurotoxicity risk associated with wildtype VSV, a chimeric virus variant VSV-GP was generated, in which the glycoprotein was exchanged with a GP protein derived from the arenavirus LCMV [18]. Just like the parental virus [19], VSV-GP is also a strong inducer of interferon stimulated genes and other pro-inflammatory molecules making it a strong immune-activating agent. Expressing foreign antigens, VSV-GP can also induce very potent target-specific immune responses [20].

This chapter presents protocols describing tumor tissue analysis after VSV-GP-based virotherapy. However, these methods are generally applicable for assessing immune-related effects of other oncolytic viruses as well. Minor adaptations, such as time of tumor harvest in the course of treatment might be necessary in respect to the agent's biology and kinetics. We provide step-by-step protocols for proper immune cell characterization, selective transcriptome analysis and cytokine quantification. The assignment of protocol steps for the individual methods is structured as follows:

- FACS for tumor-infiltrating immune cells (Subheadings 2.1, 2.2, 2.5, 3.1–3.9).
- RNA and Nanostring (Subheadings 2.1, 2.3, 2.6, 3.1, 3.10–3.14).
- Cytokines and Legendplex (Subheadings 2.1, 2.4, 2.7, 3.1, 3.15–3.17).

2 Materials

2.1 Blood Collection and Tumor Harvest

1. Isoflurane vaporizer.
2. Isoflurane.
3. 70% ethanol.
4. RPMI1640 medium.
5. Sterile Petri dishes.
6. Sterile surgical scalpels.
7. Forceps.
8. Scissors.
9. Liquid nitrogen.
10. Glass micro-hematocrit capillary tubes.
11. 2-ml EDTA blood collection tubes.
12. 1.5 ml screw-cap tubes.
13. RNA stabilization solution.

2.2 Isolation of Tumor-Infiltrating Lymphocytes

1. GentleMACS™ Octo Dissociator with Heaters (Miltenyi Biotec).
2. GentleMACS™ C-tubes (Miltenyi Biotec).
3. 70 μm cell strainers.
4. 5 ml syringe.
5. Tumor Dissociation kit, mouse (Miltenyi Biotec) containing four reagents, Enzyme D, Enzyme R, Enzyme A and Buffer A.
6. Lympholyte® M (Cedarlane): bring to room temperature before use.

2.3 Tumor Processing for RNA Analysis

1. Biosafety level 2 (BL2) hood and BL2 waste container.
2. P:C:I solution: 50% (v/v) phenol, 48% (v/v) chloroform, 2% (v/v) isoamyl alcohol saturated with 100 mM Tris pH 8.0, ~0.1% 8-hydroxyquinoline.
3. C:I solution: 96% (v/v) chloroform, 4% (v/v) isoamyl alcohol.

4. β-mercaptoethanol (βME).
5. Phase Lock Gel heavy 2 ml tube (Quantabio).
6. RNA extraction kit.
7. MagMAX™-96 Total RNA Isolation Kit (Ambion).
8. Mag MagMAX™ Express-96 deep well (MME-96DW) magnetic particle processor.
9. Fragment Analyzer™ Automated CE System (optional).
10. Homogenization tubes containing beads (*see* **Note 1**).
11. Bead-based tissue homogenization system.
12. DNase.
13. MME-96DW plate (Ambion).
14. 96-well plate.

2.4 Tumor Processing for Cytokine Analysis

1. Biosafety level 2 (BL2) hood and BL2 waste container.
2. Tissue homogenizer.
3. Homogenization tubes containing beads (*see* **Note 1**).
4. Lysis buffer: 20 mM Tris–HCl (pH 7.5), 150 mM NaCl, 1 mM Na_2EDTA, 1 mM EGTA, 1% Triton, 2.5 mM sodium pyrophosphate, 1 mM β-glycerophosphate, 1 mM Na_3VO_4, 1 μg/ml leupeptin, 1 mM PMSF (freshly added) (*see* **Notes 2** and **3**).
5. Tabletop tissue culture centrifuge.
6. 1.5 ml reaction tubes.

2.5 FACS Staining

1. FACS tubes: 5 ml polystyrene round-bottom tube, 12 × 75 mm.
2. Normal rat serum.
3. Normal hamster serum.
4. Purified rat anti-mouse CD16/CD32 (Mouse BD Fc Block™) (BD Pharmingen).
5. Antibodies: CD45.2 (clone 104), CD90.2 (clone 30-H12), CD8 (clone 53–6.7), CD11b (clone M1/70), F4/80 (clone BM8) (Biolegend).
6. True-Nuclear™ Transcription Factor Buffer Set containing 4× Fix Concentrate, Fix Diluent and 10× Perm Buffer (Biolegend).
7. LIVE/DEAD™ Fixable Dead Cell Stain Kit, available for detection in eight different channels (Thermofisher).
8. FACS buffer: Dulbecco's PBS, 1.5% FCS, 0.2% sodium azide.

2.6 Nanostring nCounter Transcriptome Analysis

1. nCounter® FLEX analysis system (NanoString Technologies).
   (a) nCounter™ prep station.
   (b) nCounter™ digital analyzer.
2. nSolver analysis software 4.0 (NanoString Technologies).
3. nCounter® CodeSet (NanoString Technologies).
4. nCounter® PanCancer Mouse Immune Profiling gene panel (NanoString Technologies).
5. nCounter® Master Kit (NanoString Technologies).
6. RNA extraction kit.
7. Thermal cycler.

2.7 Multiplex Cytokine Analysis

1. Legendplex kit for multiplex cytokine analysis (Biolegend, USA) (*see* **Note 4**). Containing premixed capture beads, detection antibodies, standard, streptavidin-PE, assay buffer, matrix A, wash buffer, 96-well v-bottom plate, plate sealers, setup beads, and data analysis software dongle.
2. *Optional:* Additional 96-multi-well v-bottom polypropylene plates.
3. Centrifuge with 96-multi-well plate capability.
4. Flow cytometer equipped with: 488 nm blue laser, 633/635/640 nm red laser, automated multiwell-plate acquisition device (*see* **Note 5**).
5. Standard tabletop vortex mixer, standard tabletop plate shaker, aluminum foil, and absorbent pads or paper towels.

3 Methods

3.1 Blood Collection and Harvest of Subcutaneous Tumors

Before performing the experiments described herein, investigators should apply for the appropriate animal and biosafety approvals according to the institutional policies and national regulations. Perform all the experiments according to approved animal care policies and biosafety protocols. For the following protocols, C57BL/6J mice bearing either modified LLC1 or B16 subcutaneous tumors were treated with VSV-GP or mock treated with PBS.

1. Anesthetize mice using the isoflurane vaporizer.
2. *Optional:* If blood is needed: Collect ≈ 700 μl blood via retro-orbital bleeding using a glass micro-hematocrit capillary tube draining into a 2 ml EDTA tube (*see* **Note 6**) and centrifuge for 10 min at 6000 × *g*. Transfer plasma into a 1.5 ml screw-cap tube and store at −80 °C.
3. Euthanize mice by an institutionally approved method.

4. Sterilize skin surface with 70% ethanol, lift the skin up using forceps and make a small incision in the skin near the tumor.
5. Remove tumor carefully using a scalpel or scissors.
6. Place the tumor on a previously tared Petri dish and weigh it (*see* **Note** 7).
7. Cut the tumor into 2–4 mm pieces using a scalpel.
8. For whole-cell tumor dissociation and flow cytometry analysis place 400–1000 mg of tumor (*see* **Note** 7) in a gentleMACS C-tube containing 1.5 ml RPMI 1640 medium. For RNA extraction place 100–200 mg of tumor (*see* **Note** 7 and **8**) in a 1.5 ml screw-cap tube containing RNA stabilization solution and store at 4 °C until use. For cytokine analysis transfer 100–300 mg of tumor (*see* **Notes** 7 and **8**) in a 1.5 ml screw-cap tube, snap freeze and store at −80 °C for subsequent processing.

3.2 Whole-Cell Tumor Dissociation Using GentleMACS

1. Reconstitute the lyophilized enzymes supplied in the kit. Reconstitute Enzyme D and Enzyme R with 3 ml and 2.7 ml RPMI 1640 (*see* **Note 9**) respectively. Reconstitute Enzyme A with 1 ml of Buffer A. Prepare 1.0125 ml enzyme mix containing 0.85 ml RPMI 1640, 100 μl Enzyme D, 50 μl Enzyme R and 12.5 μl Enzyme A per tumor sample (*see* **Note 10**).
2. Add the 1.0125 ml enzyme mix to the gentleMACS C-tubes containing tumor pieces (*see* **Notes 11** and **12**).
3. Tightly close gentleMACS C-tubes and attach upside down onto the sleeve of the gentleMACS Dissociator (*see* **Notes 13** and **14**).
4. Attach the heaters to the gentleMACS Dissociator. Select program **37C_m_TDK_1 or 37C_m_TDK_2** (*see* **Notes 15** and **16**) and press start.
5. After termination of the program, detach the C-tube from the gentleMACS Dissociator.
6. Perform a short spin of 30 s at 300 × *g* on the centrifuge to collect the sample at the bottom of the tube.
7. Resuspend sample and apply the cell suspension to a 70 μm cell strainer placed on a 50 ml Falcon tube (*see* **Note 17**).
8. Wash the cell strainer with 10 ml of RPMI 1640.
9. Centrifuge cell suspension at 300 × *g* for 10 min at 4 °C.
10. Aspirate supernatant completely.
11. Resuspend cells in RPMI 1640 and count cells.
12. Adjust cell concentration to 2×10^7 cells/ml in 5 ml RPMI 1640 and proceed to enrich lymphocytes.

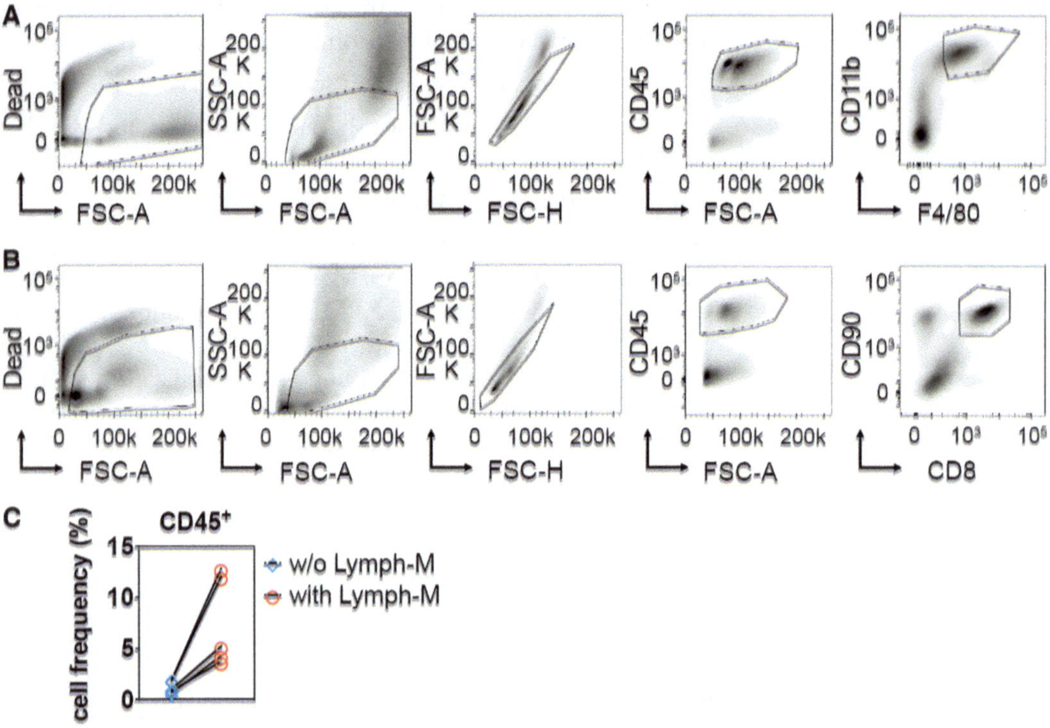

Fig. 1 Exemplary density plots depicting the gating scheme for tumor-associated macrophages and tumor-infiltrating cytotoxic T lymphocytes isolated from B16 tumors. The exclusion of dead cells is the first step followed by gating on cells based on their forward and side scatter. This is followed by gating on singlets based on forward scatter height (FSC-H) and forward scatter width (FSC-W) and on $CD45^+$ cells. These cells represent the leukocytes and can be further characterized based on other surface or intracellular markers of interest. In this example, two leukocyte sub-populations are shown: (**a**) macrophages are depicted as $CD11b^+F4/80^+$ cells as the majority of murine tissue macrophages express these two surface markers whereas (**b**) CTLs are identified by the expression of CD90 and CD8 on the cell surface. (**c**) Enrichment of $CD45^+$ leukocytes isolated from B16 tumors by density gradient centrifugation. The frequency of $CD45^+$ cells among viable cells is depicted before (w/o Lymph-M) and after enrichment with Lympholyte-M (with Lymph-M)

3.3 Density Gradient Centrifugation

The cell suspension obtained above contains tumor cells and dying cells in addition to the infiltrating immune cell fraction. Enrichment of the lymphocytes facilitates a separation and improves the quality of the FACS staining (Fig. 1c). Tumor cells and debris can be removed using discontinuous density gradient centrifugation resulting in shorter acquisition times.

1. Prepare 15 ml Falcon tubes with 5 ml Lympholyte®-M (*see* **Note 18**) and gently apply 5 ml of cell suspension on top, taking care to maintain a clear separation between the two layers (*see* **Notes 19–21**).
2. Centrifuge for 20 min at 1000–1500 × *g* without brake for 20 min at room temperature.

3. Carefully remove the cells from the interphase and transfer to a new 15 ml falcon tube.
4. Add PBS to make up the volume to 10 ml and mix well by inverting the falcon tube to make sure that there is no gradient due to residual Lympholyte M.
5. Centrifuge at 400× for 10 min at 4 °C and discard supernatant.
6. Wash the cells by resuspending them in 10 ml PBS.
7. Centrifuge at 400 × *g* for 10 min at 4 °C and discard supernatant.
8. Repeat **steps 6** and **7** before proceeding to FACS staining.

3.4 Lymphocyte FACS Staining

For optimal results, it is advisable to titrate the antibodies in advance to determine the optimal dilution for each antibody and to use the same cell number for all the samples. The protocol described below is adapted for staining in FACS tubes. However, it is possible to perform it in U-bottomed 96 well plates as well (*see* **Note 22**).

3.5 FCR Blocking

1. Prepare blocking mix of purified rat anti-mouse CD16/CD32 (diluted 1:10), rat serum (1:100) and hamster serum (1:100) in FACS buffer (*see* **Note 23**).
2. Resuspend cells in 100 μl of blocking mix.
3. Incubate 20 min at 4 °C.
4. Centrifuge cells at 400 × *g* for 5 min at 4 °C and discard supernatant.
5. Wash the cells by resuspending them in 500 μl PBS.
6. Centrifuge cells at 400 × *g* for 5 min at 4 °C and discard supernatant.
7. Repeat **steps 5** and **6** before proceeding to dead cell staining (*see* **Notes 24** and **25**).

3.6 Dead Cell Staining

1. Bring one vial of LIVE/DEAD™ Fixable Dead Cell Stain Kit to room temperature and reconstitute by adding 50 μl DMSO.
2. Prepare working solution by diluting it 1:1000 in PBS.
3. Incubate cells in 100 μl working solution for 30 min at 4 °C in the dark.
4. Centrifuge cells at 400 × *g* for 5 min at 4 °C and discard supernatant.
5. Wash the cells by resuspending them in 500 μl FACS buffer.
6. Centrifuge cells at 400 × *g* for 5 min at 4 °C and discard supernatant.
7. Repeat **steps 5** and **6** before proceeding to surface staining.

3.7 Surface Staining

1. Prepare antibody mix by diluting antibodies for surface staining in FACS buffer (CD45.2, CD8 and CD90.2 for cytotoxic T lymphocytes (Fig. 1b) or CD45.2, CD11b and F4/80 for macrophages (Fig. 1a)).
2. Resuspend cells in 100 μl antibody mix.
3. Incubate for 30 min at 4 °C.
4. Centrifuge cells at 400 × *g* for 5 min at 4 °C and discard supernatant.
5. Wash the cells by resuspending them in 500 μl FACS buffer.
6. Centrifuge cells at 400 × *g* for 5 min at 4 °C and discard supernatant.
7. Repeat **steps 5** and **6** and proceed to fixation and permeabilization. If only surface markers are to be analyzed, then proceed directly to FACS analysis.

3.8 Fixation and Permeabilization

1. Gently vortex the FACS tubes to dissociate cell pellets.
2. Prepare Transcription Factor 1× Fix solution by diluting the 4× Fix Concentrate (1 part) with the Fix Diluent (3 parts).
3. Add 1 ml of the Transcription Factor 1× Fix solution (*see* **Note 26**) to the cells and gently pipette up and down to ensure cells are fully resuspended.
4. Incubate at room temperature in the dark for 60 min.
5. During incubation prepare the Transcription Factor 1× Perm Buffer by adding 10 ml True Nuclear Perm Buffer and 90 ml double distilled water.
6. Centrifuge cells at 400 × *g* for 5 min at 4 °C and discard supernatant.
7. Resuspend cells in 2 ml Transcription Factor 1× Perm Buffer.
8. Centrifuge cells at 400 × *g* for 5 min at 4 °C and discard supernatant.
9. Repeat **steps 6** and 7 and proceed to intracellular staining.

3.9 Intracellular Staining

1. Prepare antibody mix by diluting antibodies for intracellular staining in Transcription Factor 1× Perm Buffer (*see* **Notes 27** and **28**).
2. Resuspend cells in 100 μl antibody master mix.
3. Incubate at room temperature for 30 min.
4. Centrifuge cells at 400 × *g* for 5 min at room temperature and discard supernatant.
5. Resuspend cells in 500 μl Transcription Factor 1× Perm Buffer.
6. Centrifuge cells at 400 × *g* for 5 min at room temperature and discard supernatant.

7. Repeat **steps 4** and **5**.
8. Resuspend cells in 200–300 μl FACS buffer and proceed to FACS analysis.

3.10 Tumor Homogenization for RNA Analysis

1. Prepare RLT buffer from the RNeasy Mini Kit: add 10 μl βME to 1 ml RLT-buffer (RLT-βME) and put on ice.
2. Prechill 2.0 ml reinforced screw-cap homogenization tubes, containing 1.4 mm ceramic beads, on ice.
3. Place the tumor sample in the prechilled homogenization tube.
4. Pipet 100 μl of RLT-ßME per 10 mg tissue (*see* **Note 8**).
5. Place the tubes in the tissue homogenizer and homogenize using 3 cycles of 30 s (*see* **Note 29**).
6. Centrifuge homogenate at 16000 × *g* for 2 min and 30 s.
7. Keep homogenates on ice for immediate RNA extraction or store at −80 °C until used.

3.11 RNA Extraction

1. Prepare Phase Lock Gel heavy 2 ml tube by centrifugation at 16000 × *g* for 2 min and 30 s.
2. Add homogenate (max. 700 μl) to the Phase Lock Gel heavy 2 ml tube. Add 350 μl of P:C:I to the Phase Lock Gel heavy 2 ml tube.
3. Mix for 15 s.
4. Centrifuge at 16000 × *g* for 5 min.
5. Add 350 μl C:I.
6. Mix for 15 s.
7. Incubate at room temperature for 3 min.
8. Centrifuge at 16000 × *g* for 5 min.
9. Transfer the liquid phase (ca. 700 μl) to the Mag MagMAX™ deep well block.
10. Store samples at −80 °C.

3.12 RNA Purification with MagMAX™-96 Total RNA Isolation Kit

1. Acclimatize RNA binding beads to RT.
2. Thaw RNA deep well block on ice.
3. Warm the lysis binding enhancer buffer to 37 °C until it is dissolved.
4. Prepare wash buffer 1, wash buffer 2 and RNA rebinding buffer.
5. Fill MME-96DW plates with wash buffer 1, wash buffer 2 and elution buffer as described in the manufacturer's instructions.
6. Dissolve RNase-free DNase in 550 μl H_2O.
7. Fill plate with DNase solution.

8. Add 10 µl beads to 10 µl lysis binding enhancer solution at RT and vortex shortly. Prepare a master mix for the required number of wells.
9. Add the 20 µl of bead-mix to the MME-96DW RNA plate.
10. Add 0.5 volume of isopropanol to the deep well block of the MME-96DW magnetic particle processor.
11. Transfer all plates into the robotic liquid handler according to manufacturer's instructions.
12. Start the MagMAX robotic liquid handler.
13. When program goes on hold add 100 µl of rebinding solution to the samples manually.
14. Continue with the program.
15. Take the elution plate and remove the remaining beads.
16. Determine the RNA concentration with a microvolume spectrophotometer according to the manufacturer's instructions.
17. Optional: Use an RNA fragment analyzer, based on automated capillary electrophoresis, according to manufacturer's instructions to test quality and quantity (*see* **Note 30**).

3.13 nCounter Assay

The following shows a simplified protocol for a NanoString assay. For a detailed description of the steps that need to be followed to successfully use the NanoString nCounter Gene Expression Assay, we refer to the comprehensive manuals (MAN-C0035, MAN-10023) provided by NanoString Technologies. https://www.nanostring.com/support/product-support/support-documentation#user_manuals

1. Total RNA (100 ng) is hybridized at 65 °C for 16 h in a thermo cycler.
2. Probe pairs are placed into the hybridization reaction in massive excess to target mRNA to ensure that each target finds a probe pair.
3. After hybridization, excess probes are washed away using a two-step magnetic bead-based purification on the nCounter® Prep Station.
   (a) The hybridization mixture binds to the Counter Cartridge (Streptavidin) through the capture probe.
   (b) Unbound probes and transcripts are removed during wash steps.
4. Data collection is carried out in the nCounter™ Digital Analyzer.
5. Digital images are processed by the nCounter™ Digital Analyzer and RCC files, containing barcode counts, are generated.
6. RCC files are further analyzed with the nSolver software.

3.14 Data Analysis Using the nSolver Software

The following shows a simplified process for data analysis and visualization. For a detailed description of the steps that need to be followed to successfully use the nSolver software, we refer to the comprehensive manuals (MAN-C0019, MAN-10030) provided by NanoString Technologies. https://www.nanostring.com/support/product-support/support-documentation#user_manuals

1. Profiled data are preprocessed, using nSolver, following the manufacturer's recommendations.
2. The background is subtracted using the geometric mean of the negative controls. Normalization is performed using the positive controls and housekeeping genes (*see* **Note 31**).
3. The Advanced Analysis add-on to nSolver™ can be used for cell type profiling which allows the quantification of different cell populations, based on predefined cell mRNA markers (*see* **Note 32**).
4. Heat maps of NanoString data (*see* **Notes 33** and **34**; Fig. 2a) can be generated using TreeView [21].
5. Alternatively, differentially expressed genes can be visualized in a volcano plot where individual genes are displayed with fold change and significance levels (Fig. 2b).

3.15 Tumor Lysate for Cytokine Multiplex Analysis

1. Thaw lysis buffer and add PMSF to a final concentration of 1 mM.
2. Prechill 2.0 ml reinforced screw-cap homogenization tubes, containing 1.4 mm ceramic beads, on ice.
3. Place the tumor sample in the prechilled homogenization tube.
4. Pipet 500 μl lysis buffer per 100 mg tissue (see **Note 8**).
5. Place the tubes in the tissue homogenizer and homogenize using 3 cycles of 30 s (*see* **Note 29**).
6. Centrifuge 5 min at 400 × *g* to remove tissue pieces and transfer supernatant to a prechilled 1.5 ml reaction tube.
7. Centrifuge 10 min at 16,000 × *g* and 4 °C.
8. Transfer supernatant to a new 1.5 ml reaction tube and store at −80 °C.

3.16 Multiplex Cytokine Analysis: Hybridization with Beads and Detection Antibodies

The following section describes the procedure for one 96-multiwell plate (*see* **Notes 35** and **36**). Test specific buffers are included in the Legendplex kit.

1. Prepare washing buffer, diluting 7.5 ml of the concentrated washing buffer (supplied in the kit) with 142.5 ml deionized water.
2. Reconstitute lyophilized Matrix A with 5 ml Assay Buffer. Allow to reconstitute for 15 min and vortex to mix well.

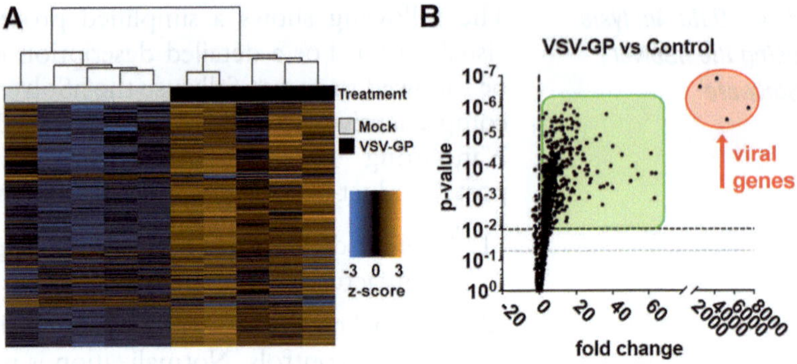

Fig. 2 Visualization of gene expression data. (**a**) Heatmap generation. Data were normalized and samples were clustered (similarity metric: Euclidean distance, Clustering method: Hierarchical clustering; Average linkage). The z-score represents the mean value subtracted from the raw value, which is then divided by the standard deviation. The measure of dissimilarity between sets of observations is displayed via Euclidean distance, represented as a straight line between two points in Euclidean space. Each column represents one tumor sample. (**b**) The volcano plot shows the fold change in gene expression, and their respective statistical significance (*p*-value), between treated and untreated tumors ($n = 5$). Individual dots depict differential gene expression between treatment groups as fold change in expression. Customized viral gene probes were added to the PanCancer Mouse Immune Profiling gene panel (shown in red circle)

3. Reconstitute lyophilized standard with 250 μl Assay Buffer. Allow to reconstitute for 10 min.
4. Label 8 1.5 ml reaction tubes as C7 to C0 for the standard curve. Pipet 37.5 μl assay buffer in the tubes C6 to C0. Pipet 50 μl of reconstituted standard in the C7 tube. Prepare fourfold serial dilutions by transferring 12.5 μl of standard from the C7 tube to the C6 tube. Proceed in the same manner performing fourfold serial dilutions for tubes C5 to C1. Assay buffer is used in C0 as background (0 pg/ml) (*see* **Note 37**).
5. Dilute tumor lysates fivefold in assay buffer at a final volume of 30 μl (*see* **Note 38**). *Optional (when analyzing EDTA-Plasma):* Dilute EDTA-Plasma samples twofold in assay buffer at a final volume of 30 μl.
6. Pipet 12.5 μl Matrix A per well in the standard wells (we recommend to run the standards in duplicates, using the first two columns of the plate, each row for one dilution step).
7. Pipet 12.5 μl Assay Buffer per well in the sample wells (we recommend to run the samples in duplicates).
8. Pipet 12.5 μl per well of standard, sample or sample dilution accordingly.

9. Vortex antibody-immobilized beads for 1 min. Add 12.5 μl beads to each well. During bead addition vortex the bead bottle from time to time to prevent bead sedimentation).
10. Seal the plate and cover with aluminum foil to protect from light. Place on a plate shaker and shake at 800 rpm at 4 °C overnight (*see* **Notes 39** and **40**).
11. Remove the aluminum foil. Briefly centrifuge the plate for 1 min at 500 × *g* to spin down the liquid attached to the plate sealer.
12. Remove the plate sealer and centrifuge the plate at 250 × *g* for 5 min.
13. Decant the supernatant into a biohazard waste container by quickly inverting the plate in one continuous forceful motion. Keeping the plate upside down, place on a stack of paper towel to drain the remaining liquid (*see* **Note 41**). The blue bead pellets might be visible at the bottom of the wells.
14. Wash by pipetting 200 μl of washing buffer into each well and incubate for 1 min. Centrifuge and decant the supernatant as detailed above.
15. Repeat the washing step once.
16. Add 12.5 μl of Detection Antibodies to each well.
17. Seal the plate and cover with aluminum foil to protect from light. Place on the plate shaker and shake at 800 rpm at room temperature for 30 min.
18. Remove the aluminum foil. Briefly centrifuge the plate for 1 min at 500 × *g* to spin down the liquid attached to the plate sealer.
19. Add 12.5 μl of Streptavidin-PE to each well.
20. Seal the plate, cover with aluminum foil and shake at 800 rpm at room temperature for 30 min.
21. Remove the aluminum foil. Briefly centrifuge the plate for 1 min at 500 × *g* to spin down the liquid attached to the plate sealer. Remove the plate sealer and wash twice as detailed above.
22. Add 150 μl of washing buffer, resuspend the beads and read on a flow cytometer.

3.17 Multiplex Cytokine Analysis: Read-Out Using the Flow Cytometer

1. Use the C0 and C7 standards to set up the flow cytometer. Transfer 75 μl of C0 and C7 standard into flow cytometry tubes containing 150 μl assay buffer. Refill the two standard wells in the plate with 75 μl to equalize the volume to the other wells (*see* **Notes 5** and **42**).
2. Use the C0 standard to set the forward and side scatter voltages so that two bead populations can be clearly identified in the

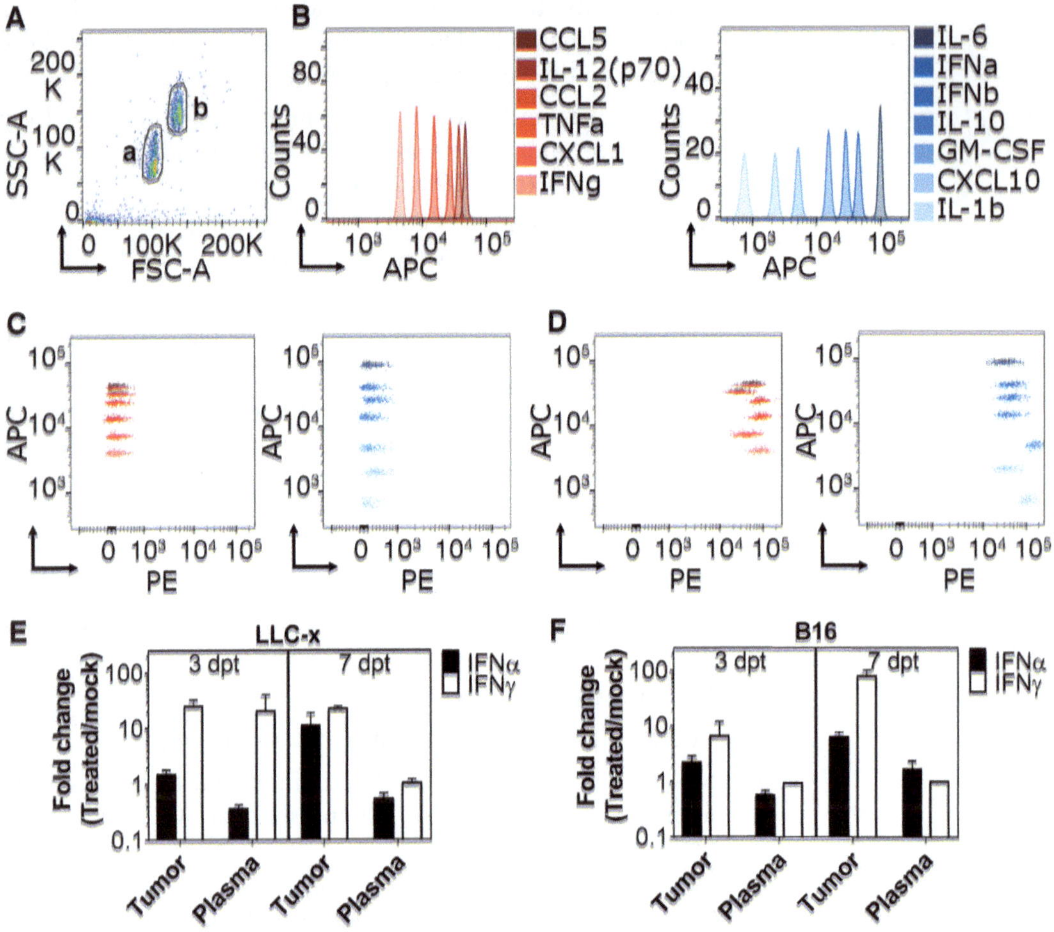

Fig. 3 Cytometer setup for Legendplex cytokine multiplex and exemplary results. (**a**) Forward and side scatter dot plot with gating of the two main bead populations. (**b**) Histogram showing the identification of the **a** (red) and **b** (blue) subpopulations determined by their intensity in the APC channel. (**c**) Dot plot of the C0 standard showing the background PE signal for each **a** and **b** subpopulation. (**d**) Dot plot of the C7 standard showing the highest PE signal for each **a** and **b** subpopulation. (**e**, **f**) Exemplary results of two cytokines (IFNα and IFNγ) measured in tumor and plasma of modified LLC1-x tumor (**e**) or B16 melanoma (**f**) bearing mice at two timepoints (3 and 7 days post treatment). The results are expressed as fold change in the treated group compared to mock treated. In tumor samples, fold change was calculated after normalizing the cytokine concentration to the concentration of total protein in the tumor lysate

middle of the dot plot as shown in Fig. 3a. Within each bead population, six (in beads A) or seven (in beads B) bead sub-populations can be distinguished according to their fluorescence intensity in the APC channel, each corresponding to one analyte (Fig. 3b).

3. Gate the two populations, A (smaller beads) and B (larger beads), and for each of them plot the APC channel in the *y*-axis and the PE channel in the *x*-axis.

4. Adjust the APC voltages so that the APC fluorescence intensities of the populations observed are between 10^2 and 10^5 (Fig. 3c). Adjust the PE voltages so that the PE fluorescence intensity is between 10^1 and 10^2 (Fig. 3c).
5. Use the C7 standard to confirm that the PE fluorescence intensity does not exceed 10^5 (Fig. 3d). If that would be the case, readjust the PE voltages.
6. Configure the automated multiwell plate acquisition device to perform three mixes of each well at a speed of 50 μl/s and a volume of 100 μl and to then measure 75 μl of the sample at a rate of 2 μl/s with a washing step of 300 μl between samples (*see* **Notes 43** and **44**).
7. After measurement, export the FCS files and use the Legendplex software to calculate the concentrations of each analyte in relation to the standard curve (*see* **Note 45**).
8. For tumor samples, it is recommended to normalize the concentration obtained to the total protein concentration in the tumor lysate. Figure 3e, f show the measurements of two exemplary cytokines measured in two different tumor types.

4 Notes

1. For homogenization, we recommend to use the specially reinforced Bertin tubes (Bertin Technologies 91-PCS-CK14 Precellys ceramic-kit 1.4 mm 50 × 2.0 ml), since the standard 2 ml screw cap Eppendorf tubes are prone to leakage and subsequent loss of sample during the homogenization process. Alternatively, for very hard tumor samples that are not thoroughly homogenized using the approach described, we recommend the use of the GentleMACS dissociator (Miltenyi Biotec) with the supplied M tubes. We have successfully used this method to obtain homogenates from subcutaneous TRAMP-C1 tumors.
2. Lysis buffer can be directly used or aliquoted and stored at −20 °C. If stocks of lysis buffer are prepared do not add PMSF. PMSF should be added fresh before use.
3. If another cell lysis buffer than the one described is used, make sure it is a nondenaturing lysis buffer.
4. Here we describe the Mouse Anti-virus response Legendplex panel using v-bottom plates. Several predefined Legendplex panels both in filter-plate and v-bottom plate format are available.
5. The use of an automated multiwell plate acquisition device is optional. The measurement with the flow cytometer can also be

performed using normal flow cytometry tubes. However, be aware that measurements taken using individual flow cytometry tubes slightly differ from measurements done using the automated multiwell-plate acquisition device. If you plan to use the automated multiwell-plate acquisition device, once the main parameters are set up, use the set-up beads supplied in the kit to perform test-measurements with the automated multiwell-plate acquisition device.

6. Since in this protocol animals are sacrificed for tumor extraction, a complete bleeding procedure from the retro-orbital bulb is described. However, 25 μl of blood are sufficient to perform the assay described herein. Alternative blood collection techniques, like collection from the tail or the facial vein can be used.
7. In case tumor weight exceeds the recommended weight, use a fraction of the tumor and make sure its weight is within the indicated ranges.
8. Avoid using less than 500 μl of buffer to prevent the homogenizing tubes from breaking during homogenization, even if the amount of tumor tissue available does not reach 100 mg.
9. DMEM can be used instead of RPMI 1640 medium for all the following steps.
10. Certain cell surface epitopes are sensitive to the enzymes present in the enzyme mix. Various epitopes have been tested by Miltenyi and the list of these epitopes is available online at: https://www.miltenyibiotec.com/upload/assets/IM0016762.PDF. The concentration of the kit components can be adjusted to mitigate the destruction of these epitopes to a certain extent.
11. To avoid exposing tumors to the enzyme mix for an extended period of time, tumors are collected in 1.5 ml RPMI 1640 and placed on ice while remaining tumors are being harvested and prepared for dissociation. Shortly before loading the tubes on the gentleMACS instrument, 1.0125 ml enzyme mix is added to the tubes. Alternatively, tumor pieces can be directly collected in gentleMACS C-tubes containing 2.35 ml RPMI 1640, 100 μl Enzyme D, 50 μl Enzyme R and 12.5 μl Enzyme A.
12. It is possible to substitute the Tumor Dissociation kit from Miltenyi with custom-made collagenase-based enzyme mixes. For instance, a mixture of 0.5 mg/ml collagenase D, 10 μg/ml DNase I and 10 mM HEPES buffer prepared in a balanced salt solution such as PBS or HBSS can be used as an alternative. The concentrations of the enzymes might have to be adjusted based on the tumor type. Collagenase D has high collagenase activity and a very low tryptic activity, which is important in order to preserve the integrity of cell surface proteins. DNase I

prevents cell clumping caused by DNA released from dying cells.

13. It is important to ensure that the sample material is located in the area of the rotor/stator.
14. The gentleMACS equipment ensures that all samples are processed in a consistent manner for high reproducibility. However, in case gentleMACS equipment is not available, the tumor pieces can be incubated in 2–4 ml enzyme mix in a shaking incubator at 37 °C for 1 h before proceeding to **step 6** (Subheading 3.2).
15. The programs **37C_m_TDK1** and **37C_m_TDK2** are suitable only when using gentleMACS Octo Dissociator with Heaters. **37C_m_TDK1** is adapted for processing of soft and medium tumors such as B16 and CT26 tumors while **37C_m_TDK2** is optimized for tumors with a harder consistency such as 4T1 and TC1 tumors.
16. If heaters are not available, select program **m_impTumor_02.** After termination of **m_impTumor_02 program**, transfer the C tube to a 37 °C incubator with continuous rotation/agitation for 40 min. Then return the C tube to the gentleMACS Dissociator. Select the program **m_impTumor_03** and press start. It is sufficient to run the program **m_impTumor_03** once for soft and medium tumors whereas it is recommended to run it twice for tumors with harder consistency.
17. If there are visible chunks of the tumor on the cell strainer, use the plunger of a 5 ml syringe to gently smash the tumor pieces against the cell strainer to get homogenous cell suspension.
18. Alternatives to Lympholyte®-M include Ficoll-Paque PLUS (GE Healthcare Life Sciences) and Histopaque-1083 (Merck) although they have slightly different densities.
19. Enrichment of lymphocytes before FACS staining is optional but it greatly improves staining quality and shortens acquisition times (Fig. 1c).
20. Instead of overlaying the cell suspension on the Lympholyte®-M, it is possible to use a Pasteur pipette to underlay the Lympholyte®-M at the bottom of the tube under the cell suspension.
21. Alternatively, leukocytes or even specific lymphocyte subsets can be enriched using magnetically labeled microbeads specific for CD45, CD8 or CD4 (Miltenyi Biotec).
22. If staining is performed in 96 well plates, use 200 μl of appropriate washing buffer and centrifuge the plates at 800 × *g* for 2 min. The same volume of 100 μl can be used for the blocking and staining steps.

23. If using antibodies raised only in mice, Purified Rat Anti-Mouse CD16/CD32 (Mouse BD Fc Block™) is sufficient to block FC receptors. However, when using antibodies raised in rats and hamster, use of normal serum from the respective host organism is recommended. Unspecific signal can be further reduced by using a dump channel to exclude other cell populations by labeling them with respective lineage markers such as NK1.1 (NK cells), CD11b (monocytes/macrophages and certain DCs), B220 or CD19 (B cells) etc.
24. The fixable dead cell dyes interact with free amines on the cells, both on the surface and inside the cells. Dead cells have a stronger staining as the dye can access the intracellular free amines due to compromised membrane integrity. It is therefore advisable to wash the cells with PBS and perform the staining in PBS. Proteins in FACS buffer can compete with free amines on the dead cells and reduce staining intensity [22].
25. In case of tetramer staining, it is advisable to start with the labeling of TCRs with pMHC-multimer before adding antibodies for surface staining as most anti-CD8 antibodies can inhibit binding of TCR to pMHC-multimers [23].
26. The choice of the buffer for intracellular staining is crucial. It depends on whether intracellular cytokines or nuclear proteins like transcription factors are stained. The True-Nuclear™ Transcription Factor Buffer Set (Biolegend) or Foxp3 Staining Buffer Set (Thermofisher) are more suitable for staining transcription factors whereas Fixation and Permeablization Wash Buffer (Biolegend) or Intracellular Fixation & Permeabilization Buffer Set (Thermofisher) are more suited for intracellular cytokines.
27. It is crucial to use the Transcription Factor 1× Perm Buffer (or the corresponding buffer if using another kit) for diluting the antibodies against intracellular targets to keep the membrane permeable.
28. When staining for intracellular cytokines or chemokines secreted by stimulated cells, it is important to add protein transport inhibitors such as brefeldin A or monensin during the last hours of stimulation to prevent their secretion.
29. Depending on the characteristics of the tumor tissue, the homogenization protocol should be optimized.
30. The Nanostring capture and reporter probes bind to relatively small RNA fragments, usable data can still be obtained from certain low quality RNA samples. When working with a tumor model exhibiting low quality RNA, we recommend running a limited number of samples to see whether usable data can still be obtained, rather than excluding this tumor model altogether.

31. The appropriate number of biological replicates in order to obtain statistically significant results will always depend on the treatments and tumor models tested. We therefore recommend to first perform a small pilot study to determine the optimal number of biological replicates for the main study. In general, the optimal amount per setting is 3–4 when analyzing cell culture samples, 6–8 for inbred animal samples and > 10 for clinical samples.
32. When analyzing murine tumor samples, many cell types in the cell type profiling option of the advanced analysis of nSolver do not have sufficient reliable cell markers for analysis. Even for the cell types with reliable RNA markers, we recommend confirming the obtained mRNA Nanostring results with protein cell markers using other experimental techniques (Nanostring Vantage 3D™ Assays, immunofluorescence, immunohistochemistry, etc.).
33. In order to exclude the possible introduction of bias during processing and analysis of the Nanostring samples, it is recommended to include processing parameters (scan date, run number, plate location, etc.) into the initial advanced analysis (nSolver) of the experiment. Possible biases will then become apparent when the samples cluster together according to one of these processing parameters.
34. For similar reasons as in **Note 33**, we also recommend to randomize the samples during processing and analysis.
35. The procedure has been optimized to use the half amount of the reagents recommended by the manufacturer. This allows to extend the use of one kit from one 96-multiwell plate to at least 2.5 plates. One v-bottom plate is supplied with the kit. If additional plates are ordered, make sure they are polypropylene plates. Do not use polystyrene plates.
36. Unused buffers, antibody-immobilized beads, detection antibodies and streptavidin-PE conjugates can be stored for 1 month at 4 °C after opening. Unused standard and protein background (matrix A) can be aliquoted and stored at −80 °C.
37. The concentration of the top standard varies from lot to lot and can be found in the certificate of analysis.
38. Total protein and analyte concentrations can vary depending on several factors (tumor type, treatment, etc). Furthermore, the concentration of each individual analyte within one sample can vary compared to the other analytes. Therefore, it is recommended to run two dilutions of each sample (e.g., pure and 1:5 diluted) to reduce the risk of having analytes below or above the detection limits.

39. Shaking is a critical step. Since there might be small variations among shakers, it is recommended to always use the same shaker once the shaking speed has been optimized. A speed should be set that the fluid shakes vigorously in the well without sprinkling the plate sealer. Sub-optimal shaking can result in a second bead population with unbound analyte.
40. This incubation step can be shortened to 2 h at room temperature. However, this can also reduce sensitivity of the assay.
41. To prevent loss of beads be sure to dump the supernatant by quickly inverting the plate in one continuous forceful motion. Remaining liquid will be removed by blotting the plate on paper towels. Do not try to remove the remaining liquid by shaking or flicking the plate on the waste container, since this will cause loss of beads.
42. Alternatively, the set-up of the flow cytometer can be performed using the set-up beads supplied in the kit.
43. Measuring with the automated multiwell plate acquisition device at higher speed might reduce the quality of the measurement. Make sure to run a test if you plan to increase the sample flow rate.
44. Measuring 75 μl of sample ensures enough counts of each analyte to obtain a meaningful measurement. The volume of sample measured can be reduced but no less than a total of 3000 beads should be measured.
45. Detailed tutorials on how to use the Legendplex analysis software can be found on Biolegend's web page (https://www.biolegend.com/legendplex).

Acknowledgments

We thank Ellen Richter, Erika Mueller, and Manuel Salvatore for excellent technical assistance. This study was supported by a grant from the Christian Doppler Research Association.

References

1. Melcher A, Parato K, Rooney CM, Bell JC (2011) Thunder and lightning: immunotherapy and oncolytic viruses collide. Mol Ther 19 (6):1008–1016. https://doi.org/10.1038/mt.2011.65
2. Bommareddy PK, Shettigar M, Kaufman HL (2018) Integrating oncolytic viruses in combination cancer immunotherapy. Nat Rev Immunol 18(8):498–513. https://doi.org/10.1038/s41577-018-0014-6
3. Lichty BD, Breitbach CJ, Stojdl DF, Bell JC (2014) Going viral with cancer immunotherapy. Nat Rev Cancer 14(8):559–567. https://doi.org/10.1038/nrc3770
4. de Graaf JF, de Vor L, Fouchier RAM, van den Hoogen BG (2018) Armed oncolytic viruses: a kick-start for anti-tumor immunity. Cytokine Growth Factor Rev 41:28–39. https://doi.org/10.1016/j.cytogfr.2018.03.006

5. Todo T (2008) "Armed" oncolytic herpes simplex viruses for brain tumor therapy. Cell Adhes Migr 2(3):208–213
6. Cesano A (2015) nCounter((R)) PanCancer immune profiling panel (NanoString technologies, Inc., Seattle, WA). J Immunother Cancer 3:42. https://doi.org/10.1186/s40425-015-0088-7
7. Galluzzi L, Chan TA, Kroemer G, Wolchok JD, Lopez-Soto A (2018) The hallmarks of successful anticancer immunotherapy. Sci Transl Med 10(459). https://doi.org/10.1126/scitranslmed.aat7807
8. Gujar S, Pol JG, Kim Y, Lee PW, Kroemer G (2018) Antitumor benefits of antiviral immunity: an underappreciated aspect of oncolytic virotherapies. Trends Immunol 39 (3):209–221. https://doi.org/10.1016/j.it.2017.11.006
9. Bentzen AK, Marquard AM, Lyngaa R, Saini SK, Ramskov S, Donia M, Such L, Furness AJ, McGranahan N, Rosenthal R, Straten PT, Szallasi Z, Svane IM, Swanton C, Quezada SA, Jakobsen SN, Eklund AC, Hadrup SR (2016) Large-scale detection of antigen-specific T cells using peptide-MHC-I multimers labeled with DNA barcodes. Nat Biotechnol 34(10):1037–1045. https://doi.org/10.1038/nbt.3662
10. Macnamara C, Eftimie R (2015) Memory versus effector immune responses in oncolytic virotherapies. J Theor Biol 377:1–9. https://doi.org/10.1016/j.jtbi.2015.04.004
11. Kerkar SP, Restifo NP (2012) Cellular constituents of immune escape within the tumor microenvironment. Cancer Res 72 (13):3125–3130. https://doi.org/10.1158/0008-5472.CAN-11-4094
12. Trujillo JA, Sweis RF, Bao R, Luke JJ (2018) T cell-inflamed versus non-T cell-inflamed tumors: a conceptual framework for cancer immunotherapy drug development and combination therapy selection. Cancer Immunol Res 6(9):990–1000. https://doi.org/10.1158/2326-6066.CIR-18-0277
13. Geiss GK, Bumgarner RE, Birditt B, Dahl T, Dowidar N, Dunaway DL, Fell HP, Ferree S, George RD, Grogan T, James JJ, Maysuria M, Mitton JD, Oliveri P, Osborn JL, Peng T, Ratcliffe AL, Webster PJ, Davidson EH, Hood L, Dimitrov K (2008) Direct multiplexed measurement of gene expression with color-coded probe pairs. Nat Biotechnol 26(3):317–325. https://doi.org/10.1038/nbt1385
14. Lyons YA, Wu SY, Overwijk WW, Baggerly KA, Sood AK (2017) Immune cell profiling in cancer: molecular approaches to cell-specific identification. NPJ Precis Oncol 1(1):26. https://doi.org/10.1038/s41698-017-0031-0
15. Kulkarni MM (2011) Digital multiplexed gene expression analysis using the NanoString nCounter system. Curr Protoc Mol Biol Chapter 25:Unit 25B 10. doi:https://doi.org/10.1002/0471142727.mb25b10s94
16. Galivo F, Diaz RM, Wongthida P, Thompson J, Kottke T, Barber G, Melcher A, Vile R (2010) Single-cycle viral gene expression, rather than progressive replication and oncolysis, is required for VSV therapy of B16 melanoma. Gene Ther 17(2):158–170. https://doi.org/10.1038/gt.2009.161
17. Simovic B, Walsh SR, Wan Y (2015) Mechanistic insights into the oncolytic activity of vesicular stomatitis virus in cancer immunotherapy. Oncolytic Virother 4:157–167. https://doi.org/10.2147/OV.S66079
18. Muik A, Stubbert LJ, Jahedi RZ, Geibeta Y, Kimpel J, Dold C, Tober R, Volk A, Klein S, Dietrich U, Yadollahi B, Falls T, Miletic H, Stojdl D, Bell JC, von Laer D (2014) Re-engineering vesicular stomatitis virus to abrogate neurotoxicity, circumvent humoral immunity, and enhance oncolytic potency. Cancer Res 74(13):3567–3578. https://doi.org/10.1158/0008-5472.CAN-13-3306
19. Melzer MK, Lopez-Martinez A, Altomonte J (2017) Oncolytic vesicular stomatitis virus as a viro-immunotherapy: defeating cancer with a "Hammer" and "Anvil". Biomedicines 5(1). https://doi.org/10.3390/biomedicines5010008
20. Tober R, Banki Z, Egerer L, Muik A, Behmuller S, Kreppel F, Greczmiel U, Oxenius A, von Laer D, Kimpel J (2014) VSV-GP: a potent viral vaccine vector that boosts the immune response upon repeated applications. J Virol 88(9):4897–4907. https://doi.org/10.1128/JVI.03276-13
21. Page RD (1996) TreeView: an application to display phylogenetic trees on personal computers. Comput Appl Biosci 12(4):357–358
22. Perfetto SP, Chattopadhyay PK, Lamoreaux L, Nguyen R, Ambrozak D, Koup RA, Roederer M (2010) Amine-reactive dyes for dead cell discrimination in fixed samples. Curr Protoc Cytom Chapter 9:Unit 9 34. doi:https://doi.org/10.1002/0471142956.cy0934s53
23. Wooldridge L, Hutchinson SL, Choi EM, Lissina A, Jones E, Mirza F, Dunbar PR, Price DA, Cerundolo V, Sewell AK (2003) Anti-CD8 antibodies can inhibit or enhance peptide-MHC class I (pMHCI) multimer binding: this is paralleled by their effects on CTL activation and occurs in the absence of an interaction between pMHCI and CD8 on the cell surface. J Immunol 171 (12):6650–6660

Chapter 11

Immunohistochemistry for Tumor-Infiltrating Immune Cells After Oncolytic Virotherapy

Dipongkor Saha and Samuel D. Rabkin

Abstract

Immunohistochemistry (IHC) is an integral laboratory staining technique, which is used for the detection of immune cells in mouse/human tissues or tumors. Oncolytic herpes simplex virus (oHSV) treatment or virotherapy of solid tumors results in antitumor immune responses and infiltration of a variety of immune cells into the tumor. Here, we describe a step-by-step chromogen/substrate-based single- and dual-color IHC protocol to stain immune cells in formalin-fixed, paraffin-embedded mouse glioblastoma (GBM) brain tumor sections after oHSV virotherapy. Tumor sections are deparaffinized with xylene, then gradually rehydrated using ethanol, followed by heat-mediated antigen retrieval using appropriate buffers. Tumor sections are incubated with primary antibodies, which detect a specific immune cell antigen, then incubated with peroxidase- or phosphatase-labeled secondary antibodies, followed by incubation with a color-producing substrate and color visualization (of immune cells) by light microscopy. The protocol described herein is also applicable to detect immune cells in other mouse and human tumors or organs after other forms of immunotherapy.

Key words Immunohistochemistry, Substrate-based immunohistochemistry, Dual-color immunohistochemistry, Oncolytic herpes simplex virus, Virotherapy, Immune cells, Antigen–antibody reaction, Antigen retrieval

1 Introduction

Oncolytic viruses are a distinct class of anticancer agents, which selectively replicate in and kill cancer cells without harming normal tissue, and often induce antitumor immune responses [1, 2]. Oncolytic herpes simplex virus (oHSV) has been genetically engineered for oncolytic activity and safety [3]. For example, T-Vec (talimogene laherparepvec) was recently approved as the first oncolytic virus in the USA for the treatment of advanced melanoma [3]. oHSV treatment induces an inflammatory reaction and attracts a variety of immune cells to infiltrate into the treatment site signifying an antitumor immune response [4, 5]. Tumor inflammation and immune cells in a tumor section can be detected or visualized by immunohistochemistry (IHC), a commonly used

Christine E. Engeland (ed.), *Oncolytic Viruses*, Methods in Molecular Biology, vol. 2058,
https://doi.org/10.1007/978-1-4939-9794-7_11,

and integral laboratory staining technique [6, 7]. This permits analysis of the distribution of labeled cells and quantification of cell numbers. The basic principle of IHC is that a reaction occurs between an immune cell antigen and an antibody conjugated with an enzyme (peroxidase, alkaline phosphatase, etc.) that can catalyze production of a colored substrate [8, 9], which is eventually visualized by light microscopy. Alternatively, the antibody can also be tagged to a fluorophore (fluorescein, rhodamine, etc.), and the antigen–antibody reaction is visualized by detecting the fluorophore under fluorescence microscopy (immunofluorescence) [10]. Fluorescence-based IHC is not stable for several months or years, since fluorophores continuously fade over time. Chromogen/substrate-based IHC has a major advantage over fluorescence-based IHC since staining quality can be preserved for years at room temperature without a fading problem. Herein, we describe a step-by-step chromogen-based single−/dual-color immunohistochemical protocol for staining tumor-infiltrating immune cells in a mouse glioblastoma (GBM) brain tumor model treated with oHSV-based immunotherapy. The strategies described herein are also applicable for other mouse/human tumors in the brain and in the periphery.

2 Materials

The materials required for single/dual IHC are listed below. Few specific materials/reagents that are exclusively used for double IHC in this review are indicated by asterisk (∗). All steps with flammable reagents, such as xylene/ethanol, must be done inside a fume hood. All water for buffers/solutions should be ultrapure, MilliQ, or distilled.

1. Slide staining set, preferably from polyoxymethylene (POM), including slide staining dishes, slide staining racks, and a power-coated metal housing.
2. Slide staining dish set rack containing a staining dish, lid, and a stainless steel rack.
3. Xylenes (Certified ACS).
4. ∗Xylene-free clearing solution.
5. 100% absolute ethanol.
6. 90% ethyl alcohol. Mix 450 mL absolute ethanol and 50 mL water.
7. 70% ethyl alcohol. Mix 350 mL absolute ethanol and 150 mL water.
8. Sodium citrate buffer. 10 mM sodium citrate, 0.05% Tween 20, pH 6.0. Dissolve 2.94 g Tri-sodium citrate (dihydrate) in

1 L water. Adjust pH with 1 N HCl. Add 0.5 mL Tween 20 and mix well. Store at 4 °C.

9. 1 mM EDTA, pH 8.0. Dissolve 0.37 g EDTA in 1 L water. Adjust pH with sodium hydroxide (NaOH). Store at 4 °C.
10. Tris–EDTA buffer. 10 mM Tris base, 1 mM EDTA solution, 0.05% Tween 20, pH 9.0. Dissolve 1.21 g Tris base and 0.37 g EDTA in 1 L water. Adjust pH with NaOH. Add 0.5 mL Tween 20 and mix well. Store at 4 °C.
11. 2 L glass beaker.
12. Plastic wrap.
13. 1× DPBS solution. Dissolve 95.5 g DPBS (Dulbecco's Phosphate-Buffered Saline) powder in 1 L water. Prepare 1× DPBS working solution for use by mixing 100 mL 10× DPBS and 900 mL water.
14. 1× DPBS/0.1% Tween 20. Add 1 mL Tween 20 in 1 L of 1× DPBS.
15. PAP Pen Liquid Blocker.
16. 3% H_2O_2 solution. Mix 20 mL 30% (w/w) hydrogen peroxide (H_2O_2) solution containing stabilizer and 180 mL water.
17. *Endogenous Peroxidase and Alkaline Phosphatase Blocking Solution.
18. Universal blocking buffer (5% BSA solution). Dissolve 1 g lyophilized bovine serum albumin (BSA) powder in 19 mL water. Store at 4 °C.
19. 10% horse serum blocking solution. Mix 1 mL normal horse serum and 9 mL water. Store at 4 °C.
20. 10% goat serum blocking solution. Mix 1 mL normal goat serum and 9 mL water. Store at 4 °C.
21. Primary antibodies for tumor-infiltrating immune cells (Table 1).
22. Humidified chamber.
23. HRP- and AP-conjugated secondary antibodies (Table 2).
24. DAB Substrate Chromogen System.
25. Red Alkaline Phosphatase (AP) substrate.
26. Blue AP substrate.
27. Hematoxylin solution. Mix 100 mL hematoxylin and 300 mL water.
28. Cover Glasses.
29. Xylene-based permanent mounting medium.
30. Xylene-free permanent mounting medium.

Table 1
List of primary antibodies with their sources and dilutions

Primary antibodies	Source	Cat. no.	Dilution	Blocking buffer
Rabbit anti-CD3	Abcam	ab5690	1:100	Horse serum
Rat anti-CD4	eBioscience	14-9766-80	1:200	Goat serum
Rat anti-CD8a	eBioscience	14-0808-80	1:100	Goat serum
Rabbit anti-granzyme B	Abcam	ab4059	1:150	Horse serum
Rabbit anti-CD68	Abcam	ab125212	1:100	Horse serum
Rabbit anti-Ki67 (for proliferating immune cells by double IHC)	Abcam	ab16667	1:100	Horse serum
Rabbit anti-PD-L1	Abcam	ab205921	1:400	Horse serum
Rabbit anti-cleaved caspase-3 (Asp175) (for apoptotic immune cells by double IHC)	Cell Signaling	9661	1:100	Horse serum
Rabbit anti-phospho-stat1 (Tyr701)	Cell Signaling	9167	1:100	Horse serum
Rabbit anti-iNOS	Abcam	ab15323	1:100	Horse serum
Mouse anti-T-bet/Tbx21	Abcam	ab91109	1:100	Horse serum
Rat anti-MHC class II	Abcam	ab25333	1:150	Goat serum
Rabbit anti-F4/80	Abcam	ab111101	1:100	Horse serum
Rat anti-CD34	Abcam	ab8158	1:150	Goat serum

Table 2
List of secondary antibodies with their sources and dilutions

Secondary antibodies	Source	Cat. no.	Dilution
HRP anti-rat IgG	Vector Lab	MP-7444	One drop/section
HRP anti-rabbit IgG	Vector Lab	MP-7401	One drop/section
HRP anti-mouse IgG	Vector Lab	MP-7402	One drop/section
AP anti-rabbit IgG	Vector Lab	MP-5401	One drop/section

3 Methods

The protocol described below is designed to perform all steps at room temperature, except incubating brain tumor sections with primary antibody overnight at 4 °C in a humidified chamber. All steps with flammable reagents, such as xylene and ethanol, must be performed inside a fume hood.

3.1 Single Immunohistochemistry (Chromogen-/Substrate-Based) for Formalin-Fixed Paraffin-Embedded Tumor Sections

1. Deparaffinize sections in xylene twice for 10 min (*see* **Note 1**).
2. Rehydrate sections with gradually decreasing concentrations of ethyl alcohol: 100%, 90% and 70% (5 min with each concentration) (*see* **Note 1**).
3. Leave the slides in water for 5 min.
4. For antigen retrieval, microwave sections in 10 mM sodium citrate buffer for 15 min using a microwaveable vessel or a glass beaker (*see* **Notes 2** and **3**).
5. After 15 min, cool tumor sections at room temperature for 20–25 min (*see* **Note 4**).
6. Rinse tumor sections in 1× DPBS twice for 5 min.
7. While running **step 6**, circle each tumor section with water-resistant Liquid Blocker or PAP pen (*see* **Note 5**).
8. Block endogenous peroxidase activity by dipping slides in 3% H_2O_2 solution for 5 min (*see* **Note 6**).
9. Rinse slides in 1× DPBS twice for 5 min.
10. Block nonspecific antigen–antibody binding by incubating tumor sections with 30–50 μL universal blocking buffer (5% BSA solution) for 30 min, followed by antibody-specific blocking buffer, i.e. 10% horse or goat serum, for another 30 min (*see* **Note** 7).
11. Incubate tumor sections with an appropriate primary antibody (optimally diluted in an appropriate blocking buffer; use 30–50 μL/section) overnight at 4 °C in a humidified chamber (*see* **Notes 8** and **9**).
12. Wash sections in 1× DPBS/0.1% Tween 20 three times for 5 min (*see* **Note 10**).
13. Incubate sections with appropriate secondary antibodies, such as HRP-conjugated anti-rabbit, anti-rat, or anti-mouse IgG (depends on the host origin of the primary antibody) for 30 min (*see* **Notes 11** and **12**).
14. Repeat washing as in **step 12**.
15. Incubate sections with a drop of DAB, prepared by adding 1 drop of chromogen in 1 mL of DAB substrate, until brown color develops (*see* **Note 13**).
16. Rinse slides in water for 1 min (*see* **Note 14**).
17. For counterstaining, dip slides in hematoxylin solution for 20 s (*see* **Note 15**).
18. Wash sections in water for 1 min, then in running tap water for another 5 min (*see* **Note 16**).

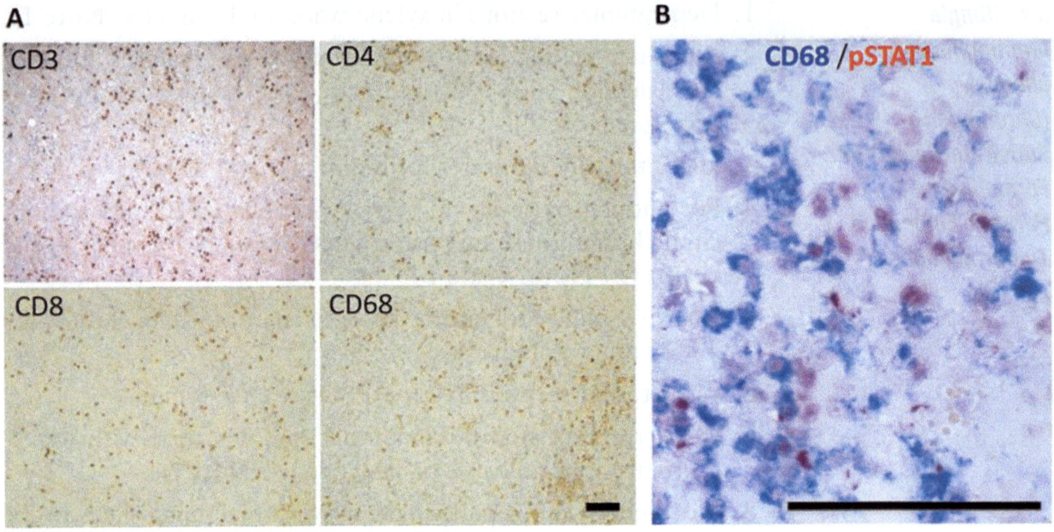

Fig. 1 (**a**) Single IHC staining of tumor infiltrating immune cells in mouse GBM. Representative images (10× magnification) are shown. Brown indicates positivity (anti-CD4, -CD8, -CD68+20% brightness correction). Scale bar = 100 μm. (**b**) Dual-color IHC staining of total macrophage (CD68; blue color) and M1-like macrophages (pSTAT1; red color) in mouse GBM. A representative image (20× magnification) is presented. Scale bar = 100 μm

19. Dehydrate sections using gradually increasing concentrations of ethyl alcohol: 70%, 90%, and 100%, 5 min with each concentration (*see* **Note 17**).
20. Dip sections in xylene twice for 10 min (*see* **Notes 17** and **18**).
21. Take the slides out of xylene (one-by-one). Place the slides on an even surface in a fume hood (*see* **Note 19**).
22. Place one drop of xylene-based mounting medium on a glass cover slip (*see* **Note 20**).
23. Apply the cover slip on a tumor section and distribute the mounting medium throughout the section/slide by applying finger pressure (*see* **Note 21**).
24. Air-dry slides in the fume hood for at least an hour before light microscopy (*see* **Note 22**). See single IHC staining of immune cells in brain tumor sections in Fig. 1a.

3.2 Double Immunohistochemistry (Substrate-Based) for Formalin-Fixed Paraffin-Embedded Sections

Substrate-based double immunohistochemistry is an easy process if both primary antibodies originated from different hosts (e.g., rat and rabbit). In most cases, both primary antibodies come from the same host (e.g., rabbit anti-CD68 for total macrophage and rabbit anti-pSTAT1 for M1-like macrophage; *see* **Note 8**; Fig. 1b), which makes this process more complicated since the same secondary antibody (e.g., HRP- or AP-conjugated anti-Rabbit IgG) is used. To solve this problem, we described previously a substrate-based double IHC protocol to stain both rabbit anti-CD68 and antipSTAT1 or rabbit anti-CD3 (total T cells) and anti-Ki67 (proliferating T cells) in the same tumor section [4]. Because it is a dual-color IHC, two distinct colors should be chosen to visually separate one color from the other by light microscopy. Here, we used blue color for CD68 or Ki67 and red color for pSTAT1 or CD3 for visualization by light microscopy. Briefly, the brain sections were incubated sequentially with primary antibodies (rabbit anti-pSTAT1 or -CD3; antibody), then secondary antibodies (AP-conjugated anti-rabbit IgG), followed by the development of red color using red alkaline phosphatase substrate. Next, the same brain sections were incubated with primary antibodies (rabbit anti-CD68 or -Ki67 antibody), then secondary antibodies (AP-conjugated anti-rabbit IgG), followed by the development of blue color using blue alkaline phosphatase substrate. The step-by-step substrate-based double IHC protocol is described below. Avoid xylene-based clearing agents/mounting media/alcoholic solutions while performing the following protocol (*see* **Note 23**).

1. Deparaffinize brain tumor sections twice for 10 min using xylene-free clearing solution.
2. Follow **steps 2–7** as in single IHC.
3. To block endogenous alkaline phosphatase, add one drop blocking solution per section and incubate for 5 min (*see* **Note 24**).
4. Rinse slides in 1× DPBS twice for 5 min.
5. Block nonspecific antigen–antibody binding by following **step 10** in the single IHC protocol.
6. Incubate tumor sections with an appropriate primary antibody (e.g., rabbit anti-pSTAT1 or anti-CD3 antibody diluted in horse serum; use 30–50 μL/section) overnight at 4 °C in a humidified chamber (*see* **Note 8**).
7. Wash sections in 1× DPBS/0.1% Tween 20 three times for 5 min (*see* **Note 10**).
8. Add one drop AP-conjugated anti-rabbit Ig per section and incubate sections for 30 min at room temperature in a humidified chamber (*see* **Note 12**).
9. Follow **step** 7.
10. Add 50 μL red substrate per section and incubate until red color has developed (*see* substrate preparation in **Note 25**).
11. Wash sections in 1× DPBS/0.1% Tween 20 for 5 min (*see* **Note 10**).
12. Incubate tumor sections with an appropriate primary antibody (e.g., anti-CD68 or Ki67; rabbit antibody diluted in horse serum; use 30–50 μL/section) overnight at 4 °C in a humidified chamber (*see* **Note 8**).

13. Follow **step 7**.
14. Add one drop AP-conjugated anti-rabbit IgG per section and incubate sections for 30 min at room temperature in a humidified chamber (*see* **Note 12**).
15. Follow **step 7**.
16. Add 50 μL blue substrate per section until blue color has developed (*see* substrate preparation in **Note 26**).
17. Wash sections in 1× DPBS/0.1% Tween 20 for 5 min.
18. Wash sections in water for 1 min.
19. Dehydrate sections using gradually increasing concentrations of ethyl alcohol: 70%, 90% and 100% (5 min per concentration) (*see* **Note 27**).
20. Dip sections in xylene-free clearing solution twice for 10 min (*see* **Note 27**).
21. Take the slides out of clearing solution (one-by-one). Place the slides on an even surface in a fume hood (*see* **Note 19**).
22. Place one drop of xylene-free mounting medium on a glass cover slip (*see* **Note 20**).
23. Apply the cover slip on a tumor section and distribute the mounting medium throughout the slide by applying finger pressure (*see* **Note 21**).
24. Air-dry slides in the fume hood for at least an hour before light microscopy (*see* **Note 22**). See dual-color IHC staining of immune cells in brain tumor sections in Fig. 1b.

4 Notes

1. Place slides containing sections in the slide rack, then dip the slide rack into xylene/ethanol in a staining dish. Xylene/ethanol solutions can be stored in airtight glass containers at room temperature and reused many times for months/years. If dirt appears, filter them before use. Xylene/ethanol steps must be performed in a fume hood.
2. Antigen retrieval step: This is a critical step for the success of this protocol. Sodium citrate buffer, 1 mM EDTA and Tris–EDTA buffer are the three most popular antigen retrieval solutions. Which one to use depends on the source of the primary antibody. It is advised to follow vendor instructions first regarding the selection of antigen retrieval solutions. However, sodium citrate buffer can be used as a default buffer solution for almost all primary antibodies (listed in Table 1) described in this protocol.

3. Antigen retrieval step: Use a 2 L microwaveable glass beaker (*see* Subheading 2) for boiling the sections. Do not fill the beaker with more than 600–800 mL retrieval solution. Before placing slides into retrieval buffer solution, first prewarm the solution by microwaving for 3 min. Afterward, place the rack holding the slides into a prewarmed buffer solution, cover the glass beaker with plastic wrap (make a few holes in the wrap to allow for evaporation during boiling), then boil for at least 15 min. Be sure to monitor for evaporation, watch out for boiling over during the procedure, and do not allow the slides to dry out.
4. The cooling process at room temperature can take time. To speed up the process, place tap water in a plastic bucket with some ice, and place the beaker in ice-cold tap water. This may take around 10 min for cooling.
5. Circling sections with liquid blocker is critical. If one slide has more than one section, this prevents spillover of primary/secondary antibodies/other reagents applied on one section to an adjacent section. Even if there is only one section/slide, circling the sections with a liquid blocker is needed to hold the antibodies/reagents on the tumor sections.
6. Endogenous peroxidase may react with the substrate during the color development stage (*see* **step 8**) and can evoke some unnecessary background. To avoid this, it is necessary to block endogenous peroxidase activity. However, blocking solution should not be applied for more than 5 min. Some protocols may require blocking alkaline phosphatase instead of endogenous peroxidase (*see* dual IHC protocol).
7. The type of blocking buffer to be used (to reduce nonspecific reactions between antigens and antibodies, and eventually unwanted background) depends on the source of secondary antibodies. For example, secondary antibodies anti-rabbit IgG and anti-mouse IgG are produced in horse, so 5–10% horse serum should be used as blocking buffer. Similarly, secondary antibody anti-Rat IgG is produced in goat, so 5–10% goat serum should be the blocking buffer. Most IHC secondary antibody kits come with ready-to-use 2.5% blocking buffer, which can be directly applied (one drop/section) on tumor sections. However, it is always better to have regular horse or goat serum at hand, so that users can prepare appropriate dilutions (e.g., 5–10%) as necessary. If a secondary antibody kit is obtained from a particular vendor, horse or goat serum should also be obtained from the same vendor.
8. Primary antibodies are listed in Table 1 with their appropriate dilution in appropriate blocking buffer. All antibodies listed detect mouse immune cell antigens (Fig. 1), and several of

them are also reactive to other species. For more information, check antibody datasheets available at the vendor's site using the listed catalogue numbers.

9. Although many primary antibodies may perform well if incubation is done at room temperature for **30–60** min, it is usually better to incubate overnight at 4 °C. Duration of overnight incubation can vary from 12 h to 24 h, and this variability does not affect antibody performance or staining quality.
10. The wash time is arbitrary. Each wash time can be extended from 5 min to 10–15 min depending on the level of nonspecific staining background that may develop during color development (*see* **step 15**). Washing should be gentle (no rocking or movement is required) in order to avoid losing sections from the slide.
11. See the list of HRP-labeled secondary antibodies in Table 2. If the primary antibody is made in rabbit, rat, or mouse, then HRP-conjugated anti-rabbit, anti-rat, or anti-mouse IgG, respectively, must be used as secondary antibodies.
12. Apply one drop/section, however, if intensity of the staining (i.e. background) is high, it can be reduced by diluting secondary antibodies 1:1 in PBS.
13. Color development either by DAB or other chromogenic system needs careful observation to minimize the level of background. For some antibodies (anti-CD3, anti-CD4, anti-CD8, anti-CD68, anti-Ki67, etc. listed in **Note 9**), color development may take 20–60 s. This can be observed by the naked eye by placing the slide on a white background. As soon as color appears, the chromogenic reaction has to be stopped immediately by dipping the slides in water in order to avoid excessive background (*see* **step 16** in Single IHC). For other antibodies (e.g., anti-Cleaved Caspase 3), color development can take 2–5 min. Sometimes color cannot be seen by the naked eye (e.g., if very low number of immune cells present in the tumor, thus less chromogenic reaction or a good choice of primary antibody and low number of immune cells present in the tumor resulting in no background reaction), so slides can be placed under a light microscope to observe color development.
14. Slides can be kept in water as long as 1 h if necessary, and this should not affect the staining quality.
15. Counterstaining is done in order to better visualize the stained tumor-infiltrating immune cells. If sections have stronger background color (brown), hematoxylin staining can be done for as long as 1 min or even undiluted hematoxylin can be used for 20 s.

16. While washing slides with running tap water, place the sections facing opposite to the running water to avoid directly hitting the sections.
17. Use the same alcohol/xylene solutions applied in **steps 1** and **2**. Use a fume hood.
18. Sections can be kept in xylene longer than 10 min. Keeping sections in xylene longer during the second wash may be required while performing the mounting step (**step 21** in single IHC).
19. Never let the sections dry in the xylene step, which can lead to fragile brain tissue sections before mounting. Finish mounting quickly within 30 s–1 min.
20. The size of the glass cover slip depends on the number of sections present on a slide and the distance between the sections. In general, a standard microscope slide (75 × 25 mm) containing three or four sections can be easily covered by a 50 × 50 mm size cover glass. A 25 × 25 mm size cover slip can easily cover a single mouse brain tumor section.
21. Bubble removal: while covering the section with a cover slip containing Cytoseal XYL, bubbles may form, which can be removed by applying finger pressure.
22. These permanent mount slides can be stored at room temperature for years while preserving the staining quality.
23. Blue substrate is partially soluble in xylene. It is used in **step 16** of dual-color IHC protocol.
24. A blocking solution that can block both endogenous peroxidase and alkaline phosphatase activity was chosen for this protocol, since alkaline phosphatase (AP)-conjugated anti-rabbit IgG secondary antibody is included in the double IHC protocol. H_2O_2 can be replaced by this solution in single IHC in **step 8**.
25. Prepare red substrate working solution preparation according to the manufacturer's instructions (50 μL/section). Prepare just before use and mix well. Discard the leftover solution after use. Color development can take 20–30 min of incubation at room temperature.
26. Prepare blue substrate working solution according to the manufacturer's instructions (50 μL/section): Prepare just before use and mix well. Discard the leftover solution. Color development may take 20–30 min of incubation at room temperature.
27. Alcohol and clearing solutions can be reused. Follow **Notes 17** and **18**.

Acknowledgments

This work was supported in part by grants from NIH (R01CA160762) and the Thomas A. Pappas Chair in Neuroscience to SDR.

References

1. Saha D, Ahmed SS, Rabkin SD (2015) Exploring the antitumor effect of virus in malignant glioma. Drugs Future 40(11):739–749. https://doi.org/10.1358/dof.2015.040.11.2383070
2. Saha D, Wakimoto H, Rabkin SD (2016) Oncolytic herpes simplex virus interactions with the host immune system. Curr Opin Virol 21:26–34. https://doi.org/10.1016/j.coviro.2016.07.007
3. Bommareddy PK, Peters C, Saha D, Rabkin SD, Kaufman HL (2018) Oncolytic herpes simplex viruses as a paradigm for the treatment of cancer. Annu Rev Cancer Biol 2 (1):155–173. https://doi.org/10.1146/annurev-cancerbio-030617-050254
4. Saha D, Martuza RL, Rabkin SD (2017) Macrophage polarization contributes to Glioblastoma eradication by combination immunovirotherapy and immune checkpoint blockade. Cancer Cell 32(2):253–267 e255. https://doi.org/10.1016/j.ccell.2017.07.006
5. Saha D, Wakimoto H, Peters CW, Antoszczyk SJ, Rabkin SD, Martuza RL (2018) Combinatorial effects of VEGFR kinase inhibitor axitinib and oncolytic virotherapy in mouse and human glioblastoma stem-like cell models. Clin Cancer Res 24(14):3409–3422. https://doi.org/10.1158/1078-0432.CCR-17-1717
6. Hofman FM, Taylor CR (2013) Immunohistochemistry. Curr Protoc Immunol 103:Unit 21 24. https://doi.org/10.1002/0471142735.im2104s103
7. Goldstein M, Watkins S (2008) Immunohistochemistry. Curr Protoc Mol Biol:Chapter 14: Unit 14 16. https://doi.org/10.1002/0471142727.mb1406s81
8. Ramos-Vara JA (2005) Technical aspects of immunohistochemistry. Vet Pathol 42 (4):405–426. https://doi.org/10.1354/vp.42-4-405
9. Ward JM, Rehg JE (2014) Rodent immunohistochemistry: pitfalls and troubleshooting. Vet Pathol 51(1):88–101. https://doi.org/10.1177/0300985813503571
10. Donaldson JG (2015) Immunofluorescence staining. Curr Protoc Cell Biol 69(4.3):1–7. https://doi.org/10.1002/0471143030.cb0403s69

Chapter 12

Detection of Tumor Antigen-Specific T-Cell Responses After Oncolytic Vaccination

Jonathan G. Pol, Byram W. Bridle, and Brian D. Lichty

Abstract

Oncolytic vaccines, which consist of recombinant oncolytic viruses (OV) encoding tumor-associated antigens (TAAs), have demonstrated potent antitumor efficacy in preclinical models and are currently evaluated in phase I/II clinical trials. On one hand, oncolysis of OV-infected malignant entities reinstates cancer immunosurveillance. On the other hand, overexpression of TAAs in infected cells further stimulates the adaptive arm of antitumor immunity. Particularly, the presence of tumor-specific $CD8^+$ T lymphocytes within the tumor microenvironment, as well as in the periphery, has demonstrated prognostic value for cancer treatments. These effector $CD8^+$ T cells can be detected through their production of the prototypical Tc1 cytokine: IFN-γ. The quantitative and qualitative assessment of this immune cell subset remains critical in the development process of efficient cancer vaccines, including oncolytic vaccines. The present chapter will describe a single-cell immunological assay, namely the intracellular cytokine staining (ICS), that allows the enumeration of IFN-γ-producing TAA-specific $CD8^+$ T cells in various tissues (tumor, blood, lymphoid organs) following oncolytic vaccination.

Key words Oncolytic virus, Cancer vaccine, Tumor antigen, IFN-γ, Intracellular cytokine staining (ICS), Flow cytometry

1 Introduction

Tumor cells infected by oncolytic viruses undergo an immunogenic cell death that triggers antitumor immunity of both innate and adaptive nature [1–6]. In order to further stimulate the adaptive arm of cancer immunosurveillance, we and others have introduced some transgenes encoding tumor-associated antigens (TAA) within the viral genome [7–26]. The resulting so-called oncolytic vaccines have proven enhanced preclinical efficacy in comparison to their unarmed parental strain, either as a monotherapy or in combination treatment. Data were essentially collected in immunocompetent mice bearing various syngeneic tumors (melanoma, prostate and ovarian cancers, lung carcinoma, or glioma) treated with oncolytic vaccines based on the vesicular stomatitis virus (VSV), Maraba

Christine E. Engeland (ed.), *Oncolytic Viruses*, Methods in Molecular Biology, vol. 2058,
https://doi.org/10.1007/978-1-4939-9794-7_12,

virus, vaccinia virus, herpes simplex virus, or Semliki forest virus [8–13, 17, 18, 20, 22–25, 27]. Improved antitumor activity associated with oncolytic vaccination was largely relying on the activation/expansion of tumor-specific $CD8^+$ T cells secreting the interferon (IFN)-γ (i.e., Tc1 response). Interestingly, an exception has been reported upon treatment of murine melanoma and prostate cancer with a set of oncolytic VSVs encoding a library of TAAs. Indeed, the antitumor activity appeared instead to depend on cancer-specific $CD4^+$ T cells secreting the interleukin (IL)-17 (i.e., Th17 response) [13, 15, 18, 20, 22]. Still, the efficacy of this experimental approach benefited from the coadministration of either IL-2, a cytokine characteristic of both Tc1 and type 1 helper (Th1) $CD4^+$ T lymphocytes, or of immune checkpoint blockade, that aims at relieving effector T cells from their coinhibitory signals [18, 22].

Oncolytic vaccinotherapy is currently under clinical investigation in three Phase I/II clinical trials in cancers positive for the melanoma-associated antigen A3 (MAGE-A3), including skin, non-small cell lung, breast, and gastroesophageal tumors. Precisely, the approach consists of a heterologous prime-boost strategy in which vaccine initiation relies on the intramuscular delivery of a replication-deficient adenoviral vaccine (AdMA3) followed by a recall with the MG1 strain of the oncolytic Maraba virus administered systemically (MG1MA3). This prime-boost oncolytic vaccination is evaluated either on its own or in combination with cyclophosphamide and/or an anti-PD-1 (clinicaltrials.gov references: NCT02285816; NCT02879760; NCT03773744) [3, 26, 28]. A fourth clinical trial is recruiting patients suffering from tumors positive for the human papillomavirus (HPV) in order to evaluate the efficacy of the same approach but targeted toward the HPV-E6 and E7 antigens (Ad/MG1-E6E7 oncolytic prime-boost) in combination with an anti-PD-L1 (NCT03618953) [3, 23, 28].

Detection of $CD8^+$ T-cells infiltrating the tumor microenvironment has demonstrated prognostic significance in cancer patients [29–33]. Moreover, the presence of TAA-specific $CD8^+$ T lymphocytes in the periphery (i.e., blood and lymphoid organs) positively correlated with tumor outcome upon immunotherapies in preclinical models [7, 34, 35]. CTLs susceptible to react against malignant entities are generally detected through immunostaining of (1) their T cell receptor (TCR) [36], (2) their cytolytic granule content (i.e., the pore-forming protein, perforin, and the proapoptotic serine proteases, granzymes) [37–39] or, more frequently, of (3) their Tc1-related cytokines (i.e., IFN-γ, IL-2, and tumor necrosis factor [TNF]-α) [40, 41]. Tetramers of class I major histocompatibility complex (MHC-I) molecules can be synthesized, labeled with a fluorochrome and coupled to one epitope peptide of a TAA. Staining with such MHC-I tetramers can identify tumor-specific $CD8^+$ T cells by binding to their surface TCR [36]. The application of MHC tetramers to the field of cancer therapy has been limited so far due to intrinsic limitations. Mainly, they demonstrate a weak interaction with the TCR of cognate T cells thus resulting in a lack of sensitivity [36]. Perforin and granzyme B are stored inside membrane-bound secretory lysosomes and are readily detectable in resting CTLs by flow cytometry following intracellular staining with fluorescent antibodies [37–39]. However, monitoring degranulation requires prior T lymphocyte restimulation in the presence of protein transport inhibitors and of a fluorescent antibody targeting CD107a. The latter marker, also known as lysosomal-associated membrane protein 1 (LAMP-1), is inserted in the lipid bilayer that surrounds granule cores and gets surface-exposed following CTL degranulation and thus amenable to anti-CD107a staining [42]. Notwithstanding, the production of IFN-γ by $CD8^+$ T cells remains the prototypical indicator of the Tc1 response. IFN-γ is a pleiotropic cytokine which displays anticancer activities [43]. First, it upregulates MHC-I expression in malignant cells thus enhancing their antigenicity and favoring their recognition and elimination by CTLs. Then, IFN-γ inhibits cell proliferation and stimulates apoptosis. Finally, it reduces tumor angiogenesis and promotes vasculature shutdown by acting on some nontransformed cells constitutive of the neoplastic microenvironment [43]. Practically, detection of IFN-γ is performed either by intracellular cytokine staining (ICS) or by enzyme-linked immunospot (ELISpot) and informs on the magnitude of the CTL response. These two immunological assays require prior T cell restimulation, respectively with and without blockade of the secretory pathway. Considering its strong sensitivity, ELISpot is preferred over ICS for measuring weak Tc1 responses (e.g., against poorly immunogenic tumor antigens). Nevertheless, ICS allows the phenotyping of polyfunctional tumor-specific $CD8^+$ T cells coproducing IL-2 and/or TNF-α, thus providing a qualitative input.

The present chapter will detail the procedures to monitor tumor-specific Tc1 responses in tissues relevant for prognosticating the efficacy of oncolytic vaccination (i.e., blood, draining lymph nodes, spleen, and tumor) by using ICS. The described method will focus on mice but has been transposed to macaque and human blood samples after minor adjustments (e.g., epitope peptides, antibodies for immunophenotyping) [26, 44]. Overall, this immunological assay has proven its reliability in the clinical evaluation of various interventions such as prophylactic vaccines against infectious diseases or therapeutic cancer vaccines [40, 45, 46].

2 Materials

Oncolytic vaccination should be performed in tumor-bearing mice of at least 6 weeks of age to warranty the presence of a mature immune system and the establishment of immune memory

[47]. Females are commonly preferred over males for their ease of hosting (e.g., rare aggressive behavior) but gender may be imposed by the malignant indication studied (e.g., genital cancers). Considering the immunocompetent status of the host, syngeneic tumors must be engrafted (a nonexhaustive list has been reported elsewhere: [48]). Mice are housed in compliance with the government directives. Animals are maintained in specific pathogen-free conditions, in a temperature-controlled environment, with a 12-h light/12-h dark daily cycle. Animals have ad libitum access to food and water. Each cage can host a maximum of 5 littermates and contains enrichments (e.g., cotton nest, shelter tubes).

2.1 Tissue Collection

1. Vaccinated tumor-bearing mice.
2. Anesthesia machine with 2.5% isoflurane vaporizer.
3. Goldenrod 5 mm animal lancet.
4. Heparin solution: 1.8 μg heparin/mL saline. Store at 4 °C.
5. Heparinized capillary tubes.
6. Gauze pads.
7. Sterile injectable saline. Store at room temperature.
8. 1 mL syringe +25–30 G needle.
9. Operating scissors.
10. Micro tweezers.
11. Ice bucket.

2.2 Tissue Sampling and Processing

1. Cell culture hood.
2. Centrifuge.
3. Polypropylene centrifuge tubes (15 and 50 mL).
4. Polystyrene round-bottom tubes (5 mL; FACS tubes).
5. FACS tubes with 35 μm pore size cell strainer snap caps.
6. Microcentrifuge tubes (1.5 mL).
7. Rack(s) for 50, 15, 5, and 1.5 mL tubes.
8. Pipet-controller and serological pipettes (5 mL).
9. Micropipettes (0.2–2 μL, 1–20 μL, 20–200 μL, 200–1000 μL) and tips.
10. Polypropylene Petri dishes.
11. Microscope glass slides (*see* **Note 1**).
12. GentleMACS™ Octo Dissociator.
13. GentleMACS™ C tubes.
14. 30 μm pore size cell strainers (MACS SmartStrainers).
15. Standard RPMI: Standard Roswell Park Memorial Institute medium (RPMI). Store at 4 °C.

16. Complete RPMI: Standard RMPI supplemented with serum (e.g., 10% v/v fetal bovine serum [FBS]), amino acids (e.g., 2 mM glutamine), and/or antibiotics (e.g., 100 IU/mL penicillin G and 100 mg/mL streptomycin). Store at 4 °C (*see* **Note 2**).
17. 1× Phosphate-buffered saline (PBS) at pH 7.4. Store at 4 °C.
18. 0.4% Trypan blue solution. Store at room temperature.
19. 1× Red blood cell (RBC) lysis buffer. Store at 4 °C (*see* **Note 3**).
20. Mouse tumor dissociation kit (Miltenyi Biotec).

2.3 T Lymphocyte Restimulation and Immunophenotyping

1. Incubator allowing standard cell culture conditions (37 °C, 5% CO_2).
2. Laminar fume hood.
3. Polystyrene round-bottom 96-well culture plates.
4. Polypropylene reservoirs.
5. Multichannel pipette (20–200 μL) and tips.
6. Aluminum foil.
7. Flow cytometer (*see* **Note 4**).
8. Computer with a software for flow cytometry data analysis (e.g., FlowJo by TreeStar).
9. Synthetic epitope peptides derived from the TAA targeted by the oncolytic vaccine (*see* Table 1 and **Note 5**). Store at −80 °C at 10 mg/mL in dimethyl sulfoxide (DMSO) except if the peptide contains methionine, cysteine or tryptophan residues (*see* **Note 6**).
10. Phorbol 12-myristate 13-acetate (PMA). Store at −20 °C at 1 mg/mL in DMSO.
11. Ionomycin. Store at −20 °C at 1 mg/mL in DMSO.
12. Brefeldin A-containing reagent (BD GolgiPlug™, BD Biosciences). Store at 4 °C (*see* **Note 7**).
13. 1× permeabilization/fixation reagent (BD Cytofix/Cytoperm™, BD Biosciences). Store at 4 °C (*see* **Note 8**).
14. 1× permeabilization/wash reagent (BD Perm/Wash™, BD Biosciences). Store at 4 °C (*see* **Note 8**).
15. FACS buffer: 1× PBS pH 7.4 containing 0.5% bovine serum albumin (BSA), filter-sterilized through a 0.22 μm pore filter nylon membrane. Store at 4 °C.
16. Fixable viability dye (LIVE/DEAD® Fixable Yellow Dead Cell Stain Kit, ThermoFisher Scientific). Store at −20 °C (*see* **Note 9**).
17. Purified anti-mouse CD16/CD32 (clone 2.4G2, BD Fc Block™, BD Biosciences). Store at 4 °C (*see* **Note 10**).
18. Fluorochrome-conjugated antibodies. Store at 4 °C (*see* **Notes 4** and **11** and Table 2).

Table 1
Nonexhaustive list of MHC-I-restricted epitopes of murine tumor antigens

Tumor antigen	Epitope location	Epitope sequence	Tumor cell line(s)	Cancer type	MHC-I haplotype	Host murine strain	References
DCT/TRP2	aa 180-188	SVYDFF VWL	B16	Melanoma	H-2K^b	C57Bl/6	[16, 17, 21, 49, 50]
Endogenous MuLV gp70[a]	aa 423-431 (AH1 peptide)	SPSYVYH QF	4T1 B16 CT26	Breast cancer Melanoma Colon cancer	H-2L^d H-2K^b H-2L^d	Balb/c C57Bl/6 Balb/c	[35] [51] [51, 52]
Endogenous MuLV p15E	aa 604–611	KSPWF TTL	B16 MC-38 MCA205	Melanoma Colon cancer Fibrosarcoma	H-2K^b H-2K^b H-2K^b	C57Bl/6 C57Bl/6 C57Bl/6	[51, 53] [53] [53]
gp100/ PMEL	aa 25-33	EGSRN QDWL	B16	Melanoma	H-2D^b	C57Bl/6	[54]
HER-2/neu	aa 66-74	TYVPANA SL	TUBO	Breast cancer	H-2K^d	Balb/c	[5, 55, 56]
HPV E6	aa 48-57	EVYDFAF RDL	TC1	Lung cancer	H-2K^b	C57Bl/6	[23, 57]
HPV E7	aa 49-57	RAHYNI VTF	TC1	Lung cancer	H-2D^b	C57Bl/6	[23, 58]
P1A	aa 35-43	LPYLG WLVF	P815	Mastocytoma	H-2L^d	DBA/2	[59]
PSCA	aa 29-37	AQMNN RDCL	TRAMP-C2	Prostate cancer	H-2D^b	C57Bl/6	[60]
STEAP	aa 327-335	VSKIN RTEM	TRAMP-C2	Prostate cancer	H-2D^b	C57Bl/6	[24]

aa amino acid, *DCT* dopachrome tautomerase, *gp70* endogenous MuLV glycoprotein 70, *gp100* glycoprotein 100, *HER-2* human epidermal growth factor receptor-2, *HPV human papillomavirus, MHC-I* class I major histocompatibility complex, *MuLV* murine leukemia virus, *p15E* endogenous MuLV transmembrane protein 15E, *PMEL* premelanosome protein, *PSCA* prostate stem cell *Antigen, STEAP* six-transmembrane epithelial antigen of prostate, *TRP2* tyrosinase related-protein 2

[a]MuLV gp70 is expressed in a wide range of murine tumor cell lines as reported by Scrimieri et al. [61]

3 Methods

The magnitude, kinetics and quality of the tumor-specific Tc1 immunity is expected to vary according to the nature, dose and delivery route of the oncolytic vaccine and its potential co-treatments. The following procedures aim at characterizing by ICS the CD8$^+$ T cell response raised against TAAs in the neoplastic tissue as well as in the blood and lymphoid organs. Ultimately, the

Table 2
Suggested panel of primary antibodies to detect Tc1 cytokines by ICS in mice

		Antibody				
Targeted marker	Marker location	Clone	Suggested fluorochrome	Supplier	Final concentration[a] (μg/mL)	References
CD45	Surface	30-F11	APC-Vio770	Miltenyi biotec	2	[62]
CD3		17A2	V450	BD biosciences	0.5	[62]
CD8		53-6.7	FITC	BD biosciences	0.25	[23, 62]
CD4		RM4-5	PerCP-Cy5.5	ThermoFisher scientific	0.25	[23]
IFN-γ	Intracytoplasmic	XMG1.2	APC	BD biosciences	0.125	[23, 37, 62]
TNF-α		MP6-XT22	PE	BD biosciences	0.125	[23, 37]
IL-2		JES6-5H4	PE-Dazzle594	BioLegend	0.125	[23, 37]

IFN-γ monostaining is sufficient to detect effector CD8+ T cells. However, co-staining of TNF-α and IL-2 will contribute to characterize their polyfunctional phenotype
[a]Antibody titration remains strongly advised to improve stain index

information gathered should facilitate the optimization of the oncolytic immunization schedule. This same process was successfully applied for the development of the Ad/MG1 oncolytic prime-boost vaccine platform currently investigated in the clinic [17, 23, 24, 26].

3.1 Blood Collection

If no other tissues than blood are required for immune analysis, mice can be kept alive and blood draws repeated over time to measure the kinetics of the TAA-specific T cell response within the same vaccinated animal (*see* **Note 12**).

1. In the lab, prepare the blood collection tubes by pipetting 200 μL of heparin solution in microcentrifuge tubes (one tube per vaccinated mouse to be bled, *see* **Note 13**).
2. In the animal facility, draw five drops of blood (i.e., ≈ 200 μL) from each mouse by puncturing the submandibular (or submental) vein with a 5 mm Goldenrod animal lancet [63]. Typically, such a volume of blood is enough to perform three conditions of T cell restimulation such as (1) no restimulation (negative control), (2) restimulation with one single TAA epitope peptide and (3) restimulation with PMA + ionomycin (positive control) (*see* **Notes 12** and **14**).
3. If the animal is kept alive, proceed to the subcutaneous injection of 300 μL of saline using a syringe with a 25–30 G needle

to compensate the volume withdrawn and speed up mouse recovery.

4. Place the microcentrifuge tubes containing the blood samples on ice.

3.2 Collection of Spleen and Lymph Nodes

1. In the lab, prepare the tissue collection tubes by adding 2 mL of standard RPMI in 15 mL polypropylene centrifuge tubes (one tube per vaccinated mouse; spleen and lymph nodes of the same animal being pooled in one single sample container).
2. In the animal facility, euthanize the vaccinated mouse by cervical dislocation.
3. Using micro tweezers and operating scissors, dissect the tumor draining lymph nodes and/or the spleen.
4. Transfer the tissues into the sample container and keep on ice.

3.3 Tumor Collection

1. In the lab, prepare the tissue collection tubes by pipetting 1 mL of standard RPMI in GentleMACS™ C tubes (one tube per vaccinated mouse).
2. Weigh each tube and record the value as "tube weight before tumor imput" (W_{bt}).
3. In the animal facility, euthanize the vaccinated mouse by cervical dislocation.
4. Using micro tweezers and operating scissors, dissect the tumor.
5. Transfer the tissues into the GentleMACS™ C tubes and keep on ice.

3.4 Whole Blood Processing

1. Using micropipettes, calculate the exact volume of blood collected (V_b). It consists of the total volume measured in the blood sample container (V_T) minus the 200 μL of heparin solution: V_b (μL) = V_T–200.
2. Transfer each blood sample into a 5 mL FACS tube.
3. Add 2 mL of 1× RBC lysis buffer. Incubate at room temperature for 5 min.
4. Add 2 mL of standard RPMI.
5. Centrifuge at 300 × *g* for 5 min at room temperature. Discard the supernatant.
6. Repeat a second time the **steps 3–5** if red blood cells are still present in the pellet.
7. Resuspend the pellet of white blood cells in a final volume of standard RPMI (V_R) corresponding to 6 times the actual volume of blood collected: $V_R = 6 \times V_b$. For each condition of T cell restimulation, 200 μL out of V_R will be required.

3.5 Lymph Node Processing

1. Transfer the lymph node(s) into one Petri dish containing 0.5 mL of standard RPMI.
2. Crush the lymph node(s) between two microscope glass slides above the Petri dish (*see* **Note 1**).
3. Clean the slides with 2 mL of standard RPMI above the Petri dish.
4. Transfer the 2.5 mL of cell suspension collected in the Petri dish into a 5 mL FACS tube with 35 μm pore size cell strainer snap cap. The cell suspension should go through the cap to get rid of large debris of tissue.
5. Rinse the Petri dish with 2 mL of standard RPMI and pipet into the FACS tube through the cell strainer snap cap.
6. Centrifuge at 300 × *g* for 5 min at room temperature. Discard the supernatant.
7. Resuspend the pellet of cells in a final volume of 600 μL per lymph node collected. For each condition of T cell restimulation, 200 μL will be required (1/3 of a lymph node).

3.6 Spleen Processing

1. Transfer the spleen to a Petri dish containing 2 mL of standard RPMI.
2. Crush the spleen between two microscope glass slides above the Petri dish (*see* **Note 1**).
3. Clean the slides with 5 mL of standard RPMI above the Petri dish.
4. Transfer the 7 mL of cell suspension collected in the Petri dish into a 15 mL centrifuge tube equipped with a 30 μm pore size MACS SmartStrainers. The cell suspension should go through the cell strainer to get rid of large tissue debris.
5. Rinse the Petri dish with 5 mL of standard RPMI and pipet into the 15 mL tube.
6. Centrifuge at 300 × *g* for 5 min at room temperature. Discard the supernatant.
7. Add 5 mL of 1× RBC lysis buffer. Incubate at room temperature for 5 min.
8. Add 5 mL of standard RPMI.
9. Centrifuge at 300 × *g* for 5 min at room temperature. Discard the supernatant.
10. Resuspend the splenocytes in a final volume of 2 mL. For each condition of T cell restimulation, 40 μL of the cell suspension will be required (1/50 of the spleen).

3.7 Tumor Processing

1. Weigh each tube and record the value as "tube weight after tumor input" (W_{at}). Calculate the actual tumor weight (W_T): $W_T = W_{at} - W_{bt}$.
2. Using operating scissors, cut tumors into small pieces (<3 mm on the side) directly into the GentleMACS™ C tubes.
3. Prepare an enzymatic mix in standard RPMI by using Miltenyi Biotec's mouse tumor dissociation kit and by following the supplier's instructions.
4. Add 2.5 mL of the enzymatic mix into each GentleMACS™ C tube.
5. Run each GentleMACS™ C tube on the GentleMACS™ Octo Dissociator following the supplier's instructions.
6. Incubate the tubes for 45 min under shaking at 37 °C.
7. Run each GentleMACS™ C tube on the GentleMACS™ Octo Dissociator following the supplier's instructions.
8. Add 5 mL of standard RPMI to each tube.
9. Centrifuge the GentleMACS™ C tubes at 300 × *g* for 5 min at room temperature. Discard the supernatant.
10. Resuspend the cell pellet in 10 mL of standard RPMI.
11. Transfer the cell suspension into a 50 mL centrifuge tube by passing through a 30 μm pore size MACS SmartStrainers. This step will eliminate large tumor debris.
12. Rinse each GentleMACS™ C tube with 10 mL of standard RPMI and pipet into the 50 mL centrifuge tube through the cell strainer.
13. Centrifuge each 50 mL centrifuge tube at 300 × *g* for 5 min at room temperature. Discard the supernatant.
14. Add 30 mL of 1× PBS to wash the cell pellet.
15. Centrifuge at 300 × *g* for 5 min at room temperature. Discard the supernatant.
16. Repeat the **steps 14** and **15**.
17. Considering the initial weight of the tumor (W_T), resuspend each tumor cell pellet at 250 mg/mL in standard RPMI. For each condition of T cell restimulation, 200 μL of the tumor cell suspension will be required.

3.8 Restimulation of TAA-Specific CD8⁺ T Lymphocytes and ICS

1. For each of the blood, tumor, lymph node, and spleen cell suspensions, transfer respectively 200 μL, 200 μL, 200 μL, and 40 μL into as many wells of round-bottom 96-well culture plates as there will be T cell restimulation conditions to achieve.

   At minimum, the three following conditions are recommended:

(a) *Restimulation 1* = no restimulation (negative control).

(b) *Restimulation 2* = restimulation with one peptide corresponding to the MHC-I restricted immunodominant epitope of the TAA targeted by the oncolytic vaccine.

(c) *Restimulation 3* = restimulation with PMA + ionomycin (positive control).

Figure 1 illustrates the resulting plates for 2 groups (untreated versus treated with the oncolytic vaccine) of 5 mice each.

2. In three reservoirs, prepare the following mixes to restimulate T lymphocytes (calculations are done for 60 samples):

(a) *Restimulation mix 1* (No restimulation) = 12 mL of complete RPMI +12 μL of BD GolgiPlug.

Plate 1	Tissues	Treatment 1 : Untreated					Treatment 2 : Oncolytic vaccination				
		Mouse 1	Mouse 2	Mouse 3	Mouse 4	Mouse 5	Mouse 1	Mouse 2	Mouse 3	Mouse 4	Mouse 5
Restim. cond° 1: Unstimulated (*negative control*)	Blood	Sample 1	Sample 5	Sample 9	Sample 13	Sample 17	Sample 21	Sample 25	Sample 29	Sample 33	Sample 37
	Spleen	Sample 2	Sample 6	Sample 10	Sample 14	Sample 18	Sample 22	Sample 26	Sample 30	Sample 34	Sample 38
	Lymph node	Sample 3	Sample 7	Sample 11	Sample 15	Sample 19	Sample 23	Sample 27	Sample 31	Sample 35	Sample 39
	Tumor	Sample 4	Sample 8	Sample 12	Sample 16	Sample 20	Sample 24	Sample 28	Sample 32	Sample 36	Sample 40
Restim. cond° 2: Epitope peptide	Blood	Sample 1	Sample 5	Sample 9	Sample 13	Sample 17	Sample 21	Sample 25	Sample 29	Sample 33	Sample 37
	Spleen	Sample 2	Sample 6	Sample 10	Sample 14	Sample 18	Sample 22	Sample 26	Sample 30	Sample 34	Sample 38
	Lymph node	Sample 3	Sample 7	Sample 11	Sample 15	Sample 19	Sample 23	Sample 27	Sample 31	Sample 35	Sample 39
	Tumor	Sample 4	Sample 8	Sample 12	Sample 16	Sample 20	Sample 24	Sample 28	Sample 32	Sample 36	Sample 40

Plate 2	Tissues	Treatment 1 : Untreated					Treatment 2 : Oncolytic vaccination				
		Mouse 1	Mouse 2	Mouse 3	Mouse 4	Mouse 5	Mouse 1	Mouse 2	Mouse 3	Mouse 4	Mouse 5
Restim. cond° 3: PMA + ionomycin (*positive control*)	Blood	Sample 1	Sample 5	Sample 9	Sample 13	Sample 17	Sample 21	Sample 25	Sample 29	Sample 33	Sample 37
	Spleen	Sample 2	Sample 6	Sample 10	Sample 14	Sample 18	Sample 22	Sample 26	Sample 30	Sample 34	Sample 38
	Lymph node	Sample 3	Sample 7	Sample 11	Sample 15	Sample 19	Sample 23	Sample 27	Sample 31	Sample 35	Sample 39
	Tumor	Sample 4	Sample 8	Sample 12	Sample 16	Sample 20	Sample 24	Sample 28	Sample 32	Sample 36	Sample 40

Fig. 1 Example of two 96-well plates for one ICS assay. Wells are represented by white background table cells. The present example includes 40 samples originating from four tissues (i.e., blood, spleen, lymph node, tumor) collected from five mice of two treatment groups (i.e., with or without oncolytic vaccination). Each sample will undergo three conditions of T lymphocyte restimulation (i.e., no restimulation versus restimulation with either a TAA-related peptide or PMA + ionomycin). Stim. cond., stimulation condition

   (b) *Restimulation mix 2* (Restimulation with one TAA epitope peptide) = 12 mL of complete RPMI + 12 μL of BD GolgiPlug + 2.4 μL of peptide at 10 mg/mL.
   (c) *Restimulation mix 3* (Restimulation with PMA + ionomycin) = 12 mL of complete RPMI + 12 μL of BD GolgiPlug + 0.6 μL of PMA at 1 mg/mL + 12 μL of ionomycin at 1 mg/mL.

3. Centrifuge the plates at 300 × *g* for 5 min at room temperature. Discard the supernatant.
4. Using a multichannel pipette, add 200 μL of each restimulation mix into the corresponding wells (*see* Fig. 1). Restimulate for 6 h at 37 °C in a cell culture incubator (*see* **Note 15**).
5. In a reservoir, prepare the following mix required to stain for dead cells (calculations are done for 60 × 3 = 180 samples):
   (a) *LIVE/DEAD staining mix* = 18 mL of 1× PBS + 36 μL of LIVE/DEAD® Fixable Yellow dye.
6. Centrifuge the plates at 300 × *g* for 5 min at room temperature. Discard the supernatant.
7. Add 100 μL of the LIVE/DEAD staining mix to each well of the plates. Incubate for 15 min at 4 °C.
8. Add 100 μL of 1× PBS to each well of the plates to top the volume up to 200 μL.
9. In a reservoir, prepare the following mix required to block Fcγ receptors (*see* **Note 10**; calculations are done for 180 samples):
   (a) *Fcγ receptor blocking mix* = 18 mL of FACS buffer + 90 μL of BD Fc Block™.
10. Centrifuge the plates at 300 × *g* for 5 min at room temperature. Discard the supernatant.
11. Add 100 μL of the Fcγ receptor blocking mix in each well of the plates. Incubate for 10 min at 4 °C.
12. Add 100 μL of FACS buffer in each well of the plates to top the volume up to 200 μL.
13. In a reservoir, prepare the following mix to stain for surface markers (calculations are done for 180 samples):
   (a) *Surface staining mix* (*see* Table 2) = 17.7 mL of FACS buffer + 240 μL of anti-CD45-APC-Vio770 + 45 μL of anti-CD3-V450 + 7.2 μL of anti-CD8-FITC + 22.5 μL of anti-CD4-PerCp-Cy5.5.
14. Centrifuge the plates at 300 × *g* for 5 min at room temperature. Discard the supernatant.
15. Add 100 μL of the surface staining mix in each well of the plates. Incubate for 25 min at 4 °C.

16. Add 100 μL of FACS buffer in each well of the plates to top the volume up to 200 μL.
17. Centrifuge the plates at 300 × *g* for 5 min at room temperature. Discard the supernatant.
18. Add 200 μL of FACS buffer in each well of the plates to wash any leftover of antibodies.
19. Centrifuge the plates at 300 × *g* for 5 min at room temperature. Discard the supernatant.
20. Add 200 μL of 1× BD Cytofix/Cytoperm™ in each well of the plates to permeabilize and fix the cells. Incubate for 15 min at 4 °C (*see* **Note 16**).
21. Centrifuge the plates at 300 × *g* for 5 min at room temperature. Discard the supernatant.
22. Add 200 μL of 1× BD Perm/Wash™ in each well of the plates to wash the excess of the fixative reagent.
23. Centrifuge the plates at 300 × *g* for 5 min at room temperature. Discard the supernatant.
24. Repeat the **steps 22** and **23**.
25. In a reservoir, prepare the following mix to stain for intracytoplasmic markers (calculations are done for 180 samples):
    (a) *Intracellular staining mix* (*see* Table 2) = 18 mL of 1× BD Perm/Wash™ + 11.25 μL of anti-IFN-γ-APC + 11.25 μL of anti-TNF-α-PE + 11.25 μL of anti-IL-2-PE-Dazzle594.
26. Add 100 μL of the intracellular staining mix in each well of the plates. Incubate for 25 min at 4 °C.
27. Add 100 μL of 1× BD Perm/Wash™ in each well of the plates to top the volume up to 200 μL.
28. Centrifuge the plates at 300 × *g* for 5 min at room temperature. Discard the supernatant.
29. Add 200 μL of 1× BD Perm/Wash™ in each well of the plates to wash any leftover of antibodies.
30. Centrifuge the plates at 300 × *g* for 5 min at room temperature. Discard the supernatant.
31. Repeat the **steps 29** and **30**.
32. Add 200 μL of FACS buffer in each well of the plates.
33. Resuspend each cell pellet and transfer the 200 μL into a FACS tube.
34. Add 200 μL of FACS buffer into each FACS tube to top the volume up to 200 μL. Protect the tubes from the light with aluminum foil.
35. Acquire the samples through a flow cytometer (*see* **Note 5**).

36. Using a flow cytometry data analysis software, evaluate the magnitude and quality of the TAA-specific Tc1 response raised upon oncolytic vaccination by enumerating LIVE/DEAD$^-$ CD45$^+$ CD3$^+$ CD4$^-$ CD8$^+$ T cells that are positive for one or several of the cytokines IFN-γ, TNF-α and IL-2 (*see* **Note 17**).

4 Notes

1. Lymphoid organs can be dissociated mechanically, for instance by crushing the tissues between two microscope glass slides. To increase lymphocyte recovery from the lymph nodes or spleen, enzymatic dissociation can be performed (e.g., Miltenyi Biotec adequate dissociation kits).
2. Serum-free media (e.g., Immunospot CTL-Test™ PLUS) may be preferred over complete RPMI for enhanced experimental standardization.
3. The RBC lysis buffer can be purchased (e.g., BioLegend 10× RBC lysis buffer) or homemade as follows: 150 mM NH_4Cl, 10 mM $KHCO_3$ and 0.1 mM Na_2EDTA in double-distilled H_2O, pH-adjusted to 7.2–7.4 and filter-sterilized through a 0.22 μm pore filter nylon membrane.
4. The described ICS procedure involves the detection of up to 8 different fluorescent signals: 1 viability dye + 7 markers (*see* Table 2). Depending on your cytometer setup (i.e., lasers and filters), you may want to reduce the number of markers analyzed and/or choose alternative fluorochromes. We recommend staining at minimum the following markers: CD3, CD8, and IFN-γ.
5. Synthetic peptides corresponding to immunodominant epitopes of an antigen expressed by the tumor and targeted by the oncolytic vaccine are mandatory to measure cancer-specific Tc1 response mounted upon treatment. To identify TAA of a given malignant tissue, immunopeptidomics analysis can be performed [64–66]. To identify immunodominant epitopes of a given TAA, we first recommend to synthesize a library of overlapping peptides covering the full-length TAA. Then, mice should be immunized against this TAA (through direct or vector-based delivery of the recombinant TAA eventually accompanied by adjuvants). Seven to 12 days later, harvest the spleen, restimulate splenocytes with each individual peptide of the synthesized TAA library, and identify the antigenic determinants by detecting IFN-γ production by ICS or ELISpot assay. Alternatively, epitopes can be predicted in silico based on the TAA amino acid sequence. Algorithms can be found for instance at http://metagenomics.iiserb.ac.in/

proinflam/, http://sysbio.unl.edu/SVMTriP/prediction.php or https://www.iedb.org/. The latter website gives access to an immune epitope database listing additional MHC-I-restricted T cell epitopes from various tumor antigens across multiple species.

6. The type of solvent used for resuspension of lyophilized synthetic peptides depends on their amino acid sequence. Useful recommendations and tools can be found at http://www.sigmaaldrich.com/life-science/custom-oligos/custom-peptides/learning-center/solubilizing-peptides.html and https://www.genscript.com/ssl-bin/site2/peptide_calculation.cgi.
7. Brefeldin A inhibits protein transport from the endoplasmic reticulum to the Golgi complex favoring intracellular accumulation of cytokines (e.g., IFN-γ, TNF-α, IL-2), thus facilitating their detection by fluorescent immunostaining and flow cytometry. Monensin is another inhibitor of the secretory pathway which prevents protein transport from the medial to trans cisternae of the Golgi apparatus. Brefeldin A can be substituted or combined with monensin. However, it is not recommended to use monensin alone for TNF-α detection [67].
8. Permeabilization/fixation reagents typically contain 0.1% saponin and 4% paraformaldehyde (PFA). Saponin transiently permeabilizes cells by removing cholesterol from membranes, leaving holes through which fluorochrome-labeled antibodies can access their intracytoplasmic targets [68]. PFA fixation mechanism relies on the cross-linking of proteins through their nitrogen atoms (e.g., NH_2 function of the lateral chain of lysine residues) [69]. Permeabilization/wash reagents typically contain 0.1% saponin and 1–10% FBS.
9. A wide range of fixable viability dyes with distinct excitation/emission properties are commercially available and detectable by flow cytometry.
10. Anti-CD16/CD32 aims at preventing the interaction between the Fcγ receptors CD16 and CD32, expressed at the surface of numerous myeloid and lymphoid cells, and the fluorochrome-conjugated antibodies required for the immunophenotyping; such binding would generate unspecific fluorescent signals on the cells that are positive for Fcγ receptors.
11. If required, phenotyping of cancer-specific Tc1 cells can be complemented with the detection of the cytolytic granule content and of the degranulation phenomenon using fluorochrome-conjugated antibodies binding to perforin, granzyme B or CD107a (*see* Table 3).
12. If the volume of blood collected within 24 h exceeds 1% of the animal body weight or 1.5% over a 2-week period, ethical

Table 3
Suggested primary antibodies for detecting cytolytic granules and degranulation of murine Tc1 lymphocytes by flow cytometry

Targeted marker	Marker location	Antibody clone	Antibody final concentration[a] (μg/mL)	References
CD107a	Surface[b]	1D4B	2	[70]
Granzyme B	Intracytoplasmic	GB11	10	[37, 62]
Perforin		eBioOMAK-D	5	[37]

The fluorochromes conjugated to the primary antibodies listed have not been suggested as these markers could be considered individually
[a]Antibody titration remains strongly advised to improve stain index
[b]CD107a is associated to cytolytic granules inside the cytoplasm of CTLs but is exposed at the surface of the cell following degranulation

committee guidelines advise to euthanize it unless justified and authorized otherwise.

13. A volume of 200 μL of heparin solution is sufficient to collect up to 1 mL of blood (i.e., terminal bleeding of mice).
14. If a blood volume larger than 200 μL is required (for instance if restimulations with various individual or pools of TAA epitope peptides need to be tested), proceed as follows:
    (a) Anaesthetize the vaccinated mice with 2.5% isoflurane (one cage at a time).
    (b) Draw blood through the retroorbital sinus using heparinized capillary tubes.
    (c) Press on the eye with a gauze pad for a few seconds to stop the bleeding.
    (d) Apply some lubricant on both eyes to prevent their irritation during the time the mouse is asleep.
    (e) Proceed to the subcutaneous injection of 300 μL of saline using a syringe with a 25–30 G needle to compensate the volume withdrawn and speed up mouse recovery.
15. The duration of the T cell restimulation in the presence of a protein transport inhibitor directly impacts the sensitivity of detection of the cytokines by ICS [71]. Following 6 h of restimulation, the assay should detect the three Tc1-associated cytokines IFN-γ, TNF-α, and IL-2. However, if the signals remain weak even in the positive controls (PMA + ionomycin restimulation), longer periods of restimulation could be tested.
16. Fixative reagents can dissociate tandem fluorochromes such as APC-Cy7, thus affecting their spectral properties. As a

consequence, it is recommended not to exceed 15 min of fixation and/or to prevent the use of such tandems. Fixative resistant surrogates such as APC-Vio770 or APC-Fire750 should be prioritized while designing the panel of antibodies to be used for the ICS assay.

17. Positive events detected in the unstimulated samples should be subtracted from the total number of events detected after restimulation with the TAA epitope peptide as they are likely not specific for the tumor antigen.

Acknowledgments

J.G.P. was supported by the Seerave Foundation. B.W.B. was supported by the Terry Fox Research Institute. B.D.L. was supported by the Terry Fox Foundation, the Ontario Institute for Cancer Research, BioCanRx, and Turnstone Biologics.

Declaration of Interests: J.G.P., B.W.B., and B.D.L. share ownership of patents for cancer vaccination involving oncolytic vaccines. B.D.L. is cofounder and equity holder of and is on the board of directors of Turnstone Biologics.

References

1. Gujar S, Pol JG, Kim Y, Lee PW, Kroemer G (2018) Antitumor benefits of antiviral immunity: an underappreciated aspect of Oncolytic virotherapies. Trends Immunol 39 (3):209–221
2. Gujar S, Pol JG, Kroemer G (2018) Heating it up: oncolytic viruses make tumors 'hot' and suitable for checkpoint blockade immunotherapies. Oncoimmunology 7(8):e1442169
3. Pol JG, Atherton MJ, Bridle BW, Stephenson KB, Le Boeuf F, Hummel JL, Martin CG, Pomoransky J, Breitbach CJ, Diallo JS, Stojdl DF, Bell JC, Wan Y, Lichty BD (2018) Development and applications of oncolytic Maraba virus vaccines. Oncolytic Virother 7:117–128
4. Workenhe ST, Mossman KL (2014) Oncolytic virotherapy and immunogenic cancer cell death: sharpening the sword for improved cancer treatment strategies. Mol Ther 22 (2):251–256
5. Workenhe ST, Simmons G, Pol JG, Lichty BD, Halford WP, Mossman KL (2014) Immunogenic HSV-mediated oncolysis shapes the antitumor immune response and contributes to therapeutic efficacy. Mol Ther 22(1):123–131
6. Workenhe ST, Mossman KL (2013) Rewiring cancer cell death to enhance oncolytic viroimmunotherapy. Oncoimmunology 2(12): e27138
7. Diaz RM, Galivo F, Kottke T, Wongthida P, Qiao J, Thompson J, Valdes M, Barber G, Vile RG (2007) Oncolytic immunovirotherapy for melanoma using vesicular stomatitis virus. Cancer Res 67(6):2840–2848
8. Bridle BW, Boudreau JE, Lichty BD, Brunelliere J, Stephenson K, Koshy S, Bramson JL, Wan Y (2009) Vesicular stomatitis virus as a novel cancer vaccine vector to prime antitumor immunity amenable to rapid boosting with adenovirus. Mol Ther 17(10): 1814–1821
9. Bridle BW, Stephenson KB, Boudreau JE, Koshy S, Kazdhan N, Pullenayegum E, Brunelliere J, Bramson JL, Lichty BD, Wan Y (2010) Potentiating cancer immunotherapy using an oncolytic virus. Mol Ther 18 (8):1430–1439
10. Castelo-Branco P, Passer BJ, Buhrman JS, Antoszczyk S, Marinelli M, Zaupa C, Rabkin SD, Martuza RL (2010) Oncolytic herpes simplex virus armed with xenogeneic homologue of prostatic acid phosphatase enhances antitumor efficacy in prostate cancer. Gene Ther 17 (6):805–810

11. Zhang YQ, Tsai YC, Monie A, Wu TC, Hung CF (2010) Enhancing the therapeutic effect against ovarian cancer through a combination of viral oncolysis and antigen-specific immunotherapy. Mol Ther 18(4):692–699
12. Wongthida P, Diaz RM, Pulido C, Rommelfanger D, Galivo F, Kaluza K, Kottke T, Thompson J, Melcher A, Vile R (2011) Activating systemic T-cell immunity against self tumor antigens to support oncolytic virotherapy with vesicular stomatitis virus. Hum Gene Ther 22(11):1343–1353
13. Kottke T, Errington F, Pulido J, Galivo F, Thompson J, Wongthida P, Diaz RM, Chong H, Ilett E, Chester J, Pandha H, Harrington K, Selby P, Melcher A, Vile R (2011) Broad antigenic coverage induced by vaccination with virus-based cDNA libraries cures established tumors. Nat Med 17 (7):854–859
14. Rommelfanger DM, Wongthida P, Diaz RM, Kaluza KM, Thompson JM, Kottke TJ, Vile RG (2012) Systemic combination virotherapy for melanoma with tumor antigen-expressing vesicular stomatitis virus and adoptive T-cell transfer. Cancer Res 72(18):4753–4764
15. Pulido J, Kottke T, Thompson J, Galivo F, Wongthida P, Diaz RM, Rommelfanger D, Ilett E, Pease L, Pandha H, Harrington K, Selby P, Melcher A, Vile R (2012) Using virally expressed melanoma cDNA libraries to identify tumor-associated antigens that cure melanoma. Nat Biotechnol 30(4):337–343
16. Bridle BW, Chen L, Lemay CG, Diallo JS, Pol J, Nguyen A, Capretta A, He R, Bramson JL, Bell JC, Lichty BD, Wan Y (2013) HDAC inhibition suppresses primary immune responses, enhances secondary immune responses, and abrogates autoimmunity during tumor immunotherapy. Mol Ther 21 (4):887–894
17. Pol JG, Zhang L, Bridle BW, Stephenson KB, Resseguier J, Hanson S, Chen L, Kazdhan N, Bramson JL, Stojdl DF, Wan Y, Lichty BD (2014) Maraba virus as a potent oncolytic vaccine vector. Mol Ther 22(2):420–429
18. Alonso-Camino V, Rajani K, Kottke T, Rommelfanger-Konkol D, Zaidi S, Thompson J, Pulido J, Ilett E, Donnelly O, Selby P, Pandha H, Melcher A, Harrington K, Diaz RM, Vile R (2014) The profile of tumor antigens which can be targeted by immunotherapy depends upon the tumor's anatomical site. Mol Ther 22(11):1936–1948
19. Blanchard M, Shim KG, Grams MP, Rajani K, Diaz RM, Furutani KM, Thompson J, Olivier KR, Park SS, Markovic SN, Pandha H, Melcher A, Harrington K, Zaidi S, Vile R (2015) Definitive management of oligometastatic melanoma in a murine model using combined ablative radiation therapy and viral immunotherapy. Int J Radiat Oncol Biol Phys 93(3):577–587
20. Zaidi S, Blanchard M, Shim K, Ilett E, Rajani K, Parrish C, Boisgerault N, Kottke T, Thompson J, Celis E, Pulido J, Selby P, Pandha H, Melcher A, Harrington K, Vile R (2015) Mutated BRAF emerges as a major effector of recurrence in a murine melanoma model after treatment with immunomodulatory agents. Mol Ther 23(5):845–856
21. Bridle BW, Nguyen A, Salem O, Zhang L, Koshy S, Clouthier D, Chen L, Pol J, Swift SL, Bowdish DM, Lichty BD, Bramson JL, Wan Y (2016) Privileged antigen presentation in splenic B cell follicles maximizes T cell responses in prime-boost vaccination. J Immunol 196(11):4587–4595
22. Cockle JV, Rajani K, Zaidi S, Kottke T, Thompson J, Diaz RM, Shim K, Peterson T, Parney IF, Short S, Selby P, Ilett E, Melcher A, Vile R (2016) Combination viroimmunotherapy with checkpoint inhibition to treat glioma, based on location-specific tumor profiling. Neuro-Oncology 18(4):518–527
23. Atherton MJ, Stephenson KB, Pol J, Wang F, Lefebvre C, Stojdl DF, Nikota JK, Dvorkin-Gheva A, Nguyen A, Chen L, Johnson-Obaseki S, Villeneuve PJ, Diallo JS, Dimitroulakos J, Wan Y, Lichty BD (2017) Customized viral immunotherapy for HPV-associated cancer. Cancer Immunol Res 5(10):847–859
24. Atherton MJ, Stephenson KB, Tzelepis F, Bakhshinyan D, Nikota JK, Son HH, Jirovec A, Lefebvre C, Dvorkin-Gheva A, Ashkar AA, Wan Y, Stojdl DF, Belanger EC, Breau RH, Bell JC, Saad F, Singh SK, Diallo JS, Lichty BD (2018) Transforming the prostatic tumor microenvironment with oncolytic virotherapy. Oncoimmunology 7(7):e1445459
25. Atherton MJ, Stephenson KB, Nikota JK, Hu QN, Nguyen A, Wan Y, Lichty BD (2018) Preclinical development of peptide vaccination combined with oncolytic MG1-E6E7 for HPV-associated cancer. Vaccine 36 (16):2181–2192
26. Pol JG, Acuna SA, Yadollahi B, Tang N, Stephenson KB, Atherton MJ, Hanwell D, El-Warrak A, Goldstein A, Moloo B, Turner PV, Lopez R, LaFrance S, Evelegh C, Denisova G, Parsons R, Millar J, Stoll G, Martin CG, Pomoransky J, Breitbach CJ, Bramson JL, Bell JC, Wan Y, Stojdl DF, Lichty BD, McCart JA (2019) Preclinical evaluation of a MAGE-A3 vaccination utilizing the oncolytic

Maraba virus currently in first-in-human trials. Oncoimmunology 8(1):e1512329

27. Boisgerault N, Kottke T, Pulido J, Thompson J, Diaz RM, Rommelfanger-Konkol D, Embry A, Saenz D, Poeschla E, Pandha H, Harrington K, Melcher A, Selby P, Vile R (2013) Functional cloning of recurrence-specific antigens identifies molecular targets to treat tumor relapse. Mol Ther 21 (8):1507–1516
28. Pol JG, Levesque S, Workenhe ST, Gujar S, Le Boeuf F, Clements DR, Fahrner JE, Fend L, Bell JC, Mossman KL, Fucikova J, Spisek R, Zitvogel L, Kroemer G, Galluzzi L (2018) Trial watch: oncolytic viro-immunotherapy of hematologic and solid tumors. Oncoimmunology 7(12):e1503032
29. Van den Eynde M, Mlecnik B, Bindea G, Fredriksen T, Church SE, Lafontaine L, Haicheur N, Marliot F, Angelova M, Vasaturo A, Bruni D, Jouret-Mourin A, Baldin P, Huyghe N, Haustermans K, Debucquoy A, Van Cutsem E, Gigot JF, Hubert C, Kartheuser A, Remue C, Leonard D, Valge-Archer V, Pages F, Machiels JP, Galon J (2018) The link between the multiverse of immune microenvironments in metastases and the survival of colorectal cancer patients. Cancer Cell 34 (6):1012–1026 e1013
30. Pages F, Mlecnik B, Marliot F, Bindea G, Ou FS, Bifulco C, Lugli A, Zlobec I, Rau TT, Berger MD, Nagtegaal ID, Vink-Borger E, Hartmann A, Geppert C, Kolwelter J, Merkel S, Grutzmann R, Van den Eynde M, Jouret-Mourin A, Kartheuser A, Leonard D, Remue C, Wang JY, Bavi P, Roehrl MHA, Ohashi PS, Nguyen LT, Han S, MacGregor HL, Hafezi-Bakhtiari S, Wouters BG, Masucci GV, Andersson EK, Zavadova E, Vocka M, Spacek J, Petruzelka L, Konopasek B, Dundr P, Skalova H, Nemejcova K, Botti G, Tatangelo F, Delrio P, Ciliberto G, Maio M, Laghi L, Grizzi F, Fredriksen T, Buttard B, Angelova M, Vasaturo A, Maby P, Church SE, Angell HK, Lafontaine L, Bruni D, El Sissy C, Haicheur N, Kirilovsky A, Berger A, Lagorce C, Meyers JP, Paustian C, Feng Z, Ballesteros-Merino C, Dijkstra J, van de Water C, van Lent-van Vliet S, Knijn N, Musina AM, Scripcariu DV, Popivanova B, Xu M, Fujita T, Hazama S, Suzuki N, Nagano H, Okuno K, Torigoe T, Sato N, Furuhata T, Takemasa I, Itoh K, Patel PS, Vora HH, Shah B, Patel JB, Rajvik KN, Pandya SJ, Shukla SN, Wang Y, Zhang G, Kawakami Y, Marincola FM, Ascierto PA, Sargent DJ, Fox BA, Galon J (2018) International validation of the consensus immunoscore for the classification of colon cancer: a prognostic and accuracy study. Lancet 391 (10135):2128–2139
31. Nizard M, Roussel H, Diniz MO, Karaki S, Tran T, Voron T, Dransart E, Sandoval F, Riquet M, Rance B, Marcheteau E, Fabre E, Mandavit M, Terme M, Blanc C, Escudie JB, Gibault L, Barthes FLP, Granier C, Ferreira LCS, Badoual C, Johannes L, Tartour E (2017) Induction of resident memory T cells enhances the efficacy of cancer vaccine. Nat Commun 8:15221
32. Fridman WH, Zitvogel L, Sautes-Fridman C, Kroemer G (2017) The immune contexture in cancer prognosis and treatment. Nat Rev Clin Oncol 14(12):717–734
33. Ganesan AP, Clarke J, Wood O, Garrido-Martin EM, Chee SJ, Mellows T, Samaniego-Castruita D, Singh D, Seumois G, Alzetani A, Woo E, Friedmann PS, King EV, Thomas GJ, Sanchez-Elsner T, Vijayanand P, Ottensmeier CH (2017) Tissue-resident memory features are linked to the magnitude of cytotoxic T cell responses in human lung cancer. Nat Immunol 18(8):940–950
34. Mould RC, AuYeung AWK, van Vloten JP, Susta L, Mutsaers AJ, Petrik JJ, Wood GA, Wootton SK, Karimi K, Bridle BW (2017) Enhancing immune responses to cancer vaccines using multi-site injections. Sci Rep 7 (1):8322
35. Liu J, Blake SJ, Yong MC, Harjunpaa H, Ngiow SF, Takeda K, Young A, O'Donnell JS, Allen S, Smyth MJ, Teng MW (2016) Improved efficacy of neoadjuvant compared to adjuvant immunotherapy to eradicate metastatic disease. Cancer Discov 6(12):1382–1399
36. Dolton G, Tungatt K, Lloyd A, Bianchi V, Theaker SM, Trimby A, Holland CJ, Donia M, Godkin AJ, Cole DK, Straten PT, Peakman M, Svane IM, Sewell AK (2015) More tricks with tetramers: a practical guide to staining T cells with peptide-MHC multimers. Immunology 146(1):11–22
37. Klein-Hessling S, Muhammad K, Klein M, Pusch T, Rudolf R, Floter J, Qureischi M, Beilhack A, Vaeth M, Kummerow C, Backes C, Schoppmeyer R, Hahn U, Hoth M, Bopp T, Berberich-Siebelt F, Patra A, Avots A, Muller N, Schulze A, Serfling E (2017) NFATc1 controls the cytotoxicity of CD8(+) T cells. Nat Commun 8(1):511
38. Hersperger AR, Makedonas G, Betts MR (2008) Flow cytometric detection of perforin upregulation in human CD8 T cells. Cytometry A 73(11):1050–1057
39. Peters PJ, Borst J, Oorschot V, Fukuda M, Krahenbuhl O, Tschopp J, Slot JW, Geuze HJ

(1991) Cytotoxic T lymphocyte granules are secretory lysosomes, containing both perforin and granzymes. J Exp Med 173(5):1099–1109

40. Smith SG, Smits K, Joosten SA, van Meijgaarden KE, Satti I, Fletcher HA, Caccamo N, Dieli F, Mascart F, McShane H, Dockrell HM, Ottenhoff TH (2015) Intracellular cytokine staining and flow cytometry: considerations for application in clinical trials of novel tuberculosis vaccines. PLoS One 10(9): e0138042
41. Maecker HT, Rinfret A, D'Souza P, Darden J, Roig E, Landry C, Hayes P, Birungi J, Anzala O, Garcia M, Harari A, Frank I, Baydo R, Baker M, Holbrook J, Ottinger J, Lamoreaux L, Epling CL, Sinclair E, Suni MA, Punt K, Calarota S, El-Bahi S, Alter G, Maila H, Kuta E, Cox J, Gray C, Altfeld M, Nougarede N, Boyer J, Tussey L, Tobery T, Bredt B, Roederer M, Koup R, Maino VC, Weinhold K, Pantaleo G, Gilmour J, Horton H, Sekaly RP (2005) Standardization of cytokine flow cytometry assays. BMC Immunol 6:13
42. Betts MR, Koup RA (2004) Detection of T-cell degranulation: CD107a and b. Methods Cell Biol 75:497–512
43. Mojic M, Takeda K, Hayakawa Y (2017) The dark side of IFN-gamma: its role in promoting cancer immunoevasion. Int J Mol Sci 19(1). https://doi.org/10.3390/ijms19010089
44. Jonker DJ, Hotte SJ, Razak ARA, Renouf DJ, Lichty B, Bell JC, Powers J, Breitbach CJ, Stojdl DF, Stephenson KB, Bramson JL, Hummel J, Lemay CG, Cutz J-C, Wells J, Eady R, Sun X, Tu D, Dancey J (2017) Phase I study of oncolytic virus (OV) MG1 maraba/ MAGE-A3 (MG1MA3), with and without transgenic MAGE-A3 adenovirus vaccine (AdMA3) in incurable advanced/metastatic MAGE-A3-expressing solid tumours: CCTG IND.214. J Clin Oncol 35(15_suppl): e14637–e14637. https://doi.org/10.1200/ JCO.2017.35.15_suppl.e14637
45. Butterfield LH, Buffo MJ (2014) Immunologic monitoring of cancer vaccine trials using the ELISPOT assay. Methods Mol Biol 1102:71–82
46. Bagarazzi ML, Yan J, Morrow MP, Shen X, Parker RL, Lee JC, Giffear M, Pankhong P, Khan AS, Broderick KE, Knott C, Lin F, Boyer JD, Draghia-Akli R, White CJ, Kim JJ, Weiner DB, Sardesai NY (2012) Immunotherapy against HPV16/18 generates potent TH1 and cytotoxic cellular immune responses. Sci Transl Med 4(155):155ra138
47. Landreth KS (2002) Critical windows in development of the rodent immune system. Hum Exp Toxicol 21(9-10):493–498
48. Humeau J, Levesque S, Kroemer G, Pol JG (2019) Gold standard assessment of immunogenic cell death in oncological mouse models. Methods Mol Biol 1884:297–315
49. Bridle BW, Clouthier D, Zhang L, Pol J, Chen L, Lichty BD, Bramson JL, Wan Y (2013) Oncolytic vesicular stomatitis virus quantitatively and qualitatively improves primary CD8(+) T-cell responses to anticancer vaccines. Oncoimmunology 2(8):e26013
50. Zhang L, Bridle BW, Chen L, Pol J, Spaner D, Boudreau JE, Rosen A, Bassett JD, Lichty BD, Bramson JL, Wan Y (2013) Delivery of viral-vectored vaccines by B cells represents a novel strategy to accelerate CD8(+) T-cell recall responses. Blood 121(13):2432–2439
51. Kershaw MH, Hsu C, Mondesire W, Parker LL, Wang G, Overwijk WW, Lapointe R, Yang JC, Wang RF, Restifo NP, Hwu P (2001) Immunization against endogenous retroviral tumor-associated antigens. Cancer Res 61(21):7920–7924
52. Huang AY, Gulden PH, Woods AS, Thomas MC, Tong CD, Wang W, Engelhard VH, Pasternack G, Cotter R, Hunt D, Pardoll DM, Jaffee EM (1996) The immunodominant major histocompatibility complex class I-restricted antigen of a murine colon tumor derives from an endogenous retroviral gene product. Proc Natl Acad Sci U S A 93 (18):9730–9735
53. Zeh HJ 3rd, Perry-Lalley D, Dudley ME, Rosenberg SA, Yang JC (1999) High avidity CTLs for two self-antigens demonstrate superior in vitro and in vivo antitumor efficacy. J Immunol 162(2):989–994
54. Overwijk WW, Tsung A, Irvine KR, Parkhurst MR, Goletz TJ, Tsung K, Carroll MW, Liu C, Moss B, Rosenberg SA, Restifo NP (1998) gp100/pmel 17 is a murine tumor rejection antigen: induction of "self"-reactive, tumoricidal T cells using high-affinity, altered peptide ligand. J Exp Med 188(2):277–286
55. Rosenthal KS, Stone S, Koski G, Zimmerman DH (2017) LEAPS vaccine incorporating HER-2/neu epitope elicits protection that prevents and limits tumor growth and spread of breast cancer in a mouse model. J Immunol Res 2017:3613505. https://doi.org/10.1155/ 2017/3613505
56. Gall VA, Philips AV, Qiao N, Clise-Dwyer K, Perakis AA, Zhang M, Clifton GT, Sukhumalchandra P, Ma Q, Reddy SM, Yu D,

Molldrem JJ, Peoples GE, Alatrash G, Mittendorf EA (2017) Trastuzumab increases HER2 uptake and cross-presentation by dendritic cells. Cancer Res 77(19):5374–5383

57. Peng S, Ji H, Trimble C, He L, Tsai YC, Yeatermeyer J, Boyd DA, Hung CF, Wu TC (2004) Development of a DNA vaccine targeting human papillomavirus type 16 oncoprotein E6. J Virol 78(16):8468–8476
58. de Oliveira LM, Morale MG, Chaves AA, Cavalher AM, Lopes AS, Diniz Mde O, Schanoski AS, de Melo RL, Ferreira LC, de Oliveira ML, Demasi M, Ho PL (2015) Design, immune responses and anti-tumor potential of an HPV16 E6E7 multi-epitope vaccine. PLoS One 10(9):e0138686
59. Naslund TI, Uyttenhove C, Nordstrom EK, Colau D, Warnier G, Jondal M, Van den Eynde BJ, Liljestrom P (2007) Comparative prime-boost vaccinations using Semliki Forest virus, adenovirus, and ALVAC vectors demonstrate differences in the generation of a protective central memory CTL response against the P815 tumor. J Immunol 178(11):6761–6769
60. Garcia-Hernandez Mde L, Gray A, Hubby B, Klinger OJ, Kast WM (2008) Prostate stem cell antigen vaccination induces a long-term protective immune response against prostate cancer in the absence of autoimmunity. Cancer Res 68(3):861–869
61. Scrimieri F, Askew D, Corn DJ, Eid S, Bobanga ID, Bjelac JA, Tsao ML, Allen F, Othman YS, Wang SC, Huang AY (2013) Murine leukemia virus envelope gp70 is a shared biomarker for the high-sensitivity quantification of murine tumor burden. Oncoimmunology 2 (11):e26889
62. Juneja VR, McGuire KA, Manguso RT, LaFleur MW, Collins N, Haining WN, Freeman GJ, Sharpe AH (2017) PD-L1 on tumor cells is sufficient for immune evasion in immunogenic tumors and inhibits CD8 T cell cytotoxicity. J Exp Med 214(4):895–904
63. Regan RD, Fenyk-Melody JE, Tran SM, Chen G, Stocking KL (2016) Comparison of submental blood collection with the retroorbital and submandibular methods in mice (Mus musculus). J Am Assoc Lab Anim Sci 55 (5):570–576
64. Fritsche J, Rakitsch B, Hoffgaard F, Romer M, Schuster H, Kowalewski DJ, Priemer M, Stos-Zweifel V, Horzer H, Satelli A, Sonntag A, Goldfinger V, Song C, Mahr A, Ott M, Schoor O, Weinschenk T (2018) Translating immunopeptidomics to immunotherapy-decision-making for patient and personalized target selection. Proteomics 18(12): e1700284. https://doi.org/10.1002/pmic.201700284
65. Murphy JP, Konda P, Kowalewski DJ, Schuster H, Clements D, Kim Y, Cohen AM, Sharif T, Nielsen M, Stevanovic S, Lee PW, Gujar S (2017) MHC-I ligand discovery using targeted database searches of mass spectrometry data: implications for T-cell immunotherapies. J Proteome Res 16 (4):1806–1816. https://doi.org/10.1021/acs.jproteome.6b00971
66. Caron E, Kowalewski DJ, Chiek Koh C, Sturm T, Schuster H, Aebersold R (2015) Analysis of major histocompatibility complex (MHC) immunopeptidomes using mass spectrometry. Mol Cell Proteomics 14 (12):3105–3117
67. Vicetti Miguel RD, Maryak SA, Cherpes TL (2012) Brefeldin A, but not monensin, enables flow cytometric detection of interleukin-4 within peripheral T cells responding to ex vivo stimulation with chlamydia trachomatis. J Immunol Methods 384(1-2):191–195
68. Jamur MC, Oliver C (2010) Permeabilization of cell membranes. Methods Mol Biol 588:63–66
69. Thavarajah R, Mudimbaimannar VK, Elizabeth J, Rao UK, Ranganathan K (2012) Chemical and physical basics of routine formaldehyde fixation. J Oral Maxillofac Pathol 16 (3):400–405
70. Fan X, Quezada SA, Sepulveda MA, Sharma P, Allison JP (2014) Engagement of the ICOS pathway markedly enhances efficacy of CTLA-4 blockade in cancer immunotherapy. J Exp Med 211(4):715–725
71. Ai W, Li H, Song N, Li L, Chen H (2013) Optimal method to stimulate cytokine production and its use in immunotoxicity assessment. Int J Environ Res Public Health 10 (9):3834–3842

Chapter 13

Evaluation of Oncolytic Virus-Induced Therapeutic Tumor Vaccination Effects in Murine Tumor Models

Rūta Veinalde

Abstract

Oncolytic virotherapy is rapidly advancing into clinical practice as an appealing strategy for cancer therapy. Studies in the recent decades indicate that immunostimulatory properties of oncolytic viruses (OVs) are crucial for their therapeutic efficacy. The specific lysis of tumor cells and release of tumor associated antigens in the context of an OV infection ensures activation of a tumor-specific immune response. The evidence for induction of a systemic, specific antitumor immune response after OV therapy in preclinical and clinical studies allows to consider oncolytic virotherapy as a type of therapeutic cancer vaccination strategy.

This chapter describes methods to evaluate the therapeutic tumor vaccination effect of an oncolytic virus in murine tumor models. Protocols for a tumor rechallenge experiment in vivo and tumor cell specific splenocyte restimulation in vitro are included, as well as protocols for analysis of memory T cell subpopulations in tumor draining lymph nodes using flow cytometry.

Key words Oncolytic virotherapy, Therapeutic cancer vaccination, Antitumor immune memory responses, Tumor rechallenge, Splenocyte restimulation, Memory T cell analysis

1 Introduction

An increased understanding of the role of the immune system in development of malignant diseases has facilitated development of a variety of clinically relevant cancer immunotherapy strategies in the past decades [1]. Among the currently exploited immunotherapy strategies, therapeutic vaccination for cancer has always been considered an attractive approach. The appeal for therapeutic cancer vaccination is the ability to induce a specific, systemic immune effector response, allowing to target not only a primary tumor, but also metastatic lesions. Furthermore, a successful vaccination also ensures protection from tumor relapse due to a generated immune memory cell pool [2]. Use of oncolytic viruses (OVs) can also be considered a type of therapeutic cancer vaccination strategy [3]. OVs selectively target, replicate within and lyse tumor cells in a patient's body [4]. After lysis of tumor cells all of

Christine E. Engeland (ed.), *Oncolytic Viruses*, Methods in Molecular Biology, vol. 2058,
https://doi.org/10.1007/978-1-4939-9794-7_13,

the contents of the damaged cells are released, including tumor associated antigens (TAAs), viral particles and different intermediates of the virus replication cycle. The released TAAs can be accessed by dendritic cells (DCs) that take them up, process and present to T cells [5]. Of particular importance is the presence of the viral components which serve as adjuvants providing activating signals to the DCs through the pathogen recognition receptors in the context of the released TAAs [6]. Activated DCs are able to provide the necessary co-stimulatory signals to the immune effector cells required for induction of a robust immune response against the specific TAAs [7]. Evidence for the therapeutic cancer vaccination effect of OVs has been provided in preclinical as well as in clinical studies. For instance, systemic immune responses induced by an OV have been observed in the phase III study for intratumoral administration of Talimogene Laherparepvec (T-VEC) in comparison to GM-CSF alone. Although in this study the virus was administered intralesionally, a reduction in size of more than 50% was observed for 15% of visceral metastases which were all uninjected [8]. Systemic antitumor vaccination effects have been demonstrated also in a phase II clinical trial for intraperitoneal administration of oncolytic measles virus encoding sodium iodide symporter (MV-NIS) for treatment of ovarian cancer. In this study T cells isolated from peripheral blood of the treated patients demonstrated a specific IFN-γ response after in vitro stimulation with ovarian cancer specific peptides [9].

Strategies to further promote the favorable therapeutic vaccination effects of OVs are also actively investigated. This can be achieved by combination of OVs with agents that generally augment effector functions of immune cells, such as different cytokines or checkpoint blockade antibodies [10]. However, a potential drawback for these methods could be amplification of an antiviral immune response along with the antitumor immune response [11, 12]. One of the strategies to overcome this limitation is the use of OVs as vectors for additional delivery of TAAs to divert the immune response toward the tumor. The feasibility and efficacy of this approach have been demonstrated in preclinical and clinical studies with different OVs [13].

This chapter describes methods to evaluate the therapeutic tumor vaccination effect of an OV in preclinical murine tumor models. Methods to evaluate the tumor vaccination effect with secondary tumor challenge experiments and in vitro splenocyte restimulation procedures, as well as a method for quantification of memory T cells in tumor draining lymph nodes using flow cytometry analysis are described. The methods are described as they have been routinely applied in studies with oncolytic measles virus vectors [14–17]. However, these general procedures can be simply adjusted for use in preclinical studies with other OV vectors.

2 Materials

2.1 Tumor Rechallenge

1. C57BL/6J mice, female, 6–8 weeks old at the beginning of the experiment.
2. Five to ten age- and gender-matched C57BL/6J mice as naive controls.
3. T175 cell culture flask with approx. 80% confluent MC38cea (murine MC38 cell line stably expressing human CEA antigen [18]) cell monolayer.
4. 70% ethanol for disinfection.
5. 0.25% trypsin–EDTA solution.
6. Dulbecco's phosphate buffered saline (D-PBS).
7. Trypan blue.
8. Hemocytometer.
9. Electrical razor.
10. Sterile 1 ml PP/PE fine dosage syringes.
11. Sterile 1 ml syringe fitting 26 G needles.
12. Digital caliper.

2.2 Tumor- and Virus-Specific IFN-γ Memory Response Recall In Vitro

1. 70% ethanol for disinfection.
2. Sterile forceps and scissors.
3. 1.5 ml and 50 ml tubes.
4. Dulbecco's phosphate buffered saline (D-PBS).
5. Bucket with ice.
6. 100 μm cell strainers.
7. 1 ml syringes.
8. Ammonium-Chloride-Potassium (ACK) lysing buffer.
9. Trypan blue.
10. Hemocytometer.
11. T75 cell culture flasks with a confluent monolayer of cell lines (MC38cea, MC38, B16) required for restimulation.
12. 0.25% trypsin–EDTA solution.
13. Dulbecco's Modified Eagle's medium (DMEM).
14. Mitomycin-C.
15. Benchtop thermomixer.
16. Cocultivation medium: Roswell Park Memorial Institute-1640 (RPMI-1640) medium, 10% fetal calf serum (FCS), 1% Penicillin-Streptomycin (P/S), 50 IU/ml recombinant murine interleukin-2 (IL-2).

17. Activation medium: RPMI-1640, 5% FCS, 1% P/S, 5 nM phorbol 12-myristate 13-acetate (PMA), 500 nM ionomycin.
18. Flat-bottom 24-well cell culture plates for adherent cells.
19. Oncolytic measles virus.
20. Enzyme-linked immunosorbent assay (ELISA) kit for detection of IFN-γ.
21. Centrifuge.

2.3 Memory T Cell Analysis in Tumors and Lymph Nodes Using Flow Cytometry

1. Sterile scissors and curved forceps.
2. Syringe needles.
3. D-PBS.
4. Bucket with ice.
5. Digestion medium: RPMI-1640, 5% FCS, 200 U/ml Collagenase Type I.
6. Sterile scalpels.
7. 1.5 ml and 50 ml tubes.
8. 100 μm cell strainers.
9. 1 ml syringes.
10. Trypan blue.
11. Hemocytometer.
12. Mouse Fc receptor blocking solution containing antibodies against mouse CD16/CD32.
13. Antibodies for cell surface marker staining: anti-mouse-CD3-PerCP-Cy-5.5, anti-mouse-CD8-APC, anti-mouse-CD4-APC-Cy7, anti-mouse-CD44-PE, and anti-mouse-CD62L-FITC.
14. 4′,6-Diamidino-2-phenylindole (DAPI).
15. 12 × 75 mm FACS tubes.
16. Flow cytometer capable of simultaneous detection of fluorescence from all the fluorochromes used for cell surface marker staining and DAPI.

3 Methods

3.1 Experimental Design

The further described procedures are suitable for a follow-up study using mice that have experienced complete tumor remissions in an experiment assessing therapeutic efficacy of oncolytic measles vectors. A schematic overview of the time schedule for the experiment including a preceding survival experiment is depicted in Fig. 1 (*see* **Note 1**). Procedures are described using the subcutaneous MC38cea tumor model as an example. For the tumor rechallenge

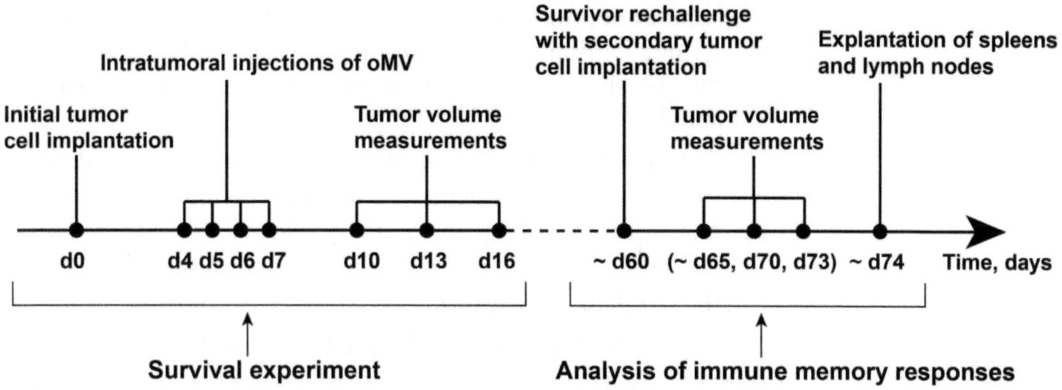

Fig. 1 Experimental schedule for assessment of efficacy of oncolytic measles virus in a survival experiment with a following study for immune memory responses. Subheadings 3.2–3.4 of this chapter describe the "Analysis of immune memory responses" part of the depicted experimental schedule. *oMV* oncolytic measles virus

experiment the same tumor cell line (MC38cea) as for the preceding survival experiment is used. For splenocyte restimulation in vitro, MC38 and B16 cells are used for preparation of specificity controls in addition to MC38cea cells. Age and gender matched naive control mice [5–10] are included in the tumor rechallenge experiment as a control group.

3.2 Tumor Rechallenge

All of the cell culture and animal handling procedures are performed in a laminar flow hood under sterile conditions.

1. Seed the MC38cea tumor cell line in cell culture flasks. On the day of rechallenge cells should be approx. 80% confluent before harvesting (*see* **Note 2**).
2. On the day of tumor rechallenge, harvest the tumor cells from culture flasks using trypsin–EDTA solution according to your routine cell culture practices.
3. Spin down the cell suspension in an appropriate size tube at 300 × *g* for 5 min at room temperature (RT).
4. Wash the pellet by resuspending in D-PBS (up to 50 ml) and spin down at 300 × *g* for 5 min at RT. Repeat this step three times in total.
5. Before the last washing step, count the cells using a hemocytometer and Trypan blue for dead cell exclusion.
6. After the last washing step, resuspend the cell pellet in D-PBS and adjust the cell concentration to 1×10^6 cells/ml (*see* **Note 3**).
7. Store the cell suspension on ice until proceeding further.
8. Shave the left flanks of the mice.

9. Aspirate approx. 600 μl of the MC38cea cell suspension using a syringe without a needle and remove air bubbles by carefully moving the plunger of the syringe up and down (*see* **Notes 4** and **5**).
10. Put a needle on the syringe with the cell suspension and adjust the aspirated cell suspension volume to 500 μl (*see* **Note 6**).
11. Pick up and restrain a mouse and inject 100 μl of the cell suspension subcutaneously in the previously shaved region on the left flank (*see* **Note 7**).
12. Repeat the injections for all mice (*see* **Note 8**).
13. Monitor the tumor growth every third day by palpation and tumor volume measurements using a digital caliper.
14. Measure the longest and the shortest diameter of the tumor using a digital caliper and calculate the volume using the formula $\frac{\text{Shortest diameter}^2 \times \text{Longest diameter}}{2}$.

3.3 Tumor and Virus Specific IFN-γ Memory Response Recall with Splenocytes

3.3.1 Harvest of Murine Spleen

1. Euthanize the mouse by cervical dislocation in a laminar flow hood.
2. Lay the mouse on the right side with head pointing left.
3. Wipe the side of the mouse with 70% ethanol.
4. Grab the skin of the animal with forceps and use clean scissors to make an approx. 1.5 cm long incision in the middle of the left flank of the animal and in parallel to the backbone. The incision should penetrate the peritoneal cavity and reveal the organs including the spleen which should now be visible on top.
5. Gently grab one side of the spleen with forceps, pull it out and cut it free from ligaments at both ends.
6. Transfer the spleen to a 1.5 ml tube with D-PBS. Close the tube and store it on ice until further processing (*see* **Note 9**).

3.3.2 Preparation of Single Cell Suspension of Murine Splenocytes

1. Place a 100 μm cell strainer in a Petri dish and fill the dish with 10 ml D-PBS.
2. Open the tube with the murine spleen and transfer the spleen onto the 100 μm cell strainer in the Petri dish.
3. Completely mesh the spleen through the cell strainer using a plunger of a 1 ml syringe (*see* **Note 10**).
4. Wash the cell strainer by pipetting the D-PBS through it 2–3 times to acquire as many splenocytes as possible.
5. Transfer the splenocyte suspension from the Petri dish to a 50 ml tube.
6. Spin down the tube for 5 min at 300 × *g* at RT.
7. Discard the supernatant.

8. Resuspend the pellet in 1 ml ACK red-blood-cell lysis buffer by pipetting up and down several times. Incubate the suspension for 10 min at RT.
9. Add 9 ml D-PBS to the suspension and spin down the tube for 5 min at 300 × *g* at RT (*see* **Note 11**).
10. Discard the supernatant.
11. Resuspend the pellet in 10 ml D-PBS and count the cells using a hemocytometer and Trypan blue for dead cell exclusion.

3.3.3 Tumor Cell Preparation for Cocultures

1. The respective target tumor cells (example for MC38cea) for splenocyte restimulation need to be expanded in cell culture in a sufficient amount for the day of the coculture preparation.
2. Harvest the MC38cea, MC38, and B16 tumor cell lines from culture flasks using trypsin–EDTA solution according to your routine cell culture practice (*see* **Note 12**).
3. Spin down the required amount of the tumor cells in 2 ml tubes and resuspend the pellets in DMEM + 20 μg/ml Mitomycin-C to achieve a cell concentration of 5×10^6 cells/ml.
4. Incubate the cells on a tabletop thermomixer at 37 °C for 2 h with shaking at approx. 800 rpm.
5. After incubation, spin down the tubes for 5 min at 300 × *g* at RT and discard the supernatants.
6. Wash the cells three times by adding 1–2 ml D-PBS per pellet and spinning down for 5 min at 300 × *g* at RT (*see* **Note 13**).
7. After the last washing step, resuspend the cell pellet in Cocultivation medium and adjust the cell concentration to 1×10^6 cells/ml.
8. Store the cells on ice until further processing.

3.3.4 Splenocyte Restimulation with Tumor Cells and Oncolytic Measles Virus

1. Prepare cocultures with splenocytes and tumor cells in 24-well cell culture plates in a total volume of 500 μl Cocultivation medium with a splenocyte to tumor cell ratio 10:1 (1×10^6 splenocytes: 1×10^5 tumor cells):

   Add 300 μl Cocultivation medium in a sufficient number of wells in 24-well cell culture plates for all the experimental samples and controls where splenocytes and tumor cells will be added.

   Transfer 1×10^6 splenocytes in 100 μl Cocultivation medium per spleen to a respective well in the prepared 24-well plates (*see* **Note 14**).

   Add 1×10^5 tumor cells in 100 μl Cocultivation medium to the respective wells with splenocytes.

   Incubate the prepared coculture plates at 37 °C and 5% CO_2 for 48 h.

2. Prepare cocultures with splenocytes and oncolytic measles virus in 24-well cell culture plates in a total volume of 500 µl Cocultivation medium with 1×10^6 splenocytes and 5×10^5 ciu of virus per well (*see* **Note 15**). Incubate the prepared coculture plates at 37 °C and 5% CO_2 for 48 h.
3. Prepare positive controls with splenocytes (*see* **Note 16**):

   Prepare fresh Activation medium according to the recipe in the Subheading 2.

   Resuspend 1×10^6 splenocytes in 500 µl Activation medium.

   Transfer the splenocyte suspension to a respective well in a 24-well plate and incubate at 37 °C and 5% CO_2 for 48 h.

3.3.5 Harvest of Supernatants and IFN-γ ELISA

1. Harvest the medium from each well in a 1.5 ml tube (*see* **Note 17**). Continue directly with IFN-γ ELISA or store the samples at −80 °C until further processing.
2. Before proceeding to IFN-γ ELISA analysis spin down the samples at $6000 \times g$ for 5 min at RT to remove cells and debris which may interfere with results of the analysis.
3. Perform IFN-γ ELISA using reagents and following instructions from a manufacturer of your choice. Exemplary IFN-γ results of the described assay are shown in Fig. 2.

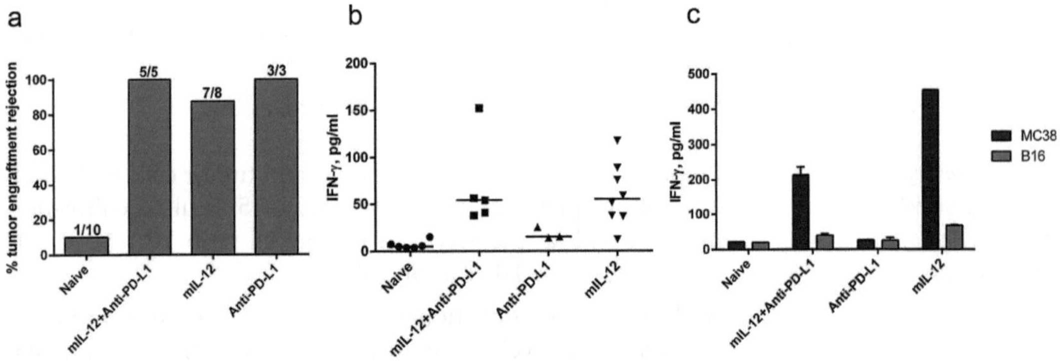

Fig. 2 Exemplary results for a tumor rechallenge experiment and splenocyte restimulation assay. Animals that experienced complete tumor remission after treatment with oncolytic measles vectors encoding the transgenes indicated on the x axis were rechallenged with subcutaneous secondary implantation of MC38cea cells. (**a**) Tumor rejection rates for animals in the rechallenge experiment are shown. (**b, c**) IFN-γ ELISA results for splenocyte restimulation assay with MC38cea cells. Results for restimulation with MC38cea cells are shown in **b**. Dots representing results from individual mice and median in the group are shown in **b**. Results for restimulation with MC38 and B16 cells (Specificity controls), as well as negative controls are shown in **c**. Average values and standard errors of the mean from two replicate measurements are shown in **c**. Figure is adapted from [22]

3.4 Memory T Cell Analysis in Tumor Draining Lymph Nodes

3.4.1 Harvest of Murine Inguinal Lymph Nodes

1. Euthanize the mouse by cervical dislocation in a laminar flow hood directly before proceeding to isolation of the lymph node.
2. Lay the mouse on a surgical pad on the side so that the flank with the tumor is exposed.
3. Stretch the top hind leg of the mouse back and the top front leg forward and fix both legs on the surgical pad using syringe needles.
4. Grab the skin with forceps and make an approx. 0.5 cm long, superficial incision with sharp sterile scissors close to the hind limb in direction to the abdomen.
5. The inguinal lymph node should now be visible as a light and slightly transparent structure with a diameter of 2–3 mm (*see* **Note 18**).
6. Take sterile curved forceps, grab the lymph node and pull it out of the tissue.
7. Directly transfer the isolated lymph node to a tube with D-PBS. The lymph node should sink to the bottom of the tube (*see* **Note 19**). Store the sample on ice until further processing.

3.4.2 Preparation of a Single Cell Suspension from Lymph Nodes

1. Prepare fresh Digestion medium (5 ml per sample) according to the recipe in Subheading 2 (*see* **Note 20**).
2. Prepare a Petri dish with 5 ml Digestion medium.
3. Transfer the lymph node from the tube with D-PBS to the Petri dish with Digestion medium.
4. Cut the lymph node open with a clean scalpel.
5. Transfer the Digestion medium with the lymph node pieces into a suitable vessel and incubate it for 30 min at 37 °C in a water bath. During incubation, gently shake the tubes every 10 min.
6. Prepare a clean 50 ml tube with a fitting 100 μm cell strainer on top.
7. After incubation, aspirate the Digestion medium with the lymph node pieces and pipette it onto the cell strainer.
8. Mesh the lymph node pieces through the cell strainer using a plunger from a 1 ml syringe.
9. Spin down the lymph node samples for 5 min at 300 × *g* at RT.
10. Discard the supernatant and resuspend the pellet in 5 ml D-PBS.
11. Count the cells using a hemocytometer and Trypan blue for dead cell exclusion.
12. Store samples on ice until further processing.

3.4.3 Cell Surface Marker Staining for Flow Cytometry Analysis

1. Prepare 1.5 ml tubes with 2×10^6 lymphocytes from lymph nodes in 100 µl D-PBS.
2. Add anti-mouse CD16/32 antibodies for Fc receptor blocking (amount as suggested by the manufacturer) to the samples and incubate for 5 min on benchtop at RT.
3. Add the appropriate amount (as suggested by the manufacturer) of anti-mouse-CD3-PerCP-Cy-5.5, anti-mouse-CD8-APC, anti-mouse-CD4-APC-Cy7, anti-mouse-CD44-PE, anti-mouse-CD62L-FITC to each of the samples and incubate for 30 min in the dark at RT.
4. Add 1 ml D-PBS to each sample and spin down for 5 min at $300 \times g$ at RT.
5. Resuspend each pellet in 500 µl D-PBS with 0.2 µg/ml DAPI for live/dead staining and incubate in the dark at RT for 5 min.
6. Spin down the samples for 5 min at $300 \times g$ at RT.
7. Resuspend the samples in 500 µl D-PBS and transfer to round bottom 12×75 mm FACS tubes.
8. Acquire data on a flow cytometer. An exemplary gating strategy is shown in Fig. 3 (*see* **Notes 21** and **22**).

4 Notes

1. The depicted timepoints for the experimental design in Fig. 1 should be regarded relatively. Adjustments can be required when using other tumor models and oncolytic vectors. In our experience, the "Analysis of immune memory responses" part can be successfully performed if started ca. 60–120 days after the initial tumor cell implantation. However, these experiments should not be delayed to more than 180 days after the initial tumor cell implantation due to aging of the mice and possible decline of immune cell functions.
2. Use of tumor cells that have not reached a 100% confluency in culture flasks ensures that the cells are in a logarithmic growth phase and will be actively proliferating when implanted in mice.
3. Prepare approximately twice the amount of the cell suspension that would be required for cell implantation in all mice to avoid shortage due to a loss of cell suspension during aspiration in syringes before implantation.
4. It is important to aspirate the cell suspension in the syringe without a needle to avoid damaging the cells before implantation.
5. The cells will settle down from the suspension while being stored on ice before implantation. It is crucial to resuspend

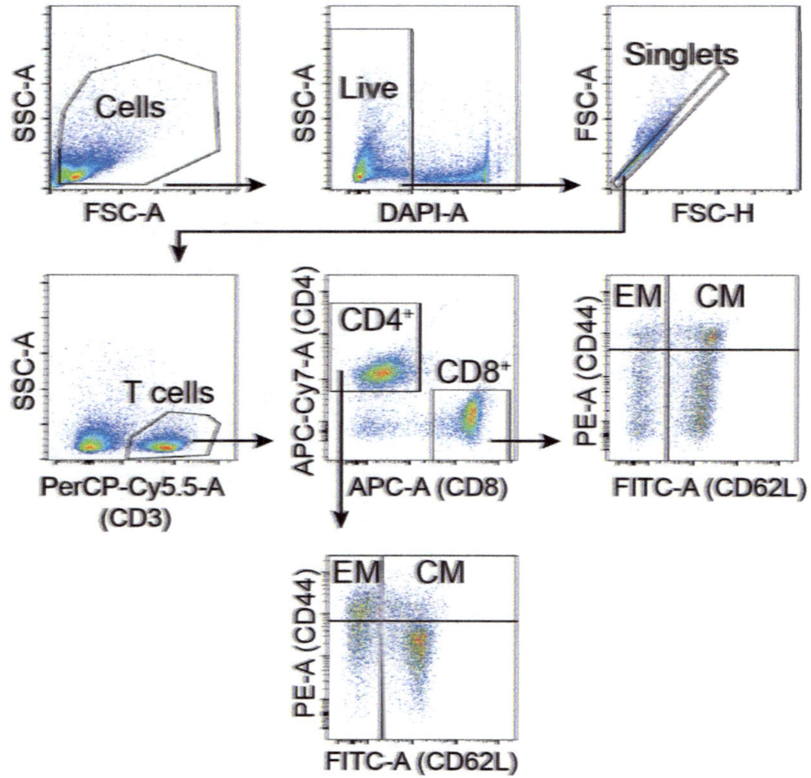

Fig. 3 Exemplary gating strategy for memory T cell analysis in tumors and lymph nodes using flow cytometry. Effector memory (EM) T cells are defined as $CD62L^{-}CD44^{high}$. Central memory (CM) T cells are defined as $CD62L^{+}CD44^{high}$

the cells before aspirating the suspension for implantation. Avoid vigorous vortexing to not impair cell viability. The tube can be gently inverted several times or the suspension can be pipetted up and down until there are no visible cell clumps. Failure to ensure an even cell suspension for implantation will result in an uneven growth of the tumors in mice.

6. We routinely cohouse five female C57BL/6J mice per cage. Therefore, the suggested cell suspension volume 500 μl per 1 ml syringe will be sufficient for cell implantation in all mice in one cage. Preparation of larger tumor cell suspension volumes per syringe for implantation is discouraged because of cells settling out of the suspension in the syringe during prolonged handling of the mice. Also when preparing syringes with only 500 μl of cell suspension, it should be ensured that no cells are settling out of the suspension in the syringe. In our experience inverting of the syringe 2–3 times between handling of each mouse is sufficient to ensure that cells remain in suspension.
7. For tumor rechallenge the cells should be implanted in the opposite flank of the mice as used in the preceding survival experiment. This will help to distinguish the secondary tumor growth after rechallenge from possible belated outgrowth of the primary tumor in rare cases.
8. After implantation of the cells in all mice it is advisable to seed tumor cells from the overleft suspension in cell culture plates in their standard culture medium and cultivate at 37 °C 5% CO_2. Monitor these plates for at least 48 h after seeding to control tumor cell growth and possible contamination.
9. The storage time of spleens on ice should be kept to minimum to avoid loss of splenocyte viability. If it is not possible to finish the whole experiment in 1 day, samples can be stored at 4 °C after preparation of single cell suspension as described in Subheading 3.3.2.
10. In alternative protocols the spleen is often meshed through the cell strainer which is on top of a 50 ml tube filled with 10 ml D-PBS. This practice is not advisable as the cell strainer is not immersed in D-PBS and it cannot be ensured that splenocytes remain in D-PBS directly after meshing through the cell strainer. Such practice may impair the viability of the splenocytes.
11. The cell pellet obtained following the centrifugation step after erythrocyte lysis with ACK buffer should be white. Redness in the pellet is associated with an incomplete lysis of erythrocytes. In such case the erythrocyte lysis step with the ACK lysis buffer should be repeated.
12. Restimulation with MC38 cells is prepared to validate that the response is not restricted to the human CEA antigen in the

MC38cea cell line. Restimulation with B16 cells is prepared to control for a response to an unrelated cell line. Restimulation with both MC38 and B16 cells must not be prepared for all samples of the experiment. It is sufficient if these controls are prepared for up to two mice per group.

13. Washing steps after incubation with Mitomycin-C are crucial to remove the cytostatic agent. Insufficient washing may lead to carry-over of Mitomycin-C to cocultures which may interfere with the splenocyte responses.
14. The splenocyte pellet before the resuspension in Cocultivation medium is too small to be directly seen in the tube. After resuspending the pellet and transferring to the coculture plate it is advisable to verify the presence of the splenocytes under a microscope.
15. It must be taken into account that the concentration of the IL-2 in the Cocultivation medium will be reduced when virus suspension is added. This must be calculated and adjusted for depending on the virus titer used for the experiment.
16. The positive controls with PMA and ionomycin are prepared to assess baseline stimulation capacity for the used splenocytes. These controls are useful in case the IFN-γ signals in the subsequent ELISA analysis are very weak or inconclusive. If the IFN-γ signals are low also in the positive controls this indicates a problem with splenocyte viability and general responsiveness in the given sample.
17. It is advisable to prepare at least two supernatant aliquots from each coculture well for storage. The second aliquot can be used as a back-up in case there are any technical problems with the subsequent analysis.
18. When using subcutaneous tumor models as described in this chapter in some cases the tumor can be growing directly over the inguinal lymph node. Care must be taken to accurately locate the lymph node if tumor explantation is also required.
19. In some cases the explanted lymph node may float on top of the D-PBS due to still attached fat tissue. The fat tissue can be carefully cut off the lymph node using a clean scalpel. The lymph node will sink to the bottom of the tube when most fat is removed.
20. Single cell suspensions from lymph nodes can be prepared also without enzymatic digestion. In our experience, enzymatic digestion of the lymph nodes increases the cell yield in comparison to mechanical dissociation alone. In some cases, if other cell surface markers are analyzed, the enzymatic digestion may cleave the marker off the cell surface. To exclude enzymatic digestion from the protocol, follow the procedure as for

preparation of single cell suspension from a murine spleen (Subheading 3.3.2).

21. Single stain, fluorescence-minus-one and isotype controls are necessary for analysis of the acquired flow cytometry data. We routinely prepare all the required controls using splenocytes prepared according to the procedure described in Subheading 3.3.2. Use single stain controls to perform fluorescence compensation. Use fluorescence minus one and isotype controls to determine the gating boundaries. For in depth information about flow cytometry experiment design, performance and data analysis, refer to [19].

22. Markers CD62L and CD44 are used in the described staining panel for analysis of memory T cell populations. CD44 expression is associated with T cell activation and its expression is maintained on memory T cells [20]. A simplified functional distinction of the $CD44^+$ memory T cell pool is further possible by analyzing CD62L expression. Central memory (CM) T cells are defined as expressing high levels of CD62L which is associated with capacity for homing to lymphoid organs. Effector memory (EM) T cells are defined as CD62 negative population which have lost the capability to migrate to lymph nodes and are therefore mainly found in peripheral tissues [21].

Acknowledgments

The author acknowledges Dr. C. E. Engeland for support and consultation for development of the described protocols. This work was supported by the European Regional Development Fund project no. 1.1.1.2/VIAA/2/18/292 (1.1.1.2/16/I/001).

References

1. Galluzzi L, Vacchelli E, Bravo-San Pedro J-M, Buqué A, Senovilla L, Baracco EE et al (2014) Classification of current anticancer immunotherapies. Oncotarget 5(24):12472–12508
2. Guo C, Manjili M, Subjeck J, Sarkar D, Fisher P, Wang X (2013) Therapeutic cancer vaccines: past, present and future. Adv Cancer Res 119:421–475
3. Russell SJ, Barber GN (2018) Oncolytic viruses as antigen-agnostic cancer vaccines. Cancer Cell 33(4):599–605
4. Bell JC, McFadden G (2014) Viruses for tumor therapy John. Cell Host Microbe 15 (3):260–265
5. Escors D (2014) Tumour immunogenicity, antigen presentation, and immunological barriers in cancer immunotherapy. N J Sci 2014:734515
6. Sánchez-Paulete DAR, Teijeira A, Cueto FJ, Garasa S, Pérez-Gracia JL, Sánchez-Arráez A, Sancho D, Melero I (2017) Antigen cross-presentation and T-cell cross-priming in cancer immunology and immunotherapy. Ann Oncol 28(Suppl_12):44–55
7. Prestwich FRJ, Harrington KJ, Pandha HS, Vile RG, Melcher AA, Errington F (2008) Oncolytic viruses: a novel form of immunotherapy. Expert Rev Anticancer Ther 8 (10):1581–1588
8. Andtbacka RHI, Kaufman HL, Collichio F, Amatruda T, Senzer N, Chesney J et al (2015) Talimogene laherparepvec improves

durable response rate in patients with advanced melanoma. J Clin Oncol 33(25):2780–2788

9. Galanis E, Atherton PJ, Maurer MJM, Knutson KL, Dowdy SC, Cliby WA et al (2014) Oncolytic measles virus expressing the sodium iodide symporter to treat drug-resistant ovarian cancer. Cancer Res 75(1):22–30
10. Lichty BD, Breitbach CJ, Stojdl DF, Bell JC (2014) Going viral with cancer immunotherapy. Nat Rev Cancer 14(8):559–567
11. Sobol PT, Boudreau JE, Stephenson K, Wan Y, Lichty BD, Mossman KL (2009) Adaptive antiviral immunity is a determinant of the therapeutic success of oncolytic virotherapy. Mol Ther 19(2):335–344
12. Li X, Wang P, Li H, Du X, Liu M, Huang Q et al (2017) The efficacy of oncolytic adenovirus is mediated by T-cell responses against virus and tumor in Syrian hamster model. Clin Cancer Res 23(1):239–249
13. Aitken A, Roy D, Bourgeois-Daigneault M-C (2017) Taking a stab at cancer; oncolytic virus-mediated anti-cancer vaccination strategies. Biomedicine 5(1):3
14. Engeland CE, Grossardt C, Veinalde R, Bossow S, Lutz D, Kaufmann JK et al (2014) CTLA-4 and PD-L1 checkpoint blockade enhances oncolytic measles virus therapy. Mol Ther 22(11)
15. Grossardt C, Engeland CE, Bossow S, Halama N, Zaoui K, Leber MF et al (2013) Granulocyte-macrophage colony-stimulating factor-armed oncolytic measles virus is an effective therapeutic cancer vaccine. Hum Gene Ther 24(7):644–654
16. Speck T, Heidbuechel J, Veinalde R, Jaeger D, von Kalle C, Ball CR, Ungerechts G, Engeland CE (2018) Targeted BiTE expression by an oncolytic vector augments therapeutic efficacy against solid tumors. Clin Cancer Res 24 (9):2128–2138
17. Veinalde R, Grossardt C, Hartmann L, Bourgeois-Daigneault M-C, Bell JC, Jäger D et al (2017) Oncolytic measles virus encoding interleukin-12 mediates potent anti-tumor effects through T cell activation. OncoImmunology 6(4):e1285992
18. Robbins PF, Kantor JA, Salgaller M, Hand PH, Fernsten PD, Schlom J (1991) Transduction and expression of the human carcinoembryonic antigen gene in a murine colon carcinoma cell line. Cancer Res 51(14):3657–3662
19. Cossarizza A, Chang H, Radbruch A, Annunziato F, Bacher P, Barnaba V et al (2017) Guidelines for the use of flow cytometry and cell sorting in immunological studies. Eur J Immunol 47(10):1584–1797
20. Puré E, Cuff CA (2001) A crucial role for CD44 in inflammation. Trends Mol Med 7 (5):213–221
21. Sallusto MF, Lenig D, Förster R, Lipp M, Lanzavecchia A (1999) Two subsets of memory T lymphocytes with distinct homing potentials and effector functions. Nature 401 (6754):708–712
22. Veinalde R (2017) Unraveling determinants of efficacy in measles immunovirotherapy. Ruprecht-Karls-Universität Heidelberg, Heidelberg

Chapter 14

Delivery of Oncolytic Reovirus by Cell Carriers

Elizabeth J. Ilett

Abstract

Oncolytic virus therapy is a rapidly expanding branch of cancer immunotherapy and represents a genuine opportunity to improve currently available treatment options. However, as single agents oncolytic viruses have shown only moderate clinical benefit and many challenges remain before their full potential is realized. Central to this is the efficient delivery of the virus to the tumor site and potentiation of the antitumor immune response. This chapter describes the loading of oncolytic reovirus onto monocytes which act as carriers for delivery of the virus to the tumor site and, as antigen presenting cells, may also thereby potentiate the development of an adaptive antitumor immune response.

Key words Cell carriage, Immune cells, Reovirus, Cancer immunotherapy, Antibodies

1 Introduction

Oncolytic viruses (OVs) are those that preferentially infect and lyse tumor cells while sparing normal tissues. They were first developed as direct cytotoxic agents [1–3] but it is now known that their antitumor activity also depends to a large extent on activation of the immune system [4, 5]. The best route of delivery for OVs is not yet known; the route designed to maximize oncolysis may not best facilitate immune-mediated tumor clearance. As of 2019, the only FDA-approved OV (T-VEC) is administered by direct injection into the tumor (melanoma). This route ensures delivery of the full viral dose to the tumor tissue but the degree of antitumor immunity generated may be submaximal; observed responses to T-VEC were 15% in visceral metastases versus 64% in directly injected lesions [6]. Perhaps more important, the use of intratumoral administration restricts the use of OV to those tumors that are readily accessible and many are not. Systemic delivery is safe, broadly applicable in a clinical setting, and more suitable for targeting visceral or widespread metastatic disease. However, virus delivered into the bloodstream faces inactivation by several mechanisms including scavenger cells in the blood or reticuloendothelial organs, and immune

Christine E. Engeland (ed.), *Oncolytic Viruses*, Methods in Molecular Biology, vol. 2058,
https://doi.org/10.1007/978-1-4939-9794-7_14, © Springer Science+Business Media, LLC, part of Springer Nature 2020

molecules such as complement and virus-specific antibodies. Thus the virus faces rapid neutralization before it can ever reach the tumor. One method that has been investigated to protect the OV within the bloodstream is to load the virus onto or into carrier cells. Various cell types have demonstrated efficacy as carriers for OV delivery to tumors including, transformed cells, endothelial cells and immune cells [7–10]. The benefit of using immune cells is the potential for a dual attack on the tumor. For example, using tumor-specific T cells for direct delivery of an oncolytic agent theoretically ensures specific delivery of the virus to the tumor target and, in addition, benefits from the cytotoxic potential of the carrier cell [11]. Although expanding endogenous tumor-specific T cells from patients is still problematic, chimeric antigen receptor (CAR) T cells represent an attractive possibility as viral carriers [12]. Alternatively, using antigen presenting cells such as dendritic cells (DC) or monocytes as the viral carriers may enhance the development of an adaptive antitumor immune response [10, 13].

More recently, it has been shown that virus-specific neutralizing antibodies are not always a barrier to systemic OV delivery and tumor therapy. Reovirus, a nonenveloped, double stranded RNA virus, is taken up more efficiently by monocytes in the presence of antiviral antibodies [14]; this is dependent on antibody receptors expressed by these cells. Furthermore, while antibody-neutralized reovirus or Coxsackievirus are unable to infect and kill tumor cells directly, when loaded onto human monocytes the neutralized viruses can be delivered to tumor cells in a functional/replicative form resulting in cell lysis [15]. The protocol for loading monocytes with reovirus or with reovirus-antibody complexes is described below. The same protocol is effective for the loading of other immune cell subtypes following their isolation by magnetic-activated cell sorting (MACS); T cells, DC and natural killer (NK) cells have all been loaded in this manner. After viral loading, the carrier cells can be cocultured with tumor targets for 72–96 h to elicit tumor cell death.

2 Materials

2.1 Preparation of Human Cells

1. Human buffy coat/leukapheresis cone/fresh blood.
2. PBS.
3. Lymphoprep density gradient medium.
4. Pasteur pipettes.
5. Human anti-CD14 MicroBeads or the Pan Monocyte isolation kit (*see* **Note 1**).
6. MACS Manual Separator.

7. Magnetic cell separation LS columns.
8. MACS buffer: PBS containing 1% (v/v) FBS and 2 mM EDTA.

2.2 Preparation of Murine Cells

1. Dissection equipment: scissors, tweezers, scalpel.
2. Sterile 70% ethanol.
3. Complete RPMI: supplement RPMI with 10% (v/v) FBS, 2 mM L-glutamine, 0.05 mM β-mercaptoethanol, penicillin 100 U/ml, streptomycin 100 μg/ml.
4. Mouse monocyte isolation kit (bone marrow).
5. MACS Manual Separator.
6. Magnetic cell separation LS columns.
7. MACS buffer: PBS containing 1% (v/v) FBS and 2 mM EDTA.

2.3 Loading Immune Cells with Reovirus

1. Complete medium as required for target cells.
2. Complete RPMI.
3. Reovirus.
4. Antireoviral antibodies (optional, for generating antibody-neutralized reovirus complexes).

3 Methods

3.1 Preparation of Cells

Carry out all steps at room temperature.

3.1.1 Preparation of Human PBMC by Density Gradient Separation

1. Dilute blood sample. Fresh blood and buffy coats should be diluted 1:1 with PBS, leukapheresis cones should be diluted with PBS to a final volume of 50 ml.
2. Layer blood onto lymphoprep at a ratio of 2:1, blood to lymphoprep. Add 15 ml lymphoprep per 50 ml Falcon tube and carefully layer 25 ml diluted blood on top.
3. Centrifuge at 800 × *g* for 25 min with minimum acceleration and brake.
4. Remove white buffy layer gently with wide tipped Pasteur pipette. Combine 2 buffy layers into 1 and add PBS to 50 ml.
5. Centrifuge at 400 × *g* for 10 min.
6. Discard supernatant and flick tube to loosen pellet. Combine cell pellets, add PBS to 50 ml and centrifuge at 300 × *g* for 5 min.
7. Discard supernatant, resuspend cell pellet in 50 ml PBS and make a 1 in 10 dilution for cell counting. Count PMBC.
8. Place the required number of PBMC into a fresh 50 ml Falcon tube, make volume up to 50 ml and centrifuge at 300 × *g* for 5 min (*see* **Note 2**).
9. Discard supernatant. Cells are now ready for selection (*see* **Note 3**).

3.1.2 Preparation of Murine Bone Marrow Cells

1. Dissect out mouse hind legs removing as much muscle as possible and place in medium, on ice.
2. Put bones in 70% ethanol for 1 min.
3. Put bones into fresh complete RPMI in 10 cm dish and separate and strip tibias and femurs, then remove both ends of the bones with a scalpel.
4. Transfer into fresh complete RPMI in a new dish and using a 21 G needle and 10 ml syringe. Pull up a few ml of medium and flush the marrow out each of bone into the dish.
5. Transfer cells and medium to a 50 ml Falcon tube and pipette up and down repeatedly to dissociate the cells. Pass through a 40 μm cell strainer into a fresh tube.
6. Add PBS to a final volume of 50 ml, take an aliquot of cells for counting and centrifuge the remainder at 400 × *g* for 5 min, minimum brake (*see* **Note 4**).
7. Discard supernatant. Cells are now ready for selection.

3.2 Selection of Immune Cells (Monocytes)

3.2.1 Human Monocytes

1. Resuspend the PBMC in the required volume of cold MACS buffer according to the manufacturer's instructions and add the appropriate volume of CD14 MicroBeads (*see* **Note 5**).
2. Incubate at 4 °C for 15 min.
3. Add 10 ml cold MACS buffer and centrifuge cells at 300 × *g* for 5 min. Discard the supernatant, resuspend the cells in an appropriate volume of MACS buffer according to the manufacturer's instructions and add to a prepared LS column (*see* **Note 6**).
4. Apply cell suspension and allow to run through.
5. Rinse through three times with 3 ml MACS buffer.
6. Add 5 ml of MACS buffer to the column, immediately remove from the magnet and flush positive cells out into a fresh 30 ml Universal tube or 15 ml Falcon tube using the plunger.
7. Count CD14 cells. Verification of cell purity should be carried out by FACS analysis. The cells are now ready for loading with reovirus (*see* **Note 7**).

3.2.2 Murine Monocytes

1. Resuspend cells in 175 μl MACS buffer per 5 × 10^7 cells.
2. Add 25 μl FcR blocking reagent and mix well.
3. Add 50 μl antibody cocktail per 5 × 10^7 cells, mix well and incubate at 4 °C for 5 min.
4. Add 10 ml MACS buffer per 5 × 10^7cells and centrifuge at 300 × *g* for 10 min.

5. Resuspend cells in 400 μl MACS buffer per 5×10^7 cells, add 100 μl microbeads per 5×10^7 cells and incubate at 4 °C for 10 min.
6. Apply this solution directly to a prepared LS column and allow to run through.
7. Rinse through with 3×3 ml MACS buffer and collect eluate containing the untouched monocytes.
8. Count the monocytes. Verification of cell purity should be carried out by FACS analysis. The cells are now ready for loading with reovirus (*see* **Note 8**).

3.3 Loading Immune Cells with Reovirus

1. Place the required number of immune cells into a 15 ml Falcon tube (*see* **Note 9**).
2. Add 10 ml of PBS to the tube, centrifuge cells at $300 \times g$ for 10 min and discard supernatant.
3. Add reovirus at the required multiplicity of infection (MOI) to the cells in the residual volume of PBS (*see* **Note 10**).
4. Incubate for 3–4 h at 4 °C.
5. Add 10 ml PBS to the tube, centrifuge at $300 \times g$ for 10 min and discard the supernatant.
6. Repeat the PBS wash twice more to ensure complete removal of unbound reovirus.
7. Gently resuspend the cell pellet in the required volume of complete RPMI and add the loaded carrier cells to the tumor targets (*see* **Note 11**).
8. Monitor tumor cell death.

3.4 Preparation and Loading of Reovirus-Antibody Complexes

Reovirus-neutralizing antibody complexes (reoNAb) have been generated using (1) whole serum from clinical trial patients receiving Reolysin® therapy or (2) reovirus-specific mouse monoclonal antibodies [15].

1. To form the reoNAb complexes, incubate the reovirus with a neutralizing volume of the serum/antibody in a 500 μl eppendorf and incubate at 37 °C for 2–3 h (*see* **Note 12**).
2. After incubation, store the reoNAb at 4 °C (*see* **Note 13**).
3. ReoNAb loading onto immune cells is performed in exactly the same manner as described above for reovirus.

4 Notes

1. If the loaded monocytes are to be processed for electron microscopy the use of negative selection to isolate untouched monocytes is preferable as this avoids the problem of bead

debris in the preparation which can produce artifacts in the microscopy images.

2. Estimate the number of PBMC required from the expected frequency of the cell subset found within the whole PBMC population; for example, monocytes make up approx. 10–20% of human PBMC, T cells up to 70% and NK cells 5–10%. DC are rare (1–2%) but myeloid-derived DC can be obtained by culturing isolated monocytes in medium containing 800 U/ml human recombinant GM-CSF and 500 U/ml human recombinant IL-4 for 5–6 days. Anticipate up to 30% cell loss to occur during the culture period. The cells will have an immature DC phenotype and can be used for viral loading or matured with a suitable stimulus prior to use. We routinely loaded 2×10^6 carrier cells with reovirus for each condition required.
3. Cell separation is best carried out using fresh samples rather than frozen. However, buffy coats and leukapheresis cones can be stored at room temperature overnight prior to PBMC preparation.
4. Typical count of total BM cells = $4–5 \times 10^7$ cells per mouse.
5. For positive selection of monocytes using CD14 MicroBeads, it is possible to obtain efficient isolation using only 5 μl of the beads per 10^7 cells rather than the recommended 20 μl. We have only verified this for Miltenyi CD14 MicroBeads.
6. 2 LS columns are required per whole buffy coat/ leukapheresis cone.
7. If required, isolated cells can be cultured overnight in RPMI/ 10% FBS/2 mM L-glutamine prior to loading with virus.
8. If DC are required the monocytes can be cultured in RPMI complete medium (no Pen/Strep) containing 20% L929 conditioned medium as a source of GM-CSF. The cells should be viewed daily and fresh medium added as required. At day 5–6 the cells will have an immature DC phenotype. They can then be used for viral loading or matured with a suitable stimulus prior to use.
9. 15 ml polypropylene Falcon tubes give more efficient retrieval of cells following viral loading than 30 ml Universal polystyrene tubes. Fewer than 2×10^6 cells per tube is not recommended as the cell pellet is easily lost during the post-loading wash steps.
10. MOI ranging from 1–10 with respect to the target cells have been used successfully [10, 13, 15]. Coxsackievirus A21 has also been loaded efficiently using this protocol [15].
11. Tumor target cells should be seeded into multiwell plates 24 h prior to the addition of the carrier cells. Routinely 3×10^5 melanoma cells were seeded into wells in 6-well plates.

Immediately prior to adding the loaded carrier cells, the medium in the wells should be removed and replaced with 2 ml fresh complete medium. Virus-loaded carrier cells were then added at a 4:1, 2:1, or 1:1 ratio, carrier cells to targets, in a volume of 500–700 μl.

12. The amount of serum/antibody required should be determined by a standard neutralization assay on L929 cells and confirmed by the absence of cell death in target cells treated directly with reoNAb. In general, a reovirus to serum ratio of 1:10 was used for the formation of reoNAb complexes; this was in excess of the volume that was found to be completely neutralizing for all sera tested.
13. Preformed reoNAb complexes can be stored at 4 °C for up to 12 h prior to use.

References

1. Stojdl DF, Lichty B, Knowles S, Marius R, Atkins H, Sonenberg N, Bell JC (2000) Exploiting tumor-specific defects in the interferon pathway with a previously unknown oncolytic virus. Nat Med 6(7):821–825
2. Brun J, McManus D, Lefebvre C, Hu K, Falls T, Atkins H, Bell JC, McCart JA, Mahoney D, Stojdl DF (2010) Identification of genetically modified Maraba virus as an oncolytic rhabdovirus. Mol Ther 18 (8):1440–1449
3. Martuza RL, Malick A, Markert JM, Ruffner KL, Coen DM (1991) Experimental therapy of human glioma by means of a genetically engineered virus mutant. Science 252 (5007):854–856
4. Kaufman HL, Kohlhapp FJ, Zloza A (2015) Oncolytic viruses: a new class of immunotherapy drugs. Nat Rev Drug Discov 14 (9):642–662
5. Prestwich RJ, Errington F, Ilett EJ, Morgan RS, Scott KJ, Kottke T, Thompson J, Morrison EE, Harrington KJ, Pandha HS, Selby PJ, Vile RG, Melcher AA (2008) Tumor infection by oncolytic reovirus primes adaptive antitumor immunity. Clin Cancer Res 14(22):7358–7366
6. Andtbacka RH, Kaufman HL, Collichio F, Amatruda T, Senzer N, Chesney J, Delman KA, Spitler LE, Puzanov I, Agarwala SS, Milhem M, Cranmer L, Curti B, Lewis K, Ross M, Guthrie T, Linette GP, Daniels GA, Harrington K, Middleton MR, Miller WH Jr, Zager JS, Ye Y, Yao B, Li A, Doleman S, VanderWalde A, Gansert J, Coffin RS (2015) Talimogene laherparepvec improves durable response rate in patients with advanced melanoma. J Clin Oncol 33(25):2780–2788
7. Power AT, Wang J, Falls TJ, Paterson JM, Parato KA, Lichty BD, Stojdl DF, Forsyth PA, Atkins H, Bell JC (2007) Carrier cell-based delivery of an oncolytic virus circumvents antiviral immunity. Mol Ther 15(1):123–130
8. Iankov ID, Blechacz B, Liu C, Schmeckpeper JD, Tarara JE, Federspiel MJ, Caplice N, Russell SJ (2007) Infected cell carriers: a new strategy for systemic delivery of oncolytic measles viruses in cancer virotherapy. Mol Ther 15 (1):114–122
9. Cole C, Qiao J, Kottke T, Diaz RM, Ahmed A, Sanchez-Perez L, Brunn G, Thompson J, Chester J, Vile RG (2005) Tumor-targeted, systemic delivery of therapeutic viral vectors using hitchhiking on antigen-specific T cells. Nat Med 11(10):1073–1081
10. Ilett EJ, Prestwich RJ, Kottke T, Errington F, Thompson JM, Harrington KJ, Pandha HS, Coffey M, Selby PJ, Vile RG, Melcher AA (2009) Dendritic cells and T cells deliver oncolytic reovirus for tumour killing despite pre-existing anti-viral immunity. Gene Ther 16(5):689–699
11. Qiao J, Wang H, Kottke T, Diaz RM, Willmon C, Hudacek A, Thompson J, Parato K, Bell J, Naik J, Chester J, Selby P, Harrington K, Melcher A, Vile RG (2008) Loading of oncolytic vesicular stomatitis virus onto antigen-specific T cells enhances the efficacy of adoptive T-cell therapy of tumors. Gene Ther 15(8):604–616
12. VanSeggelen H, Tantalo DG, Afsahi A, Hammill JA, Bramson JL (2015) Chimeric antigen receptor-engineered T cells as oncolytic virus carriers. Mol Ther Oncolytics 2:15014

13. Ilett EJ, Barcena M, Errington-Mais F, Griffin S, Harrington KJ, Pandha HS, Coffey M, Selby PJ, Limpens RW, Mommaas M, Hoeben RC, Vile RG, Melcher AA (2011) Internalization of oncolytic reovirus by human dendritic cell carriers protects the virus from neutralization. Clin Cancer Res 17 (9):2767–2776
14. Ilett E, Kottke T, Donnelly O, Thompson J, Willmon C, Diaz R, Zaidi S, Coffey M, Selby P, Harrington K, Pandha H, Melcher A, Vile R (2014) Cytokine conditioning enhances systemic delivery and therapy of an oncolytic virus. Mol Ther 22(10):1851–1863
15. Berkeley RA, Steele LP, Mulder AA, van den Wollenberg DJM, Kottke TJ, Thompson J, Coffey M, Hoeben RC, Vile RG, Melcher A, Ilett EJ (2018) Antibody-neutralized reovirus is effective in oncolytic virotherapy. Cancer Immunol Res 6(10):1161–1173

Chapter 15

In Vivo Bioimaging for Monitoring Intratumoral Virus Activity

Liesa-Marie Schreiber, Carles Urbiola, Patrik Erlmann, and Guido Wollmann

Abstract

In vivo studies are the mainstay of translational immune-oncology and virotherapy research. In general oncology, bioluminescence imaging provides a convenient and reliable tool to visualize disseminated tumors and monitor growth kinetics or treatment effects. Unique aspects of this method in the field of oncolytic viruses are tracing the process of tumor-specific targeting, assessing potential off-target replication, and visualizing intratumoral spread. In addition, the longitudinal monitoring of virus activity kinetics over time is a very powerful feature supporting the subsequent, often elaborate, preclinical biodistribution and pharmtox program. Here we present a step-by-step standard imaging protocol used in our group for both tumor and virus monitoring, along with background information and general principles that should allow the reader to modify and adapt the protocol according to their needs.

Key words In vivo imaging, (Firefly) luciferase, Luciferin, Bioluminescence, IVIS® Lumina, Virus, VSV-GP

1 Introduction

Bioimaging is a powerful and noninvasive tool for live imaging of small laboratory animals, enabling the study of real-time biological processes in vivo. It is a widely applied approach for live imaging and can be implemented more easily than for instance imaging techniques based on radioactive tracers used in PET or SPECT. The most frequently used reporters include proteins that catalyze enzymatic activation of light-generating substrates such as luciferases and fluorescent reporters such as GFP, that produce detectable signals intrinsically upon excitation [1]. As fluorescent reporters inherently display a high background-to-signal ratio and lack of deep tissue penetration, luciferases are generally the reporters of choice. Current standard luciferase enzymes originate from different organisms such as firefly (*Photinus pyralis*), Gaussia (*Gaussia princeps*), or Renilla (*Renilla reniformis*). Firefly luciferase

Christine E. Engeland (ed.), *Oncolytic Viruses*, Methods in Molecular Biology, vol. 2058,
https://doi.org/10.1007/978-1-4939-9794-7_15,

requires luciferin substrate for conversion compared to Gaussia and Renilla luciferases, which use the different substrate coelenterazine. Bioluminescent imaging (BLI) has been widely applied in studies monitoring transgene expression, tumor growth and metastasis, as well as viral infections and spread [1]. The ability to monitor the same subjects over time provides insight into the kinetics of biological processes with a reasonable macroscopic spatial resolution without the need to sacrifice study animals.

For the experimental setup described in this chapter we use an oncolytic virus VSV-GP encoding firefly luciferase (VSV-GP-Luc) to visualize viral targeting of tumor tissue, intratumoral propagation and potential off-target spread. Alternatively, the same protocol has been applied in our lab for tracing and monitoring of luciferase expressing tumor cells in orthotopic, oligometastatic, or disseminated settings [2, 3]. Dual live imaging of two different reporters is feasible with most current setups, requiring the use of luciferase variants with distinct emission spectra, for example from *Photinus pyralis* and *Luciola italica* [4]. In the context of oncolytic virotherapy, such dual systems can be employed for virus tracing to disseminated tumor lesions.

Caution has to be raised when implementing luciferase-based tumor studies—be it using reporter-expressing cells or viruses—in the context of immune-oncology. Luciferase is xenogenic to mice and as such triggers an immune response in immune-competent hosts [5]. Generally, any confounding effect might depend on the tumor model. Some studies reported no significant alteration of tumor growth kinetics and tumor microenvironment by the expression of luciferase [6]. Others observed reduction in tumor growth and metastasis formation [7]. Hence we advise to validate newly established luciferase-expressing tumor models.

2 Materials

1. IVIS® Lumina II (PerkinElmer).
2. XGI-8 Gas Anesthesia System (Caliper Life Sciences).
3. Dulbecco's Phosphate Buffered Saline (DPBS).
4. 50 mg/mL D-luciferin: Dilute (VivoGlo Luciferin in vivo grade (Promega) to 50 mg/mL in DPBS and pass through a 0.45 μm sterile filter. Prepare aliquots for one (30 μL) or up to five animals (150 μL) and store at −20 °C. Protect from light and avoid freeze-thaw cycles.
5. Isoflurane.
6. 0.5 mL 30G insulin syringe.
7. Living Image software (version 4.3.1.)
8. Optional: nonreflective transparent adhesive tape.

3 Methods

3.1 Animal and Substrate Preparation for In Vivo Luciferase Imaging

The protocol presented below has been applied for in vivo bioimaging of oncolytic virus activity in various established subcutaneous tumors in C57BL/6JRj, B6(C)/Rj-Tyr$^{c/c}$ Albino, BalbC, C3H/HeOuJ, or athymic NMRI Nude mice. Exemplary data shown for illustration purposes were generated using a highly susceptible modified mouse lung cancer tumor syngeneic for C57BL/6 hosts. Mice were treated via intratumoral (i.t.) injection of 30 μL solution containing 1×10^8 $TCID_{50}$ VSV-GP expressing Firefly luciferase.

1. Shave area of interest. Fur and skin pigmentation present a significant factor of signal absorption (*see* **Note 1** and Fig. 5).
2. Prepare D-luciferin substrate to be injected, depending on number of mice to be measured. Up to five animals can be injected per round of imaging. Mix 170 μL or 850 μL DPBS with 30 μL or 150 μL D-luciferin 50 mg/mL aliquots, respectively (equals 1.5 mg D-luciferin/mouse) (*see* **Note 2**).
3. Draw 200 μL in a 0.5 mL syringe.
4. Apply one-handed mouse restraint so the abdomen is exposed.
5. Inject 200 μL D-luciferin solution intraperitoneally into the lower (caudal) right or left quadrant of the ventral abdomen (angle of 45–60°) (*see* **Note 3**).
6. Return the animal to the cage and allow the substrate to distribute in the awake animal for about 10–15 min (*see* **Note 4**).
7. In the meantime, start the Living Image software (version 4.3.1. discussed herein). After selecting your user ID and password, the system needs to be initialized. The "IVIS Acquisition Control Panel" will open automatically. To start initialization, click on the "Initialize" button in the right corner of the control panel (a in Fig. 1). It will take 5–10 min for the checkup of the system, for the camera to cool down to −90 °C and the sample stage to heat up to 37 °C. The current temperature can be checked by clicking on the red bar at the bottom of the acquisition control panel. These setups are set as default but can be changed manually; for example, the sample stage can be set to temperatures between 20 and 40 °C. Once the system is initialized and the correct temperatures have been reached, the red bar will turn green (b in Fig. 1) and the system is ready for image acquisition.
8. To anesthetize mice undergoing imaging, a gas anesthesia system (discussed herein: XGI-8) consisting of a gas anesthesia module and an induction chamber, is utilized. Turn on the "Evacuation Pump" switch located in the upper left part of

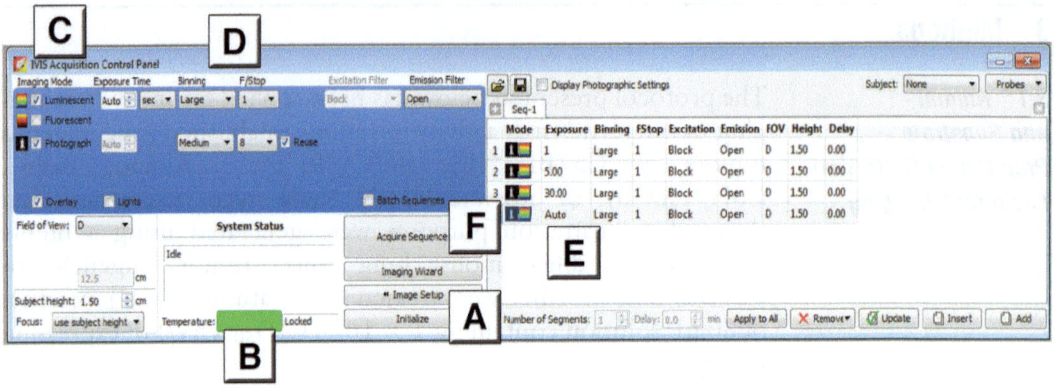

Fig. 1 Setting up Living Image software using the acquisition control panel. Prior imaging, the system needs to initialize (**a**) for 5–10 min. Once the system is ready for image acquisition, the red bar will turn green (**b**) and experimental/user-dependent modules, for example, choosing the appropriate imaging mode (**c**), camera sensitivity (**d**), and exposure times (**e**) can be enabled

the anesthesia module. The flow should be greater than 8 L/min. Enable oxygen flow from the attached oxygen bottle and turn the green "Oxygen" handle at the lower left corner to "ON." Turn on the "IVIS Manifold" and "Induction Chamber" located in the upper right part of the module using the toggle switches and adjust the flow rate to 1.5–2 L/min and 3–4 L/min, respectively. Set the main dial of the isoflurane vaporizer to 2–3. Place up to five mice into the anesthesia chamber (*see* **Note 5**).

9. After mice are fully anesthetized (2–3 min), transfer one mouse from the anesthesia chamber to the nose cone attached to the manifold in the imaging chamber to ensure continuous anesthesia during imaging. Place the mouse limbs-down-back-up close to the center of the field of view (*see* **Note 6**). Alternatively, up to five mice can be imaged simultaneously next to each other (*see* **Note 7**). Start image acquisition as described in Subheading 3.2).
10. After image generation, the animal is returned to the cage where it recovers quickly.
11. Transfer the next animal from the anesthesia chamber to the nose cone in the imaging chamber.

3.2 Image Acquisition Using Living Image Software

1. After the system has been initialized, in the control panel select "Luminescent" as imaging mode with the option for a "Photograph" overlay (c in Fig. 1). Imaging mode "Fluorescent" can be chosen for studies with respective reporter types.
2. The parameter "Binning" controls the pixel size on the CCD camera. By increasing the binning, the pixel size and the sensitivity are also increased. This improves the signal to noise ratio

for luminescent images but reduces spatial resolution (d in Fig. 1). Using of the following binning parameters depending on the specimen is recommended: 1–4 (small) for cells/tissue sections, 4–8 (medium) for in vivo imaging of specimens and 8–16 (large) for in vivo imaging of specimens with very dim light sources. We usually set it to "Large" for luminescent images and to "Medium" for the photographic image.

3. The parameter "F/Stop" (d in Fig. 1) controls the lens aperture size adjusting the amount of light detected by the CCD camera. A smaller F/Stop means larger aperture and thereby increased sensitivity while larger F/stop means smaller aperture resulting in less sensitivity but larger depth of field. Typically "F/1" is used for high sensitivity luminescence images and "F/8" or "F/16" for photographic images to gain a sharp image.
4. The parameter "Exposure Time" (e in Fig. 1) can be manually set depending on the signal intensity based on the reporter construct, the anatomical location of the emitted signal, the mouse strain, and the general experimental setup. Weak reporters, tumors hidden within the body or subcutaneous tumors covered in thick, dark skin may require longer exposure times (*see* **Notes 1** and **8**). In general, the exposure time for fluorescent images is limited to 60 s while luminescent images can theoretically be generated without a limit on maximum exposure time. For the detection of our virus-encoded luciferase signals, we routinely choose an exposure time set covering 1, 5, and 30 s in addition to auto exposure. Add the respective exposure times to a new tab and the imaging sequence will appear in a window adjacent to the control panel (e in Fig. 1).
5. To automatically save the generated images, use the drop-down menu tab "Acquisition" and enable "Automatically save to" using your storage location of choice.
6. Once all parameters are set and the mouse is positioned correctly, click the "Acquire" button (f in Fig. 1).
7. After the image acquisition has started, the "Edit Image Labels" tab will appear where information about the experiment such as time points, animal number, strain or sex, treatment, or reporter construct can be added. The generated images (Fig. 2) will appear in a new window. If several exposure times have been chosen, the individual images will be displayed as a group. If the "automatically save to" option has been enabled, there is no need to separately save the images.

3.3 Image Data Analysis

1. After the images haven been successfully acquired a "Tool Palette" window appears (Fig. 3). It is also possible to reload already generated images for analysis (*see* **Note 9**).
2. Using the tab "Image Adjust" images can be navigated and adjustments for color settings and scales are provided (a in

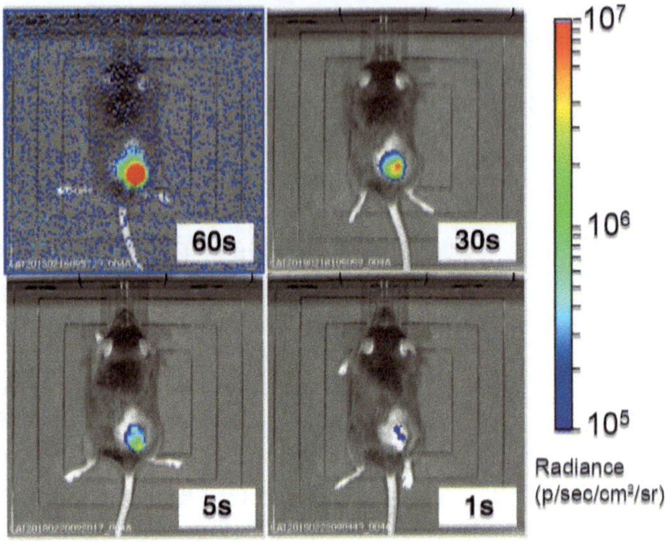

Fig. 2 Exemplary depiction of a life imaging recording detecting intratumoral virally encoded luciferase signals. All four images show the same mouse and same time point of imaging but with different exposure times ranging from 60 to 30 to 5 to 1 s. Color scale indicates quantity of the radiance and covers two logs

Fig. 3). If the luciferase signal intensity is to be compared over a defined timespan or between animals from the same treatment group, it is advisable to apply the identical color scale for all images (Fig. 4). The range of the color scale (Min to Max) can be adjusted individually, simply enable the "Manual" option in the tab "Color Scale Limits" (b in Fig. 3) and define the range (c in Fig. 3). Also consider logarithmic scales to cover a broader signal range (d in Fig. 3).

3. For quantification of bioluminescent signals, regions of interest (ROIs) can be manually defined and photons radiating from that area are measured by the software using the tab "ROI Tools." Use one of the predefined shapes or draw a region of interest manually (e in Fig. 3). By clicking on "Measure ROIs" the measurement is performed (f in Fig. 3); a table including the average intensity (photons/s/cm^2/sr) and total photon flux (photons/s) is displayed and can be saved as a text file (*see* **Note 10**).

4 Notes

1. In our studies it has been observed that bioluminescent signals originating from s.c. tumors on C57BL/6 mice have a ~1–1.5 log reduction in intensity compared to similar tumors implanted onto Albino B6 or NMRI Nude mice (Fig. 5).

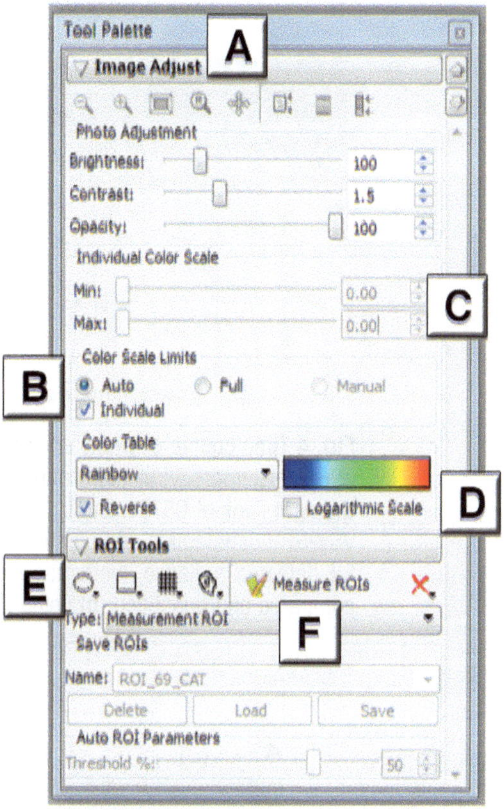

Fig. 3 Image data analysis using the Living Image tool palette. The "Tool Palette" window appears after image acquisition. This tool controls color adjustment (**a**) and scales (**b–d**) as well as the definition of regions of interest (ROIs) (**e**) for quantification (**f**) of bioluminescent signals

Tumors were imaged with intact, but shaved skin and after removal of the skin. Removal of skin significantly increases the signal indicating that the skin condition is a major factor of signal absorption. This trend was not observed in Albino B6 or NMRI Nude mice and the signal intensities were comparable in the presence of skin. We therefore recommend using otherwise genetically identical Albino B6 instead of C57BL/6 mice for live imaging studies. In either model, hair removal in the region of interest by shaving and/or chemical depilation prior to imaging is strongly advised. Variations in melanin pigmentation of the skin (dark patches) due to hair cycle progression may further reduce the signal intensity [8].

2. The amount of 1.5 mg D-luciferin per mouse used in this setup is at the lower end of the range (1.5 mg). The manufacturer (Promega) recommends using 150 mg/kg, which equals 3 mg D-luciferin for an average mouse of 20 g; 2–3 mg per mouse are commonly reported in the literature [9, 10]. It might be

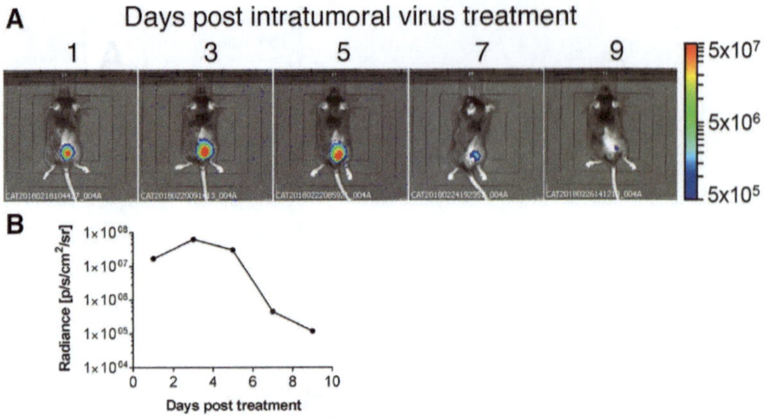

Fig. 4 Time course of virally encoded luciferase signal in subcutaneous tumors after intratumoral virus application. Mouse lung tumor cells were implanted into the right flank of C57BL/6 mice. Bioimaging was performed starting 1 day post-VSV-GP-Luc infection and continued for 9 days (**a**). Due to identical color scale for detected radiance, luciferase signal at various time points can be directly compared and quantified (**b**) using the "Region of Interest" tool

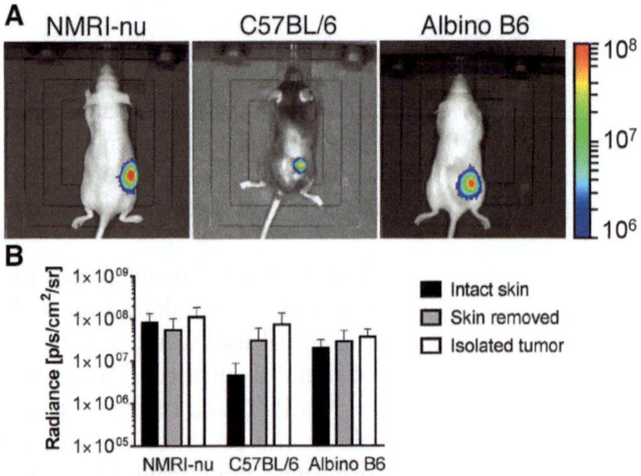

Fig. 5 Effect of skin composition on luciferase signals in subcutaneous tumors. Lung tumor cells were implanted in NMRI-nu, C57BL/6 and Albino B6 mice and imaged 24 h post-luciferase-virus treatment (**a**). Luminescence was assessed within a 15 min sequence: (1) with intact, shaved skin, (2) after removal of the skin, and (3) after removal of the tumor (**b**). Bioluminescent signal originating from subcutaneous tumors on C57BL/6 mice showed a ~1–1.5 log reduction in intensity compared to nude and albino mice. Signal alignment after removal of skin suggests the skin composition to be a significant factor in signal absorption

advisable to test different amounts in your specific approach also taking the route of substrate delivery into account.

3. The standard route for administration of luciferin is intraperitoneal (i.p.). However it is also possible to deliver luciferin via the intravenous (i.v.), retro orbital, and subcutaneous (s.c.) route. Depending on the location of photon emission (e.g., anatomical location of luciferase-expressing tumor cells or the organ to be visualized), one administration route may be more suitable than another and has to be determined for the specific experimental setup. Substrate delivery via the i.v. or retro orbital route usually allows imaging right after application as the substrate directly enters the bloodstream. I.v. substrate application offers high sensitivity and better reproducibility compared to i.p. injection [11]. In several studies it was also shown that i.p. injection led to the overestimation of signals from intraperitoneal tissues while s.c. injection was shown to be a convenient alternative for visualizing signals in and outside the peritoneal cavity [12, 13]. Moreover biodistribution studies revealed that early substrate uptake into liver, lung, myocardium and muscle were lower with i.p. than with i.v. administration [14]. There here is also evidence that bioluminescence detection of pulmonary luciferase expression is greatly enhanced by intranasal (i.n.) luciferin administration compared with i.p. [15].
4. In case of i.p. application, the substrate has to be absorbed from the peritoneal cavity; therefore the image acquisition starts 10–15 min after injection. The time-to-peak signal intensity is slower compared to i.v. application but there is a longer and more stable release of photons (plateau phase). If a different application route is chosen, be aware that these dynamics may change. For example, i.v. substrate delivery shows shorter time-to-peak intervals but more rapid drop of light emission; the rapid peak of light emission is explained by the instant delivery of luciferin to the target site via the bloodstream [11]. As the kinetics of substrate tissue biodistribution may vary for different animal models/administration routes, it is advisable to perform a kinetic study of luciferase detection to determine the time-to-peak and length of plateau phase.
5. Isoflurane is often used as the anesthetic of choice for bioimaging due to its user friendly, easy application, fast induction and recovery of the animal. However it has been shown that volatile anesthetics exhibit strong inhibitory effect on the luciferase activity in vivo [10, 16–18]. Photon emission after isoflurane application is reduced by 60% compared to unanesthetized animals with decreased sensitivity and significantly longer time-to-peak values. If applicable, one should consider the

use of alternative anesthetics (e.g., ketamine/xylazine or tribromoethanol).

6. IVIS Lumina II is equipped with an adjustable sample stage on which the mice are placed for imaging. Changing the height of the platform enables close-up images from specific body regions (e.g., the head if intracranial tumors are to be investigated). If the ventral side of the animal is to be imaged, spreading and fixation of the front limbs with nonreflective transparent adhesive tape might be required.
7. As the IVIS Lumina II comprises five nose cones within the imaging chamber, it is possible to image up to five mice at the same time. Luciferase signal intensities can vary strongly between mice, even within treatment groups. If applicable, also use light baffle dividers between mice to keep adjacent bioluminescent/fluorescent signals from interfering. Generally, if less than five mice are imaged, use the supplied black rubber stoppers for ports not in use to prevent isoflurane leakage into the imaging chamber. In case of intravenous application of luciferase encoding cells or viruses, a strong radiance signal may be observed along the tail of mice. In this scenario, it is helpful to cover the tail, for example, with a dark-colored cardboard to prevent off-target signal distortion.
8. Especially for novel models or reporter constructs it is advisable to acquire images after several exposure times. It is also helpful to include one image acquisition where the ideal exposure time is chosen by the Living Image software itself (auto mode). This way the risk of generating images with suboptimal exposure times is reduced. Although the software issues a warning for overexposed images, these should not be used for quantification of luciferase signals!
9. If images are analyzed after completion of the experiment or if mice from different time points/treatment groups should be compared, it is useful to load them as group and thereby apply adjustments to all animals simultaneously. To do so, use the "Browse" tab within the Living Image browser. Load the first image into the browser, enable "Add to list" to load more images into the sequence followed by clicking on "Load as group." The images can also be saved as a sequence which facilitates downstream processing of the images.
10. Although background signals in luminescent imaging are inherently very low, it is advisable to measure the average signal intensity in a user-specified area of the image that is considered background and subtract it from the ROI radiance.

Acknowledgments

We thank Dr. Janine Kimpel (Medical University Innsbruck) for providing VSV-GP-Luc master stock. This study was supported by a grant from the Christian Doppler Research Association.

References

1. Close DM, Xu T, Sayler GS, Ripp S (2011) In vivo bioluminescent imaging (BLI): noninvasive visualization and interrogation of biological processes in living animals. Sensors (Basel) 11(1):180–206. https://doi.org/10.3390/s110100180
2. Urbiola C, Santer FR, Petersson M, van der Pluijm G, Horninger W, Erlmann P, Wollmann G, Kimpel J, Culig Z, von Laer D (2018) Oncolytic activity of the rhabdovirus VSV-GP against prostate cancer. Int J Cancer. https://doi.org/10.1002/ijc.31556
3. Park JW, Wollmann G, Urbiola C, Fogli B, Florio T, Geley S, Klimaschewski L (2018) Sprouty2 enhances the tumorigenic potential of glioblastoma cells. Neuro-Oncology 20 (8):1044–1054. https://doi.org/10.1093/neuonc/noy028
4. Maguire CA, van der Mijn JC, Degeling MH, Morse D, Tannous BA (2012) Codon-optimized Luciola italica luciferase variants for mammalian gene expression in culture and in vivo. Mol Imaging 11(1):13–21
5. Jeon YH, Choi Y, Kang JH, Kim CW, Jeong JM, Lee DS, Chung JK (2007) Immune response to firefly luciferase as a naked DNA. Cancer Biol Ther 6(5):781–786
6. Liao JB, Ovenell KJ, Curtis EE, Cecil DL, Koehnlein MR, Rastetter LR, Gad EA, Disis ML (2015) Preservation of tumor-host immune interactions with luciferase-tagged imaging in a murine model of ovarian cancer. J Immunother Cancer 3:16. https://doi.org/10.1186/s40425-015-0060-6
7. Baklaushev VP, Kilpelainen A, Petkov S, Abakumov MA, Grinenko NF, Yusubalieva GM, Latanova AA, Gubskiy IL, Zabozlaev FG, Starodubova ES, Abakumova TO, Isaguliants MG, Chekhonin VP (2017) Luciferase expression allows bioluminescence imaging but imposes limitations on the orthotopic mouse (4T1) model of breast cancer. Sci Rep 7(1):7715. https://doi.org/10.1038/s41598-017-07851-z
8. Curtis A, Calabro K, Galarneau JR, Bigio IJ, Krucker T (2011) Temporal variations of skin pigmentation in C57BL/6 mice affect optical bioluminescence quantitation. Mol Imaging Biol 13(6):1114–1123. https://doi.org/10.1007/s11307-010-0440-8
9. Zinn KR, Chaudhuri TR, Szafran AA, O'Quinn D, Weaver C, Dugger K, Lamar D, Kesterson RA, Wang X, Frank SJ (2008) Non-invasive bioluminescence imaging in small animals. ILAR J 49(1):103–115
10. Keyaerts M, Remory I, Caveliers V, Breckpot K, Bos TJ, Poelaert J, Bossuyt A, Lahoutte T (2012) Inhibition of firefly luciferase by general anesthetics: effect on in vitro and in vivo bioluminescence imaging. PLoS One 7 (1):e30061. https://doi.org/10.1371/journal.pone.0030061
11. Keyaerts M, Verschueren J, Bos TJ, Tchouate-Gainkam LO, Peleman C, Breckpot K, Vanhove C, Caveliers V, Bossuyt A, Lahoutte T (2008) Dynamic bioluminescence imaging for quantitative tumour burden assessment using IV or IP administration of D: -luciferin: effect on intensity, time kinetics and repeatability of photon emission. Eur J Nucl Med Mol Imaging 35(5):999–1007. https://doi.org/10.1007/s00259-007-0664-2
12. Khalil AA, Jameson MJ, Broaddus WC, Chung TD, Golding SE, Dever SM, Rosenberg E, Valerie K (2013) Subcutaneous administration of D-luciferin is an effective alternative to intraperitoneal injection in bioluminescence imaging of xenograft tumors in nude mice. ISRN Mol Imaging 2013:689279. https://doi.org/10.1155/2013/689279
13. Inoue Y, Kiryu S, Izawa K, Watanabe M, Tojo A, Ohtomo K (2009) Comparison of subcutaneous and intraperitoneal injection of D-luciferin for in vivo bioluminescence imaging. Eur J Nucl Med Mol Imaging 36 (5):771–779. https://doi.org/10.1007/s00259-008-1022-8
14. Lee KH, Byun SS, Paik JY, Lee SY, Song SH, Choe YS, Kim BT (2003) Cell uptake and tissue distribution of radioiodine labelled D-luciferin: implications for luciferase based gene imaging. Nucl Med Commun 24 (9):1003–1009. https://doi.org/10.1097/01.mnm.0000090431.24184.49

15. Buckley SM, Howe SJ, Rahim AA, Buning H, McIntosh J, Wong SP, Baker AH, Nathwani A, Thrasher AJ, Coutelle C, McKay TR, Waddington SN (2008) Luciferin detection after intranasal vector delivery is improved by intranasal rather than intraperitoneal luciferin administration. Hum Gene Ther 19 (10):1050–1056. https://doi.org/10.1089/hum.2008.023
16. Ueda I, Kamaya H, Eyring H (1976) Molecular mechanism of inhibition of firefly luminescence by local anesthetics. Proc Natl Acad Sci U S A 73(2):481–485
17. Franks NP, Jenkins A, Conti E, Lieb WR, Brick P (1998) Structural basis for the inhibition of firefly luciferase by a general anesthetic. Biophys J 75(5):2205–2211. https://doi.org/10.1016/S0006-3495(98)77664-7
18. Franks NP, Lieb WR (1984) Do general anaesthetics act by competitive binding to specific receptors? Nature 310(5978):599–601

Chapter 16

Oncolytic Immunotherapy for Bladder Cancer Using Coxsackie A21 Virus: Using a Bladder Tumor Precision-Cut Slice Model System to Assess Viral Efficacy

Kate Relph, Nicola Annels, Chris Smith, Marcos Kostalas, and Hardev Pandha

Abstract

Oncolytic viruses are anticancer agents that selectively target and kill cancer cells by direct lysis, while at the same time stimulating a tumor antigen-specific adaptive immune response. These promising therapeutic agents target multiple cancers and have already proven to be an effective treatment option for solid malignancies. One such agent, T-Vec (Talimogene laherparepvec) has been licensed and is in routine clinical use for treatment of malignant melanoma.

Non-muscle invasive bladder cancer (NMIBC) is an ideal potential target for oncolytic immunotherapy as locally instilled live biological therapy using Bacille Calmette-Guerin (BCG) is already well established in the clinical setting. Coxsackievirus A21 (CVA21) is a novel intercellular adhesion molecule-1 (ICAM-1)-targeted immunotherapeutic virus. We have investigated CVA21-induced cytotoxicity in a panel of human bladder cancer cell lines, revealing a range of sensitivities largely correlating with expression of the viral receptor ICAM-1. CVA21 in combination with low doses of mitomycin-C enhanced CVA21 viral replication and oncolysis by increasing surface expression levels of ICAM-1. In addition to cell lines and an animal model a key component of our studies into oncolytic immunotherapy for bladder cancer was the use of a bladder tumor precision slice preclinical model system which represents tumor architecture, heterogeneity, and the complexity of a tumor in vitro. Results seen in cell lines were reflected in the tumor slice model whereby levels of virus protein expression and induction of apoptosis were enhanced with prior exposure to mitomycin-C. In this chapter we demonstrate the utility of the precision cut tumor slice model as a unique organotypic model to test oncolytic viruses. We will describe how to prepare and slice the tumor using a vibrating microtome together with the optimum culture and conditions for treatment.

Key words Oncolytic virus, Bladder cancer, Vibrating microtome, Immunotherapy, Tissue slice model

1 Introduction

Urothelial cancer of the bladder is the seventh most common cancer in the UK, with over 10,000 new cases diagnosed annually [1]. Superficial, non-muscle invasive bladder cancers (NMIBCs) are managed with cystoscopic transurethral resection of all visible

Christine E. Engeland (ed.), *Oncolytic Viruses*, Methods in Molecular Biology, vol. 2058,
https://doi.org/10.1007/978-1-4939-9794-7_16,

lesions followed by intravesical chemotherapy and/or immunotherapy [2]. The use of Bacillus Calmette-Guerin (BCG) as an immunotherapy for NMIBC and its proven effects of reducing recurrence and progression and improving disease-specific survival revolutionized the treatment of this malignancy in the 1970s. However, the potential for serious side effects of local and systemic BCG infection coupled with the use of a potentially hazardous live agent in the clinic as well as the fact that there is a significant (30%) group of nonresponder patients to BCG highlights the need to develop future immune-based therapies that overcome these problems [3]. The accessibility and superficial location of NMIBC allows direct intravesical administration of antitumor agents and is an ideal model for evaluation of new therapies. Alternatives to BCG have been long sought, due to its toxicity and limited efficacy and, also in recent years, the difficulty with its manufacture [4, 5]. Viruses, both nonreplicating for gene transfer and replication-competent oncolytic viruses (OVs), have been evaluated for the treatment of NMIBC. These viruses infect and selectively lyse tumor cells, leaving nontransformed cells unharmed [6, 7].

Coxsackievirus A21 (CVA21), a naturally occurring "common cold" intercellular adhesion molecule-1 (ICAM-1)-targeted RNA virus, has exhibited selective oncolytic activity in a range of solid tumors [8–12]. CAVATAK is a novel bioselected formulation of CVA21, whose oncolytic and immunotherapeutic capacity has already been clearly demonstrated in in vitro cultures, in vivo melanoma models, and several human trials where CAVATAK has been administered intratumorally alone or in combination with immune checkpoint inhibitors, resulting in significant bystander effects with reduction of distant noninjected metastases [13].

We have evaluated CVA21 as a novel oncolytic virus for the treatment of human bladder cancer [14]. Bladder cancer cell lines were assessed for surface expression of the viral receptor ICAM-1 by flow cytometry and subsequent susceptibility to viral-induced lytic infection. We showed that lytic infection could be facilitated/enhanced by treatment of bladder cancer cell lines and bladder tumor slices with mitomycin-C by increasing ICAM-1 expression on the surface of the bladder cancer cells. Furthermore, we investigated the mode of cell death induced by CVA21 and potential immunogenicity in an immunocompetent murine bladder cancer model.

The testing of sensitivity and resistance to novel drugs/drug combinations has traditionally been carried out in cell lines and using patient-derived xenografts. While these techniques have proved invaluable to understanding some aspects of the drug, they do have their failings: testing of compounds in cell lines is not predictive of their efficacy in vivo and in vivo models in animals are very expensive and time consuming. A key component of our studies into oncolytic immunotherapy for bladder cancer was the use of a bladder tumor precision slice preclinical model system which represents tumor architecture, heterogeneity, and the complexity of a tumor in situ [15]. The tumor microenvironment consists of tumor cells, stromal components and numerous immune cells, each contributing to local immune suppression or immune activation. Physical attributes of the tumor microenvironment such as interstitial pressure are also key factors not modelled in cell line-based studies. A variety of techniques have been used over the years to assess primary tumor cells in vitro including disaggregation into a cell suspension. However, while this technique is satisfactory it is often difficult to localize events within the tumor to specific areas such as malignant cells and vasculature or to understand the interplay of the heterogeneous populations within the tumor architecture. The tissue slice model system has been used for a variety of tumor types as it represents a low-throughput but high-content in vitro platform [16]. In particular its use has been recognized as crucial when testing novel treatments as the therapeutic response is not just a reflection of the molecular composition of the cells but also influenced by the tumor microenvironment [17, 18]. In this chapter we will describe how to establish the model, use it to study oncolytic virus treatment, and assess the effects of the treatment on the tissue using immunohistochemistry.

2 Materials

2.1 Slicing Tumor

1. Vibratome (e.g., Leica VT1200S). Other vibratomes are available.
2. Razor blades.
3. Paddle forceps.
4. Sterile forceps.
5. Embedding mold—6 well plate.
6. Adhesive: ethyl cyanoacrylate based glue, although any superglue will do.
7. 24 well plates.
8. Agarose solution: 4% solution. Weigh 4 g agar into 100 mL distilled water. Heat intermittently in microwave and mix until clear consistency. After use store in cold room at 4 °C.

2.2 Culturing and Treating Tumor

1. Culture medium: Dulbecco's Modified Eagle Medium (DMEM), 10% fetal bovine serum (FBS), 1% Penicillin/Streptomycin (P/S), 2 mM L-Glutamine. Add serum and antibiotics from stock concentrates to obtain final concentration of 10% FBS and 1% PenStrep (100 U/mL Penicillin and 100 μg/mL Streptomycin) and 2 mM L-Glutamine (*see* **Note 1**).

2. Slicing medium: DMEM, 2% FBS,1% P/S. Add serum and antibiotics from stock concentrates to obtain final concentration of 2% FBS and 1% P/S.
3. Six-well plate with cell culture inserts.
4. Mitomycin-C: 0.5 mg/mL stock solution. Dilute using culture medium (*see* **Note 2**).
5. Formalin: 10% neutral buffered formalin.
6. Staining reagents: Hematoxylin and eosin staining is performed using an automated IHC stainer. Manual IHC staining is performed to detect the presence of viral protein (red) and the nuclear stain TO-PRO-3 (blue) and (B) evidence of apoptosis using the TumorTACS in situ apoptosis detection kit (apoptotic cells are identified by the dark brown staining).

3 Methods

3.1 Tissue Procurement

1. Establish a source of primary tumor tissue from a local surgical team and ensure that the necessary authorization and ethical documentation is in place (*see* **Note 3**).
2. Where possible direct the surgeon to give tumor which has been cut toward the edge of the tumor to avoid central necrosis (Fig. 1b). The surgical team must act quickly and place the tumor into a sterile leak proof container within a leak proof bag. The sample must not be placed in formalin or fixative. The sample must be transported on ice and taken to the laboratory as soon as possible for sectioning (*see* **Note 4**). Where possible, obtain an adjacent section of normal tissue which should be stored and transported in the same way.

3.2 Preparing Tissue for Sectioning

1. Cut the tumor into a 0.5 cm^3 piece/core in a laminar flow hood (*see* **Note 5**).
2. Microwave to melt the 4% agarose solution. Allow the agarose to cool slightly after melting and then place into a water bath at 42 °C until needed.
3. Place the tumor tissue chunk into the middle of an embedding mold (e.g., a well of a sterile 6-well plate) using sterile forceps and then slowly pour on the agarose keeping the tumor piece in the correct orientation for cutting. Allow to set on ice for 10–20 min (Fig. 1d, *see* **Note 6**).

3.3 Preparation of the Vibratome

1. The microtome is used for cutting fixed or unfixed fresh bladder tissue. All users should receive the appropriate training and read and be familiar with the manual, SOP and the risks associated with using microtomes. All users should have and document their Hepatitis B vaccination.

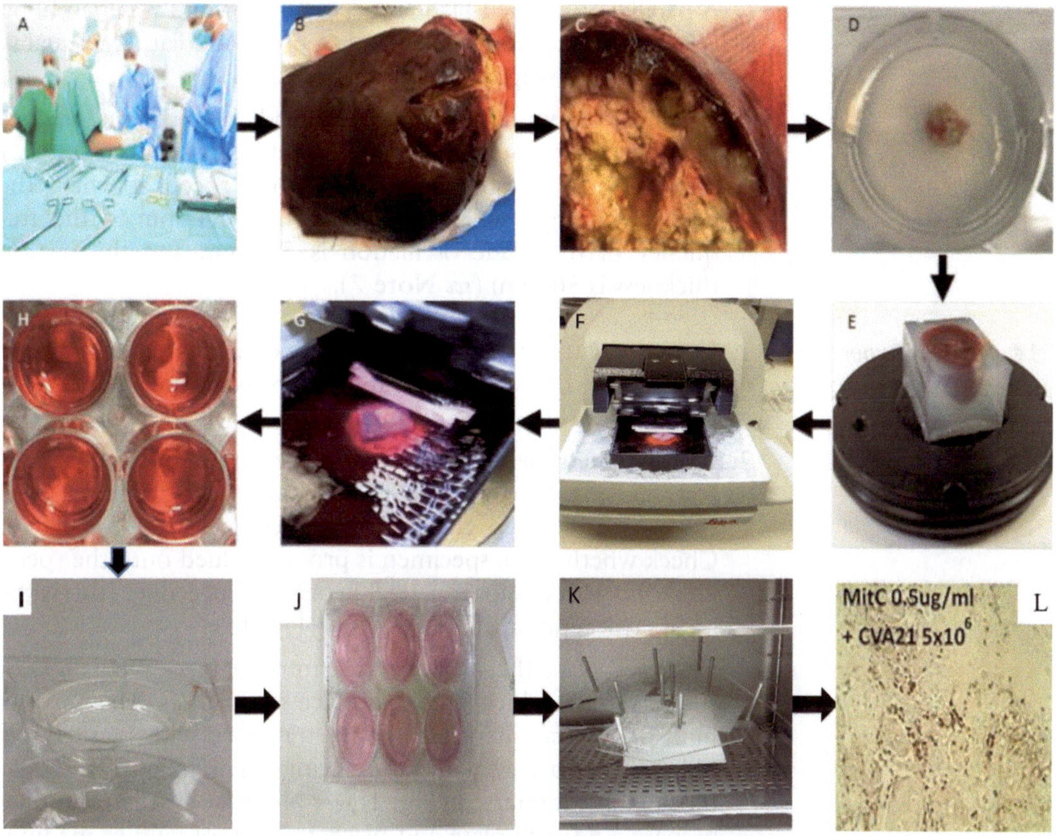

Fig. 1 Tissue slice model—From theater to experimental treatment. (**a**) Surgeon removes the tumor during routine operation. (**b**, **c**) Section of the tumor is removed for analysis. If possible an adjacent section of normal tissue is also removed as a control. (**d**) Tumor tissue is embedded in liquid agarose which is allowed to set. (**e**) A cube of agarose is cut containing the tumor and stuck onto a mounting disc from the microtome. (**f**) The mounting disc containing the agarose embedded tumor is placed into the black media tray of the vibratome which is held in place by magnets. The purpose built media tray sits within an ice tray filled with crushed ice. Fill the media tray with slicing medium until the agar block is just immersed. (**g**) As slices are cut they float off into the media and paddle forceps are used to scoop up the cut sections and transfer into a well of a 24-well plate containing culture medium (**h**). (**i**) Working within a sterile containment hood, place a cell culture insert into each well of a 6-well plate. (**j**) Add 1 mL media ± mitomycin C. Use the paddle forceps to transfer tissue from the 24 well plate onto the cell culture insert in the 6-well plate containing media ± mitomycin and ensure that it is flat on the insert. (**k**) Incubate for 1 h in a 37 °C incubator on a rotating plate. After 1 h replace media with media containing virus and incubate for 48 h. (**l**) When treatment has finished fix and stain tissue

2. All users must wear appropriate personal protective equipment (PPE), including lab coats, nitrile gloves and safety specs when preparing and using the microtome. All blades and tissue should go into sharps clinical waste containers.
3. When inserting a new razor blade, the use of the Vibrocheck is recommended to measure the vertical deflection of the blade. With the adjustment screw on the blade holder, the blade can be tilted as long as it sits perfectly parallel to the cutting plane.

A parallel positioned blade is necessary to avoid venetian blind effect on the section surface and to produce a flat, high quality section with living cells on the surface.

4. Once the optimum position of the blade has been set, remove the Vibrocheck and set the appropriate cutting variables. For sectioning the fresh bladder tumor the speed at which the microtome approaches the specimen is 0.4 mm/s, the frequency of the blade oscillation is 3.0 mm and the section thickness is 300 μm (*see* **Note 7**).

3.4 Cutting Tumor Slices

1. Once set, cut the agar into the shape of a cube containing the tumor core. Leave a good margin (approx. 5 mm) around it as this supports the tissue during slicing.
2. Remove the agar cube containing the tumor from the 6-well plate and fix onto the metallic specimen disc using an ethyl cyanoacrylate-based glue. Allow to set on ice for 5–10 min. Check whether the specimen is properly glued onto the specimen disc with cyanoacrylate glue before proceeding (this is the most critical point of preparation) (Fig. 1e).
3. Once set, place the specimen disc into the black media tray of the vibratome which is held in place by magnets. The purpose built media tray sits within an ice tray filled with crushed ice. Fill the media tray with cold slicing medium until the agar block is just immersed (Fig. 1f, *see* **Note 8**). In the meantime prepare a 24-well plate and place 1–2 mL of culture medium into each well.
4. Start slicing tissue. As slices are produced use paddle forceps to scoop up the cut sections and transfer into a well of the 24-well plate (*see* **Note 9**).
5. Work as quickly as possible and create as many tissue slices as possible.

3.5 Culturing and Treating Tumor Slices

1. Allow the tissue slices to acclimatize in the slicing medium in the 24-well plate for approximately 1 h.
2. After 1 h of acclimatization, working within a sterile culture hood, place a cell culture insert into each well of a 6-well plate containing 1 mL culture medium ± mitomycin C (Fig. 1i, *see* **Note 2** for preparation of media + mitomycin c). Use the paddle forceps to transfer tissue from the 24-well plate onto the cell culture insert and ensure that it is flat on the insert (*see* **Note 10**).
3. Add a further 700 μL of media ± mitomycin C on top of the tissue slice on the culture insert. Incubate for 1 h in a 37 °C incubator on a rotating plate (Fig. 1k, *see* **Note 11**).

4. Remove the medium containing mitomycin C and replace with 1.7 mL fresh culture medium alone or CVA21 in fresh culture medium (*see* **Note 12** *for concentrations of mitomycin and CVA21 used*) NB. Place 1 mL below the insert and 700 μL on top of the tissue in the insert. Then incubate the plate for 48 h at 37 °C on a rotating plate (*see* **Note 13**).
5. When the treatment period has ended fix each slice in 10% neutral buffered formalin.

3.6 Processing the Tumor

1. The fixed tumor slices can be processed in an automated tissue processing machine and left overnight. Following processing tissue slices can be placed in a paraffin embedding machine and can be manually embedded within cassettes.
2. The tissue can be analyzed using a variety of staining methods. Hematoxylin and eosin staining can be performed using an automated IHC stainer. Immunofluorescent staining for the presence of viral protein showed increased viral replication within the combination-treated slices as compared to the CVA21-alone-treated tumor slice (Fig. 2a). As expected, there was no evidence of virus within the slices treated with mitomycin-C alone or in the untreated slices. Furthermore, a significantly higher level of apoptotic activity within the combination-treated tissue slice as compared to CVA21- or mitomycin-C-alone-treated slices was observed. The control slice had little to no evidence of apoptotic cells, reflecting the situation in the original tumor biopsy (Fig. 2b).

4 Notes

1. To heat inactivate serum: (1) Thaw the frozen serum (companies supply FBS at −20 °C) at 37 °C for 5–6 h. (2) Incubate the thawed serum at 56–65 °C for 30 min (shake the bottle gently every 10 min). (3) Aliquot the serum into 50 mL conical tubes in a culture hood. (4) Freeze and store. Thaw aliquots in a 37 °C water bath as needed.

   It is useful to aliquot PenStrep as the half-life of antibiotics is quite short at 37 °C (~2 days for Penicillin and ~4 days for Streptomycin). Aliquot PenStrep (100 U/mL Penicillin and 100 μg/mL Streptomycin) in a culture hood into 5 mL aliquots and freeze. Thaw a 5 mL aliquot and add to 500 mL medium bottles as needed for medium preparation.

   Aliquot the 200 mM L-Glutamine stock concentration into 5 mL aliquots and freeze. When needed defrost 5 mL aliquot and add to 500 mL medium to give a 2 mM final concentration. N.B. L-Glutamine precipitates at low temperature to give a milky liquid. When you add this to the medium it dissolves.

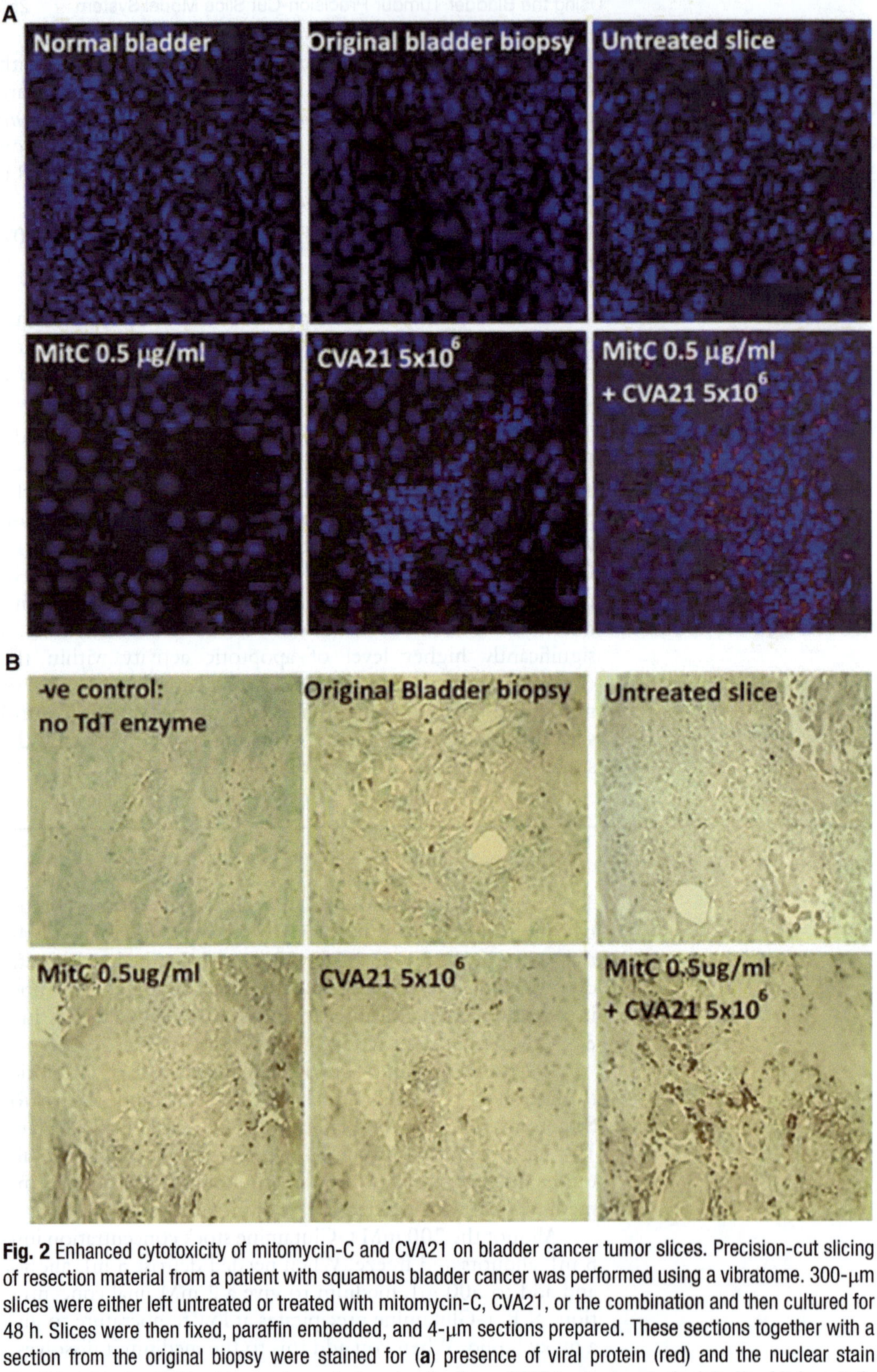

Fig. 2 Enhanced cytotoxicity of mitomycin-C and CVA21 on bladder cancer tumor slices. Precision-cut slicing of resection material from a patient with squamous bladder cancer was performed using a vibratome. 300-μm slices were either left untreated or treated with mitomycin-C, CVA21, or the combination and then cultured for 48 h. Slices were then fixed, paraffin embedded, and 4-μm sections prepared. These sections together with a section from the original biopsy were stained for (**a**) presence of viral protein (red) and the nuclear stain TO-PRO-3 (blue) and (**b**) evidence of apoptosis using the TumorTACS in situ apoptosis detection kit (apoptotic cells are identified by the dark brown staining)

2. Mitomycin C (MC) dilutions.
   STOCK mitomycin C = 0.5 mg/mL.

Stock MC	600μl	600μl	600μl	**1200μl**
	↘ ↗	↘ ↗	↘ ↗	
Media	5.4mL	5.4 mL	5.4mL	4.8mL
	↓	↓	↓	↓
Final MC conc.	50	5	0.5	0.1 μg/mL

3. We obtain tumor tissue which is surplus to requirements from routine operations to remove bladder tumors as part of a patient's routine standard of care. The patients have given their full consent and we have full ethics approval and assigned study number from the National Research Ethics Committee (NRES) and Human Research Authority (HRA).
4. A sample coordinator liaises with the surgeon to identify patients which are undergoing surgery and identifies a time at which the sample should be picked up from outside the operating theatre.
5. The use of disposable biopsy punches allows for a defined diameter core of tumor tissue for embedding.
6. A thin layer of agarose solution is added to the well before placing the tumor tissue in the correct orientation for subsequent cutting. It may be necessary to support/hold the tumor tissue in place using sterile forceps while it is being submerged in the agarose solution. Allow the agarose solution a few minutes to begin to set before carefully moving the 6-well plate onto ice. Embedding in agarose is recommended for very small and soft specimens. The embedding has no effect on the specimen itself. The embedding gives support to the specimen and avoids pushing away the specimen by the vibrating blade. To avoid heat artefacts Agarose with a low melting point of 37 °C is recommended.
7. The "Rule of thumb" for setting the cutting variables is that softer tissues require slower forward speed, higher amplitude and frequency, steeper blade angle and cold buffer. For fresh tissue a knife travel speed below 0.1 mm/s is recommended. The amplitude should be between 1 and 1.5 mm for the VT1200. The selection of the section thickness depends on the purpose of the study. For bladder tumor slices we recommend a thickness of 300 μm. The decision to slice the tissue at

300 μm intervals was based on the existing literature regarding the use of organotypic cultures as well as our experience in cutting tumors. While thinner slices can be procured, we found that the cutting of tissue cores at this thickness was less consistent.

8. Sectioning of fresh or fixed tissue is always easier when the specimen is cold by using cooled slicing medium.
9. As tumor is cut, the slices float off into the medium. Use paddle forceps to "scoop-up" the cut tissue sections to prevent damage to the fragile tissue sections. Do not try to grab the tissue sections when transferring them.
10. Both from our own experience and others it is important to cultivate slices on a filter support rather than cultivate floating tumor slices as this helps to maintain tissue morphology.
11. The use of a rotating incubation unit allows for the 6-well plates to be rotated slowly (1 rpm) on an inclined plane to alternately expose the slices to the atmosphere of the tissue culture incubator or dip them in the culture medium.
12. Mitomycin C and CVA21 concentrations and combinations that we have studied. Mitomycin C (or culture medium alone as control) was added 1 h prior to CVA21, this was then removed and then medium containing CVA21 (or medium alone as control) was added and incubated for 48 h.

Controls	CVA21 5×10^6 PFU	CVA21 1×10^6 PFU	CVA21 5×10^5 PFU	MC 50 μg/mL	MC 5 μg/mL	MC 0.5 μg/mL U	MC 0.1 μg/mL U	
Combinations	MC 50 μg/mL + CVA21 5×10^6 PFU	MC 5 μg/mL CVA21 + 5×10^6 PFU	MC 0.1 μg/mL + CVA21 5×10^6 PFU	MC 0.5 μg/mL + CVA21 5×10^6 PFU	MC 50 μg/mL + CVA21 5×10^6 PFU	MC 5 μg/mL + CVA21 5×10^6 PFU	MC 0.1 μg/mL + CVA21 5×10^6 PFU	MC 0.5 μg/mL + CVA21 5×10^6 PFU

13. If experiment requires longer incubation top up media with 200 μL every 48 h to prevent tissue becoming dry.

Acknowledgments

We would like to acknowledge and thank Thames West Unicorn and Topic of Cancer charities for funding the vibratome. We would also like to thank Guy Simpson for his editing and reviewing of the manuscript.

References

1. Cancer Research UK (2016) Bladder cancer statistics. http://www.cancerresearchuk.org/health-professional/cancer-statistics/statistics-by-cancer-type/bladder-cancer
2. Aldousari S, Kassouf W (2010) Update on the management of non-muscle invasive bladder cancer. Can Urol Assoc J 4:56–64
3. Fuge O, Vasdev N, Allchorne P, Green JSA (2015) Immunotherapy for bladder cancer. Res Rep Urol 7:65–79
4. Sanofi Pasteur (2016) Sanofi Pasteur statement on discontinuation of BCG. http://www.sanofipasteur.ca/node/50701
5. Davies B (2016) Sanofi shuts down bladder cancer drug production: inevitable drug shortage to harm patients. https://www.forbes.com/sites/benjamindavies/2016/11/17/sanofi-shuts-down-bladder-cancer-drug-production-inevitable-drug-shortage-to-harm-patients/#27d4a3abc132
6. Russell SJ, Peng KW, Bell JC (2012) Oncolytic virotherapy. Nat Biotechnol 30:658–670
7. Patel MR, Kratzke RA (2013) Oncolytic virus therapy for cancer: the first wave of translational clinical trials. Transl Res 161:355–364
8. Shafren DR, Au GG, Nguyen T et al (2004) Systemic therapy of malignant human melanoma tumours by a common cold-producing enterovirus, coxsackievirus A21. Clin Cancer Res 10:53–60
9. Au GG, Lindberg AM, Barry RD et al (2005) Oncolysis of vascular malignant human melanoma tumours by Coxsackievirus A21. Int J Oncol 26:1471–1476
10. Berry LJ, Au GG, Barry RD et al (2008) Potent oncolytic activity of human enteroviruses against human prostate cancer. Prostate 68:577–587
11. Skelding KA, Barry RD, Shafren DR (2009) Systemic targeting of metastatic human breast tumour xenografts by Coxsackievirus A21. Breast Cancer Res Treat 113:21–30
12. Au GG, Lincz LF, Enno A et al (2007) Oncolytic coxsackievirus A21 as a novel therapy for multiple myeloma. Br J Haematol 137:133–141
13. Andtbacka R, Curti B, Hallmeyer S et al (2015) Abstract CT214: phase II CALM study: changes in the tumour microenvironment induced by the immunotherapeutic agent coxsackievirus A21 delivered intratumorally in patients with advanced melanoma. Cancer Res 75:CT214
14. Annels NE, Arif M, Simpson GR et al (2018) Oncolytic immunotherapy for bladder cancer using Coxsackie A21 virus. Mol Ther Oncolytics 14(9):1–12
15. Hood CJ, Parham DM (1998) A simple method of tumour culture. Pathol Res Pract 194(3):177–181
16. Davies EJ, Dong M, Gutekunst M et al (2015) Capturing complex tumour biology in vitro: histological and molecular characterisation of precision cut slices. Sci Rep 9(5):17187
17. Gerlach MM, Merz F, Wichmann G et al (2014) Slice cultures from head and neck squamous cell carcinoma: a novel test system for drug susceptibility and mechanisms of resistance. Br J Cancer 110(2):479–488
18. Carranza-Torres IE, Guzmán-Delgado NE, Coronado-Martínez C et al (2015) Organotypic culture of breast tumour explants as a multicellular system for the screening of natural compounds with antineoplastic potential. Biomed Res Int 2015:618021

Chapter 17

Use of Liquid Patient Ascites Fluids as a Preclinical Model for Oncolytic Virus Activity

Eleanor M. Scott, Sally Frost, Hena Khalique, Joshua D. Freedman, Len W. Seymour, and Janet Lei-Rossmann

Abstract

The translational success of oncolytic virotherapies would benefit from the widespread use of clinically relevant ex vivo models. Malignant ascites, an accumulation of fluid in the peritoneum due to disseminated cancer, recapitulates many features of the tumor microenvironment, making it a valuable model for studying oncolytic virus activity. Here, we describe a method for the separation and storage of cellular and acellular components of malignant ascites, followed by flow cytometric characterization of the cellular fraction. We then outline a simple experiment using whole ascites to assess the activity of a bispecific T cell engager (BiTE)-expressing oncolytic adenovirus.

Key words Ascites, Liquid biopsies, Oncolytic virus, Ex vivo model, Flow cytometry

1 Introduction

An accumulation of fluid in the peritoneal cavity due to the intraperitoneal spread of cancer, liquid human malignant ascites comprises a mixture of tumor cells, noncancerous cells, and soluble factors. Together, these form a complex microenvironment that fosters tumor growth, spread and immune evasion, as well as promoting resistance to chemotherapy (reviewed in [1]). The most common malignancies associated with the development of ascites are ovarian, breast, gastrointestinal, and lung cancers. The presence of malignant ascites is a grave prognostic factor and associated with significant morbidity, causing abdominal pain, anorexia, nausea, vomiting, and respiratory issues [2].

To alleviate symptoms, patients are regularly drained of excessive fluid by paracentesis. Ascites collected in this manner can be up to several liters in volume and would otherwise be discarded, making it a readily available source of material for study. Primary neoplastic cells isolated from ascites, which may exist as single cells

Christine E. Engeland (ed.), *Oncolytic Viruses*, Methods in Molecular Biology, vol. 2058,
https://doi.org/10.1007/978-1-4939-9794-7_17,

or aggregates, can be propagated in vitro, acting as a model system that more faithfully resembles the patient situation than immortalized cell lines [3, 4]. Sub-populations of ascites cancer cells have been found to display cancer stem cell (CSC)-like characteristics, exhibiting an enhanced proliferative capacity and resistance to chemotherapy [5]. Ascites also contains a range of noncancerous cells such as cancer-associated fibroblasts, mesenchymal stem cells, myeloid-derived suppressor cells, M2-like macrophages and T lymphocytes, which together influence tumour cell behavior and response to therapy [1]. Meanwhile, the acellular fraction of malignant ascites is rich in proinflammatory chemokines and cytokines such as interleukin (IL)-6, IL-8, and IL-10 [6], as well as prosurvival factors, proangiogenic factors, and extracellular matrix fragments. These properties make malignant ascites an especially attractive model for studying the replication of oncolytic viruses, including those that encode a therapeutic transgene, as many of the cell subpopulations present in the tumor microenvironment can be targeted [7].

Here, we describe a method for separating and storing the cellular and acellular components of malignant ascites, including a step for removal of erythrocytes from the cellular fraction. We then describe flow cytometric characterization of the ascites cell subpopulations, followed by a simple in vitro experiment using ascites cells and supernatant to assess the activity of a bispecific T cell engager (BiTE)-armed oncolytic virus (OV).

2 Materials

2.1 Materials and Reagents

1. Ascites fluid, acquired with informed patient consent and appropriate regulatory approval.
2. Decontamination solution for clinical-grade waste.
3. Plasticware for disposal of clinical-grade waste.
4. 70% ethanol.
5. Sterile containers such as empty PBS or media bottles, or T175 flasks.
6. Sterile conical centrifuge tubes (15 mL and 50 mL).
7. Sterile stripettes (5, 10, and 25 mL).
8. Phosphate-buffered saline (PBS) without magnesium or calcium, sterile, pH 7.4.
9. Red cell lysis buffer (155 mM NH_4Cl, 12 mM $NaHCO_3$, 0.1 mM EDTA).
10. Freezing medium: 10% dimethyl sulfoxide in fetal bovine serum (FBS).

11. Cryovials suitable for archiving at −80 °C or liquid nitrogen (1 mL or 2 mL).
12. Complete RPMI: RPMI-1640 basal medium supplemented with 10% FBS.
13. Nonenzymatic cell dissociation buffer.
14. Trypan blue.
15. Low-adherence tissue culture flask of appropriate size.
16. Flat-bottom 96-well plate.
17. MACS buffer: 2 mM EDTA and 0.5% BSA in 1× PBS, sterile filtered.
18. Paraformaldehye (4%), buffered, pH 6.9.
19. Flow cytometry-compatible antibodies against cell populations of interest.
20. Ficoll-Paque (as needed).
21. Sterile 70 μm cell strainers (as needed).

2.2 Equipment

1. Biosafety cabinet.
2. Laboratory clamp stand and two clamps.
3. Nalgene cryo 1 °C freezing container.
4. Pipette gun.
5. Benchtop centrifuge with corresponding adaptors for 50 mL conical tubes.
6. Incubator set to 37 °C and 5% CO_2.
7. Suitable flow cytometer, equipped with the appropriate number of lasers and detection channels for the antibodies selected.
8. Flow cytometry analysis software.

3 Methods

All procedures should be performed at room temperature in a biosafety level 2 laboratory using a Class II biosafety cabinet, unless otherwise specified, following all local safety and disposal regulations for handling biosafety level 2 materials. It is imperative to obtain approval from the appropriate ethics committee before starting this or any other protocol involving human biopsies.

3.1 Separation of Cellular and Acellular Fractions

1. Receive freshly isolated ascites from hospital staff in a sterile drainage bag (*see* **Note 1**).
2. Assemble a clamp stand within the biosafety cabinet.

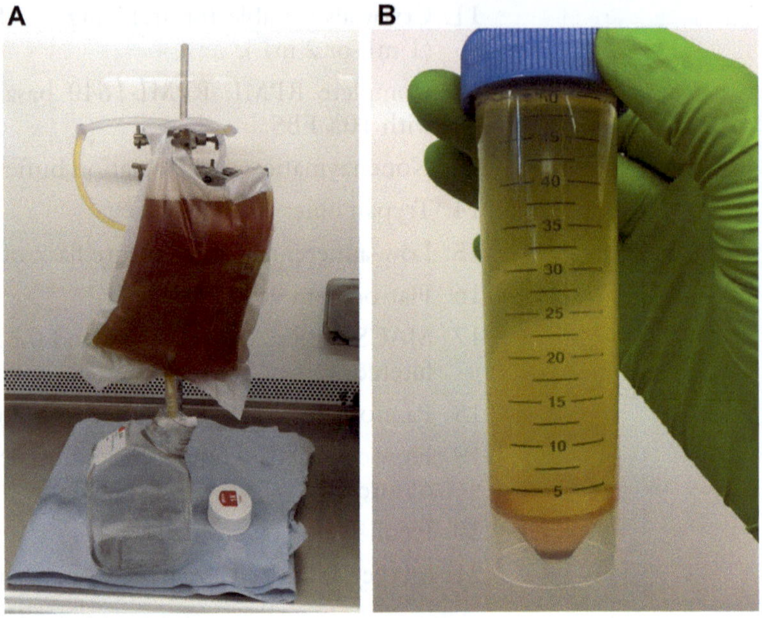

Fig. 1 Setup of ascites sample bag processing. (**a**) The sample bag should be clamped securely with the tap facing downward ready for draining into a sterile PBS bottle. (**b**) Initial centrifugation of ascites samples separates the cellular (pellet) and acellular fractions (supernatant)

3. Clean the entire bag with 70% ethanol, taking particular care to disinfect the tap, then attach the bag to the clamp stand (Fig. 1a).
4. Massage the bag to homogenize the fluid and dislodge any large lumps that may obstruct the tap.
5. Carefully open the tap and transfer the ascites into a large sterile container with a wide mouth, such as an empty PBS or media bottle (*see* **Note 2**).
6. Once the container is full, aseptically transfer the fluid into sterile 50 mL conical tubes for centrifugation, filling the centrifuge to maximum capacity (*see* **Note 3**).
7. Centrifuge at 400 × g for 10 min. For efficient processing, repeat **steps 4** and **5** during this centrifugation step with the remaining fluid.
8. Centrifugation will yield a cell pellet, which is usually red due to the presence of erythrocytes (*see* Subheading 3.2 for advice regarding erythrocyte removal), and a yellow/orange supernatant, resembling serum (Fig. 1b). Depending on the needs of the researcher, a portion of the supernatant can be retained (*see* Subheading 3.3 for sample storage recommendations), while the rest should be discarded according to local regulations for the disposal of clinical-grade waste.

9. Repeat **steps 4–8** until all of fluid has been centrifuged. Ascites drainage bags are typically around two liters in volume per sample, so expect to perform **steps 4–8** up to four times.
10. Resuspend each pellet in a small volume of PBS, then combine the cell suspensions into a small number of 50 mL centrifuge tubes and centrifuge at 400 × *g* for 10 min. Discard the supernatant.

3.2 Erythrocyte Removal by Red Cell Lysis

Erythrocytes should ideally be removed from the cellular fraction prior to conducting further experiments. The method of erythrocyte removal will depend on the nature of the pellet, which will usually be compact (containing few erythrocytes), but may also be loose (containing many erythrocytes). Dense pellets can be cleared of erythrocytes using red cell lysis buffer, as outlined below, whilst loose pellets may require density gradient centrifugation (please *see* Subheading 3.2, **step 5** for more details).

1. Resuspend each pellet in 5 mL red cell lysis buffer.
2. Incubate at 4 °C for 10 min.
3. Top up to 50 mL with PBS and centrifuge at 400 × *g* for 10 min.
4. Discard the supernatant containing lysed red blood cells, which should appear pink.
5. If erythrocytes remain in the pellet, repeat **steps 1–4** up to two more times until the pellet is cleared (*see* **Note 4**).
6. Finally, wash the cells twice by resuspending the pellet in 50 mL PBS and centrifuging at 400 × *g* for 10 min, discarding the supernatant.

3.3 Supernatant Storage

We recommend retaining at least 200 mL of fluid, which can be aliquoted into tubes or cryovials of different sizes and stored as desired.

1. To ensure complete removal of cells, centrifuge the supernatant again at 400 × *g* for 10 min.
2. Aliquot clarified supernatant into new sterile tubes or cryovials and freeze at −20 °C or −80 °C for long-term storage (*see* **Note 5**).

3.4 Cellular Fraction Storage

Following the removal of erythrocytes, determine the cell number and viability by trypan blue exclusion. Cells may be used immediately or cryopreserved for later use.

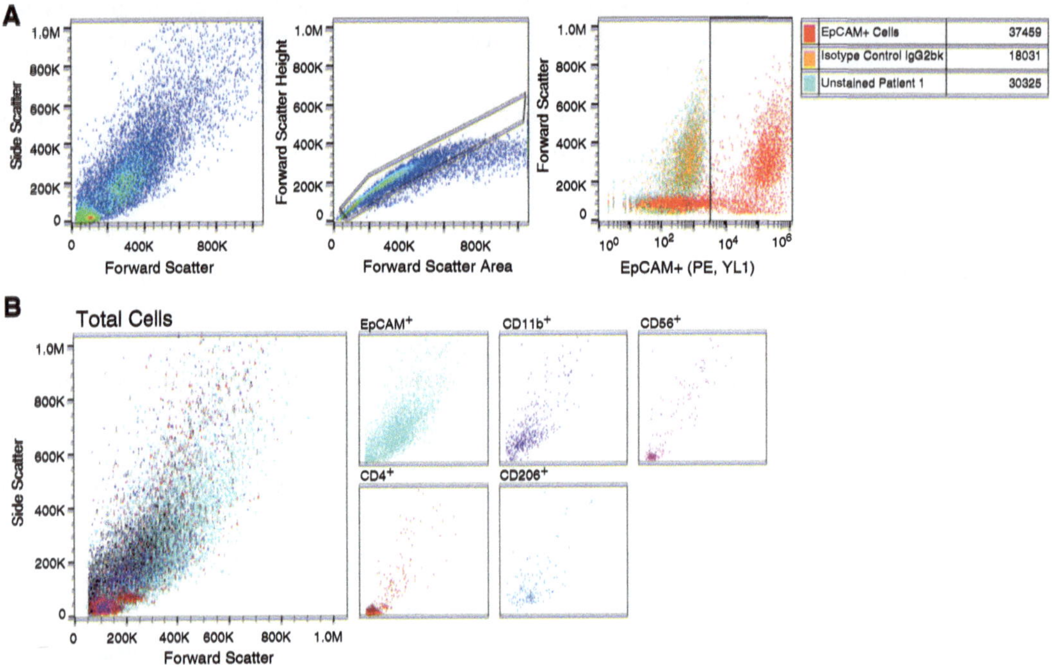

Fig. 2 Characterization of the cellular fraction of an ascites sample by flow cytometry. (**a**) Example of gating to determine EpCAM expression on cells. Total cells are gated first using FSC-H/FSC-A for single cells, then in the EpCAM channel (shown here as PE-EpCAM in channel YL1) against isotype and unstained control samples. (**b**) The total ascites cell population was stained using a multicolor panel and analyzed on an Attune NxT flow cytometer for total cells (forward scatter FSC/side scatter SSC), EpCAM (cyan), CD11b (purple), CD56 (pink), CD4 (red), and CD206 (blue)

1. To cryopreserve the ascites cells, resuspend in freezing medium at a density of 20 million cells/mL, and aliquot into cryovials. We recommend designating a small aliquot (around 5 million cells) for use in characterization by flow cytometry (*see* Subheading 3.5).
2. Transfer cryovials to a 1 °C freezing container and freeze at −80 °C, then transfer to liquid nitrogen for long-term storage.

3.5 Sample Characterization

To inform the design of subsequent experiments, it is useful to firstly identify the key cell populations present within each ascites sample. We recommend antibody staining for EpCAM, EGFR, CD3, CD4, CD8, CD11b, and PDL1, CTLA-4, followed by analysis by flow cytometry (not detailed in this protocol). An exemplary characterization experiment is displayed in Fig. 2 (*see* **Note 6**).

Here, we describe an experiment using an EpCAM-targeted bispecific T cell engager delivered by an oncolytic adenovirus. Details of the viruses used here can be found in [7].

3.6 Exemplary Experiment: Treatment of Whole Ascites with an EpCAM BiTE-Armed Oncolytic Adenovirus

1. Thaw cryopreserved ascites cells into complete RPMI pre-warmed to 37 °C, diluting the freezing medium at least 1:10 to adequately dilute the DMSO in the freezing medium.
2. Centrifuge at 400 × *g* for 10 min, discarding the supernatant.
3. Resuspend the cell pellet in complete RPMI at 2×10^6 cells/mL and rest the cells overnight at 37 °C in a low-adherence flask (*see* **Note 7**).
4. The next day, harvest the cells by collecting the supernatant (containing suspension cells) and removing the remaining adherent cells with cell dissociation buffer. Combine the two fractions of cells into the same sterile conical tube.
5. Centrifuge and resuspend the cells in complete RPMI at a density of 3×10^6 cells/mL.
6. Seed 50 μL cell suspension (1.5×10^5 cells)/well in a 96-well plate.
7. Add 50 μL complete RPMI containing EpCAM-BiTE-expressing oncolytic adenovirus or corresponding virus expressing a control-BiTE against an irrelevant antigen target.
8. Add 100 μL of either complete RPMI or autologous ascites supernatant (*see* **Note 8**).
9. Three days later, analyse T cell activation and cancer cell death by antibody staining for CD3, CD25, EGFR and live/dead staining, followed by flow cytometric analysis (*see* Fig. 3, latter two stains not shown, *see* **Notes 9** and **10**).

4 Notes

1. Isolation of ascites cells and fluids works best with freshly acquired samples at room temperature. Samples that have been at room temperature for more than 18 h should not be used, as many of the cell subpopulations will have lysed or may not be in good health. Samples should also not be placed on ice before isolation, as this may cause lipids or proteins to aggregate and precipitate, making separation of the cells and the fluid difficult.
2. The type of drainage tap on the sample will depend on the unit from which the sample bag is received. We recommend that opening and closing of the tap is tested carefully (preferably with the tap facing upward) before draining the sample into a sterile bottle to prevent large volumes of sample loss onto the sterile bench due to an inability to reclose the tap. Samples contained in a bag without a tap can be removed by cutting a small opening using sterilized scissors at the top of the bag and pipetting the fluid directly into 50 mL conical tubes.

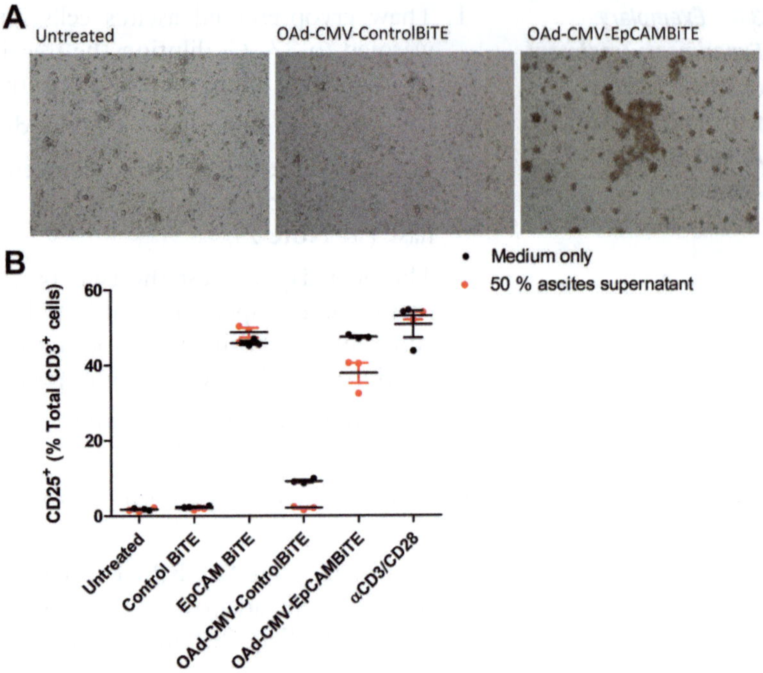

Fig. 3 Exemplary experiment showing treatment of whole ascites with an EpCAM BiTE-armed oncolytic adenovirus. (**a**) Representative brightfield images of whole ascites cells 3 days after infection with oncolytic viruses. (**b**) Flow cytometric analysis of T cell activation, assessed 3 days after virus treatment by antibody staining for CD3 and CD25, followed by flow cytometric analysis with an Attune NxT flow cytometer and FlowJo software

3. We have found that the use of larger centrifugation vessels, such as 250 mL conical tubes, reduces the yield of the cellular fraction that is recovered from the sample. Therefore, we recommend that samples are harvested as described above.
4. The number of erythrocytes present in each sample can vary greatly. Occasionally, the concentration of erythrocytes will be so high that the pellets formed after centrifugation are loose, appearing more like a small volume of concentrated blood at the bottom of the centrifuge tube than a robust pellet. For these samples, we recommend passing the cellular fraction over a Ficoll-Paque gradient as follows:
   (a) Very carefully discard the supernatant (the pellet will be lost if the supernatant is poured off too quickly).
   (b) Dilute the cells in PBS, using a volume of PBS that is at least equal to that of the pellet.
   (c) Carefully layer 35 mL of diluted cell suspension onto 15 mL Ficoll-Paque solution, ensuring that the cells and Ficoll-Paque solution do not mix. The number of tubes required will vary depending on the size of the pellets.

(d) Centrifuge at 950 × *g* for 30 min with half-maximal acceleration and braking turned off.

(e) Carefully collect the interphase containing mononuclear cells and transfer to a sterile centrifuge tube.

(f) Wash the cells three times with PBS.

(g) First wash: 950 × *g* for 10 min with maximal acceleration and half-maximal braking.

(h) Second wash: 450 × *g* for 10 min with maximal acceleration and maximal braking.

(i) Third wash: 200 × *g* for 10 min with maximal acceleration and maximal braking.

(j) If the mononuclear cell fraction contains contaminating erythrocytes, treat with red cell lysis buffer as described in Subheading 3.2.

5. The number of freeze–thaw cycles of the supernatant should be kept consistent between experiments. We recommend that frozen supernatants are thawed once, aliquoted into smaller volumes, and frozen a second time. These small aliquots of twice-frozen supernatants can then be used for experimentation.
6. The number of tumor cells varies greatly between patients and the time since last treatment. Thorough characterization of the cell populations of interest (tumor cells, lymphocytes, macrophages, etc.) is necessary to determine whether the samples contain the cell subpopulations of interest.
7. When thawing ascites cells for experimentation, we recommend resting them overnight at 37 °C in complete RPMI (or equivalent basal medium supplemented with 10% FCS) to allow them to recover from freeze–thaw shock, as this can greatly affect cell viability as measured in subsequent assays.
8. Frozen and thawed ascites fluid samples should be passed through a 70 μm cell strainer before use to remove any precipitates that may have formed during thawing.
9. This protocol outlines detection of cytotoxicity using live/dead staining in flow cytometry. While it may also be possible to use supernatant or cell-based plate assays such as MTS or lactate dehydrogenase, we have found that results for these two methods (particularly lactate dehydrogenase) can be convoluted due to the presence of existing lactate dehydrogenase in the ascites supernatant itself. We recommend using a cytometry-based method where cell death can be correlated with the specific cell population of interest.
10. We have not described antibody staining in detail here, as it is beyond the scope of this protocol. However, antibody panels should be carefully designed to minimize spectral overlap and antibody concentrations should be titrated before use.

Acknowledgments

This work was supported by grants from Cancer Research UK (C552/A17720 and C5255/A20936) and the Medical Research Council (MR/N013468/1). JLR is supported in part by the Sir Paul Nurse Junior Research Fellowship at Linacre College. The studies described in this protocol were performed under an ethics permission from the Oxford Radcliffe Biobank (16/A070).

References

1. Ahmed N, Stenvers KL (2013) Getting to know ovarian cancer ascites: opportunities for targeted therapy-based translational research. Front Oncol 3:256. https://doi.org/10.3389/fonc.2013.00256
2. Sangisetty SL, Miner TJ (2012) Malignant ascites: a review of prognostic factors, pathophysiology and therapeutic measures. World J Gastrointest Surg 4(4):87–95. https://doi.org/10.4240/wjgs.v4.i4.87
3. Dunfield LD, Shepherd TG, Nachtigal MW (2002) Primary culture and mRNA analysis of human ovarian cells. Biol Proced Online 4:55–61. https://doi.org/10.1251/bpo34
4. Golan T, Atias D, Barshack I, Avivi C, Goldstein RS, Berger R (2014) Ascites-derived pancreatic ductal adenocarcinoma primary cell cultures as a platform for personalised medicine. Br J Cancer 110(9):2269–2276. https://doi.org/10.1038/bjc.2014.123
5. Alvero AB, Chen R, Fu HH, Montagna M, Schwartz PE, Rutherford T, Silasi DA, Steffensen KD, Waldstrom M, Visintin I, Mor G (2009) Molecular phenotyping of human ovarian cancer stem cells unravels the mechanisms for repair and chemoresistance. Cell Cycle 8(1):158–166. https://doi.org/10.4161/cc.8.1.7533
6. Matte I, Lane D, Laplante C, Rancourt C, Piché A (2012) Profiling of cytokines in human epithelial ovarian cancer ascites. Am J Cancer Res 2 (5):566–580
7. Freedman JD, Hagel J, Scott EM, Psallidas I, Gupta A, Spiers L, Miller P, Kanellakis N, Ashfield R, Fisher KD, Duffy MR, Seymour LW (2017) Oncolytic adenovirus expressing bispecific antibody targets T-cell cytotoxicity in cancer biopsies. EMBO Mol Med 9 (8):1067–1087. https://doi.org/10.15252/emmm.201707567

Chapter 18

Generating Primary Models of Human Cancer to Aid in the Development of Clinically Relevant Oncolytic Viruses

Brian A. Keller, Marie-Ève Wedge, Abera Surendran, and Carolina S. Ilkow

Abstract

Oncolytic viruses (OVs) represent a rapidly advancing class of cancer immunotherapeutic that is helping to shift the treatment paradigm of cancer. Initially thought to only have direct oncolysis ability, the established anticancer mechanisms of action of OVs have rapidly expanded to incorporate a multitude of immune-activating properties. Specifically, the field is focused on harnessing OVs to train patients' immune systems to recognize and eradicate cancer. Recent scientific progress has enabled the shift in focus to clinical translation, and as such it has become increasingly important to utilize clinically relevant models in the development and testing of novel OVs. As a result of this need, some research groups have developed their own experimental models derived from clinical or surgical specimens at their home institutions. This chapter outlines methodology for the generation of three distinct patient-derived experimental models. This includes cell cultures derived from (1) malignant ascites of ovarian cancer patients, (2) direct excision of metastatic melanoma, and (3) core biopsies of pancreatic cancer. Each of these methodologies has been utilized to generate replicating cellular models and have been demonstrated as valuable tools in in vitro, in vivo, and ex vivo studies.

Key words Oncolytic virus, Cancer, Primary model, Patient derived, Cell culture, Ovarian cancer, Melanoma, Pancreatic cancer, Personalized medicine

1 Introduction

As the focus of the oncolytic virus (OV) field has changed to one in which true translational research is occurring [1, 2], it has become increasingly important to study the effects of clinical candidate OVs in models that closely replicate human cancers. Tissue culture using immortalized cell lines is an incredibly powerful tool for basic scientific studies; however, as our field moves toward one in which clinical implications are becoming commonplace, more realistic alternatives to such cell lines are welcome. As such, some groups have sought techniques to develop models directly from patient-derived tissues, and in so doing have streamlined several approaches by which these models can be both generated and utilized.

Christine E. Engeland (ed.), *Oncolytic Viruses*, Methods in Molecular Biology, vol. 2058,
https://doi.org/10.1007/978-1-4939-9794-7_18,

There are multiple disadvantages to using immortalized cell lines as experimental systems. One example has resulted from a question that has been the subject of extensive literature debate in recent years and highlights the lack of clarity regarding the origin of some immortalized cell lines. Specifically, is was found that MDA-MB-435, a commonly used model of metastatic breast cancer, has a molecular signature that actually resembles that of the melanoma cell line UCLA-SO-M14 (M14), which raised questions about the tissue of origin of each of these cell lines [3, 4]. It has also been demonstrated that various environmental pressures placed on cells in culture can result in genomic evolution [5, 6], and one can easily speculate on the degree of genomic divergence that might therefore exist amongst immortalized cell lines that have, in many cases, been in culture for several decades. It is well accepted that increasing passages in culture tends to result in cell lines that are decreasingly representative of their original disease state with respect to their metabolism [7], genome [5, 8], and epigenome [9, 10]. In fact, significant evidence exists that supports genomic instability as a necessary factor in the process of natural cellular immortalization [8]. Ultimately, the very genomic changes that promote immortalization have been shown to affect the phenotype of an experimental model. Concurrently, there is abundant evidence in the literature to support the fact that in vitro studies are rarely predictive of what will happen in vivo or in human studies [11]. Despite this knowledge that implies the contrary may be prudent, it is interesting that working with well-established and well-studied immortalized cell lines largely remains the technique of choice when studying human cancers and developing novel OVs.

In response to these model-based shortcomings, we sought to develop methodologies that enabled more clinically relevant models and ultimately improve the quality of OV therapeutic development. As a result, we chose to develop primary cell cultures from tumor samples obtained from individuals with cancer undergoing medical procedures relating to diagnosis, treatment, or symptomatic management. Specifically, we have chosen to focus on patients with diagnoses of (1) ovarian cancer who are undergoing peritoneal paracentesis for symptomatic relief of accumulating malignant ascites, (2) malignant melanoma who are undergoing surgical excision of their metastatic foci, in-transit metastases, or their primary tumor, and (3) pancreatic cancer who are undergoing endoscopic ultrasound-guided fine needle aspiration (EUS-FNA) biopsy procedures to aid in histopathologic diagnosis. We have therefore developed protocols to isolate short-term cultures of patient-derived ovarian cancer, melanoma, and pancreatic cancer that closely resemble their original sources (Keller BA et al., and Chatterjee A et al., unpublished data). The techniques presented in this chapter outline our approach to developing these short-term cultures of the aforementioned tumor types; however, these techniques can theoretically be adapted to derive models from any cancer of interest. A workflow diagram of the steps taken to establish cell cultures from all three sources of tumor tissue outlined in this chapter can be found in Fig. 1.

2 Materials

It is critical for all patient-derived specimens to be obtained following the ethics approval of local institutional research ethics committees, which includes obtaining appropriate patient consent (*see* **Note 1**). At the Ottawa Hospital, patient specimens have been obtained following ethics approval granted by Ottawa Health Science Network Research Ethics Board (OHSN-REB) protocol numbers 20120559-01 (solid tumors) and 20140075-01H (malignant ascites). The methods herein focus on the derivation of experimental models following clinical acquisition of patient specimens and do not detail the clinical acquisition techniques themselves. Likewise, animal experiments need to align with the ethics protocols set forth by local institutional guidelines for animal care.

2.1 Collection of Material Derived from the Paracentesis of Malignant Ascites

For the scientist, aspirated clinical material will arrive in a vacuum-sealed bag or container following peritoneal paracentesis by the clinician. These materials are more likely to be free of any microbial contaminants than solid tumor material that must undergo several steps of handling before arriving for experimental manipulation.

1. Direct delivery of freshly obtained malignant ascites from clinical source(s) to experimental laboratories (*see* **Note 2**).
2. 50 mL conical centrifuge tubes.
3. Benchtop centrifuge dedicated for tissue culture experiments.
4. Investigator-selected disposable tissue culture plates.
5. Tissue culture medium: Roswell Park Memorial Institute (RPMI)-1640 medium, 10% fetal bovine serum, and 1× Antibiotic-Antimycotic.
6. Single-purpose Biological Safety Cabinet for tissue culture experiments.
7. Standardized incubator for maintaining cell cultures: 37 °C, 5% CO_2.
8. Microscope for tissue culture purposes.
9. All additional standard disposable tissue culture supplies.

2.2 Collection of Material from Solid Tumors

1. Tissue collection medium for melanoma cultures: RPMI-1640 medium, 10% fetal bovine serum, 100 μg/mL Gentamicin, 0.1 mg/mL Normocin™, and 1× Antibiotic-Antimycotic.

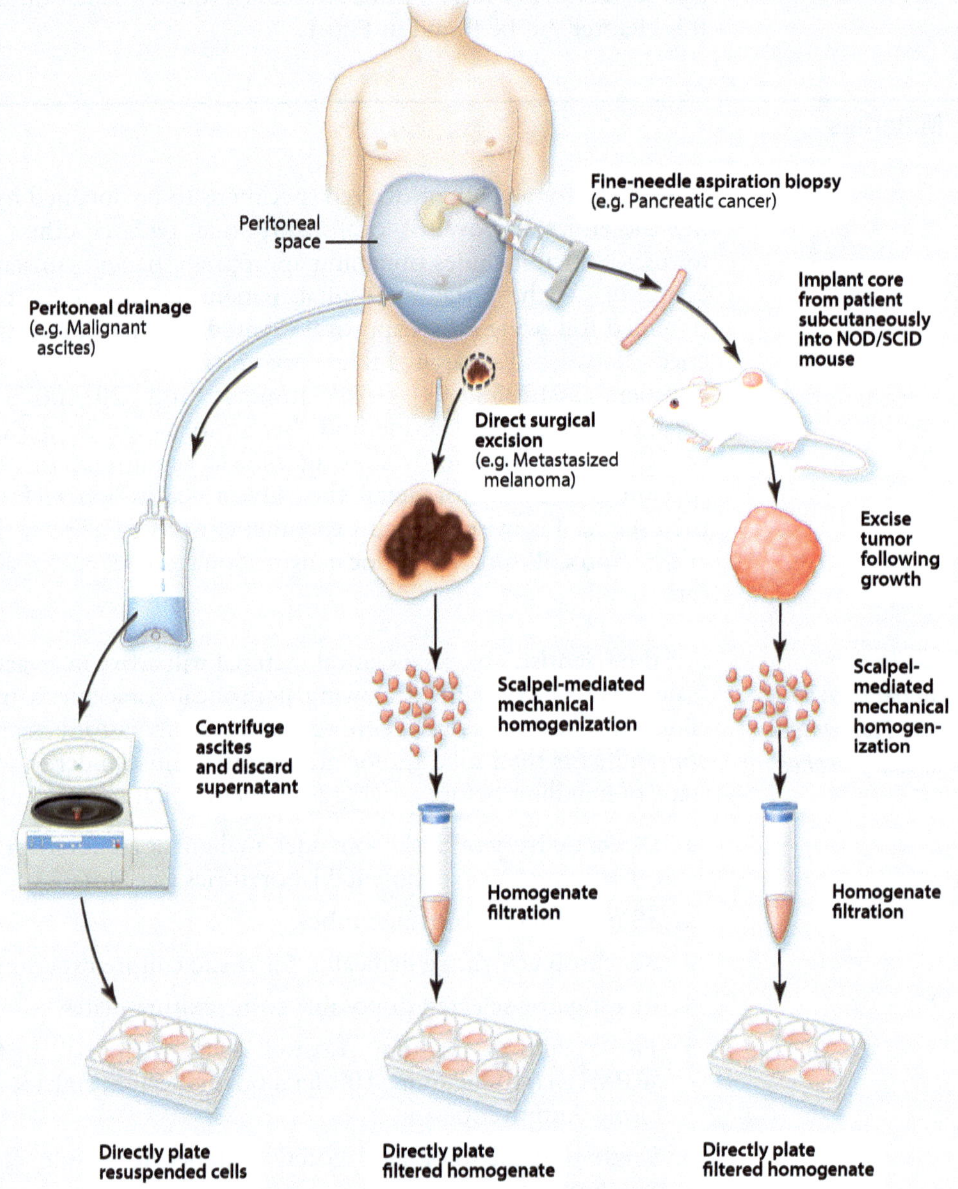

Fig. 1 Overview of the three independent pathways of primary short-term cell culture development highlighted in this chapter. Each pathway outlined in this figure schematizes the methodology used in the development of primary cultures from malignant ascites, from malignant melanoma, and from cores derived from the fine-needle aspiration biopsy of pancreatic cancer specimens

2. Tissue collection medium for pancreatic cancer cell cultures: RPMI-1640 medium, 10% fetal bovine serum, 15 mM HEPES, 0.5 mM sodium pyruvate, 2 μg/mL insulin, 5 μg/mL transferrin, 40 mg/mL hydrocortisone, 10 mg/mL hEGF, 100 μg/mL Gentamicin, 0.1 mg/mL Normocin™, and 1× Antibiotic-Antimycotic (*see* **Note 3**).
3. 1× Phosphate buffered saline (PBS).
4. Collection tubes that can be taken to operating theater and/or to gross dissection room (*see* **Note 4**).
5. Standardized surgical scalpel blades or razor blades.
6. Standardized tissue culture plates: 15 cm plates, 24-well plates, and/or preferred tissue culture flask sealed with filtered cap.
7. Tissue culture materials as described above: 50 mL conical centrifuge tubes, 70 μm nylon mesh cell strainer, single-purpose Biological Safety Cabinet for tissue culture experiments, standardized incubator for maintaining cell cultures, microscope for tissue culture purposes, and all additional standard disposable tissue culture supplies.

2.3 Diagnostic Confirmation of Early Culture Materials

Establishing patient-derived experimental models can at times seem labour-intensive, especially for inexperienced individuals working in a setting lacking streamlined protocols in place. Thus, there may be interest in performing various procedures to confirm diagnosis of isolated specimens before investing significant time and resources into establishing a model. For this reason, we have used several methods to confirm the initial diagnosis, the materials for which are outlined here.

1. Optional: punch biopsy and histology tools to aid in the diagnostic confirmation of isolated specimen by histology.
2. Optional: primers for RT-PCR or RT-qPCR of early culture material (once established) if interested in obtaining some degree of diagnostic confirmation.
3. Optional: appropriate antibodies if diagnostic confirmation is desired at the protein level.
4. Optional: access to molecular diagnostician and appropriate laboratory space if next-generation sequencing (NGS)-based diagnostic confirmation is desired.

2.4 Oncolytic Virus Development

1. Oncolytic virus platforms of choice.
2. Purpose-dedicated pipettes for working with OV candidates.
3. All tissue culture material as outlined above.

2.5 In Vivo Mouse Models of Patient-Derived Cancers (Optional)

1. Appropriate local institutional ethics protocols must be in place before implanting any human-derived surgical specimens of cancer and/or injecting any OV platform into animal models.
2. NOD/SCID mice.
3. CD1 nude mice.
4. For implantation of pancreatic cancer cores: Cultrex® high protein concentration Basement Membrane Extract.
5. 5×10^6 to 1×10^7 cells from model of interest or directly excised human tumor for implantation.
6. Large-bore needle and syringe for injection of cultured cells that have been resuspended, and/or surgical scalpel for implantation of solid specimens.
7. Clean forceps for manipulation of solid tumor specimens.

3 Methods

3.1 Generation of Cell Lines from Malignant Ascites

As mentioned, it is crucial to have appropriate ethics committee approval in place before malignant ascites is collected for research purposes. Likely, it will be important to obtain patient consent prior to the procedure, which is classically the responsibility of the clinician or a member of the clinical team. Typically, the purpose of this procedure is to provide symptomatic relief to the patient and depending on the clinical needs, the sample may not need to be analyzed and can therefore be entirely devoted to research activities. In some cases, patients will tolerate the removal of several liters of ascites, which we have found can provide an abundant source of cellular material.

1. Obtain collected paracentesis specimen from clinical source. This is usually collected into a vacuum sealed container or single-use plastic bag that is produced for this purpose and will be transferred to the laboratory from the clinic quickly after use. Since this procedure is usually performed at the bedside using aseptic protocols, there is greater chance for the researcher to obtain a sample free of microbial contaminants.
2. Place entire container into biological safety cabinet and prepare enough 50 mL conical tubes to hold the contents of fluid that will be utilized. Choice of tubes or containers can be modified to suit experimental needs if, for example, a significant volume of ascites is obtained.
3. Centrifuge fluid at $1000 \times g$ for 10 min. Aspirate off supernatant fluid and discard appropriately. Alternatively, one may consider keeping some acellular fluid following centrifugation and mixing it with medium (1:4; acellular fluid: cell culture medium) for application to initial culture once plated.

4. Resuspend pelleted material into the desired amount of tissue collection medium (or mixture of acellular fluid: cell culture medium).
5. Directly plate resuspended cells onto desired tissue culture dishes. We recommend the use of 24-well tissue culture plates, or if interested in generating a larger culture, into 15 cm tissue culture plate or tissue culture flask sealed with a filtered cap.
6. After overnight incubation, visualize degree of cell adherence using a microscope. Significant debris can be expected at this point and should be aspirated off and discarded.
7. Top up wells with fresh tissue collection medium.
8. After overnight incubation, repeat this process of removing approximately half of the medium and topping up with fresh medium for 3–5 days; this ensures that the replicating cancer cells are not completely starved of any potentially secreted growth factors while also ensuring a relatively consistent supply of fresh medium and antibiotics to prevent any residual microbial growth.

3.2 Generation of Cell Lines from Excised Melanoma Specimens

All steps prior to incubation should be carried out at room temperature. Before getting the tissue to the experimental laboratory, most institutions will not have facilities to handle the specimen in a sterile manner, so care must be taken at each step to avoid microbial contamination. It is important to understand that the time required to establish patient-derived cell cultures is quite variable and case-dependent. For some specimens, a replicating and relatively pure cell line can be obtained in as little as 3–4 days, while other specimens can take several weeks to be well established. It will be up to the end user to gain an understanding of their individual specimens.

1. Obtain surgical specimen from primary melanoma, site of in-transit metastasis, or site of lymphatic spread. This tissue can be obtained directly from the surgeon, or otherwise must be obtained from the individual assigned to gross dissection, depending on what ethics protocol allowances are in place. In all cases in which melanoma is excised from a patient, histologic confirmation of negative margins is of paramount clinical and prognostic importance and therefore acquisition of tissue for experimental purposes cannot interfere with these needs.
2. Tissue should be collected directly into a conical tube containing sufficient antibiotic-containing growth medium to submerge the specimen.
3. Place conical tube containing tissue and medium into Biological Safety Cabinet and pour entire contents of tube into 15 cm tissue culture plate.

4. With one scalpel in each hand, homogenize the tumor specimen extensively into small fragments that can be transferred via a pipette.
5. Place 70 μm cell strainer onto the top of a 50 mL conical tube.
6. Transfer entirety of homogenized specimen contents and medium from 15 cm tissue culture plate onto the top of the 70 μm cell strainer and allow passive filtration of contents into tube (*see* **Note 5**).
7. Discard the tumor specimen that did not pass through the cell strainer along with the cell strainer itself.
8. Top up the 50 mL conical tube with enough fresh medium to ensure safe centrifugation.
9. Centrifuge the contents of the 50 mL conical tube for 10 min at 1000 × *g*. Discard supernatant. Resuspend pelleted material into enough fresh tissue collection medium to meet final volume requirements for your respective experimental protocol.
10. Divide resuspended material into individual wells of 24-well tissue culture plate, or into 15 cm tissue culture plate or tissue culture flask sealed with a filtered cap for larger cultures.
11. After overnight incubation, use microscope to examine degree of cellular adherence. One can expect significant non-adherent cellular debris at this point, which can sometimes contain a rich source of patient-derived T cells that could potentially be collected and propagated in tissue culture for experimental purposes. The present protocol does not include details on this methodology; however, the medium must be supplemented with various growth factors and/or antibodies for success [12–15], and replicating tumor-infiltrating lymphocytes must then be characterized using flow cytometry (*see* **Note 6**).
12. Remove approximately half of the liquid contents of the well at this point by aspiration and top up well with fresh tissue collection medium.
13. After overnight incubation, repeat this process of removing approximately half of the medium and topping up with fresh medium for 3–5 days. This ensures that the replicating cancer cells are not completely starved of any potentially secreted growth factors and also ensures a relatively consistent supply of fresh medium and antibiotics. After 3–5 days of replenishing the culture medium, notable cell replication or early colony formation is observed and the medium should be relatively clear of any non-adherent debris (*see* **Note 7**).

3.3 Generation of Cell Lines from Fine-Needle Aspiration Biopsy of Pancreatic Cancer

The generation of cell lines from EUS-FNA biopsy materials follows a fundamentally similar protocol to that outlined above; however, it is inherently more challenging by definition due to the relative paucity of starting tissue. Like with other tissue sources, prime importance must be placed on getting appropriate ethical consent following institutional protocols, patient consent, and on not sacrificing any clinical outcome for experimental gain. We have optimized this protocol for the derivation of cell lines from pancreatic cancer; however, it could theoretically be utilized successfully whenever fine-needle aspiration biopsy is performed.

1. Obtain excised material from the clinician following the acquisition of all appropriate consent from the patient and institution.
2. Tissue should be collected directly into a conical tube containing sterile 1 × PBS and taken directly to local animal care facility for implantation into NOD/SCID mice.
3. Prepare NOD/SCID mice following local handling instructions.
4. Remove excised tumor core from conical tube and 1 × PBS and place onto tissue culture plate so that tissue can be manipulated. Cut cores into smaller lengths of approximately 5 mm and repeat this process until entire core has been divided into smaller tubular fragments of approximately 5 mm in length by 2 mm in diameter.
5. Make small sub-cutaneous incision(s) on one or both flanks of each mouse following standardized minor surgical procedures. Implant 1–2 tumor core fragments at sub-cutaneous incision site along with Cultrex® high protein concentration Basement Membrane Extract. Proceed to close incision with surgical sutures, staples, or skin glue as your normal animal care practices would dictate.
6. Provide normal nutrition and animal care to mice having received implantation of xenograft and observe intermittently for tumor growth under implanted flanks. Once tumor growth is established and has progressed to the desired size, mice can be sacrificed and tumors can be excised for manipulation. Tissue should be collected directly into a conical tube containing sufficient antibiotic-containing growth medium to submerge it fully. This tumor-containing conical tube can stay at room temperature for transferring between animal care facility and laboratory space.
7. Place conical tube containing tissue and medium into biological safety cabinet and pour entire contents of tube into 15 cm tissue culture plate.

8. With one scalpel in each hand, homogenize tumor specimen extensively into small fragments that can be transferred via pipette.
9. Filter homogenized specimen contents and medium from 15 cm tissue culture through a 70 μm cell strainer into a 50 mL conical tube.
10. Top up the 50 mL conical tube with enough fresh medium to ensure safe centrifugation.
11. Centrifuge the contents of the 50 mL conical tube for 10 min at 1000 × *g*. Discard supernatant. Resuspend pelleted material into enough fresh tissue collection medium to meet the final volume requirements for your respective experimental protocol.
12. Divide resuspended material into individual wells of 24-well tissue culture plate, or if interested in generating a larger culture, into 15 cm tissue culture plate or tissue culture flask sealed with a filtered cap.
13. After overnight incubation, use microscope to examine degree to which cellular adherence to the tissue culture plate has occurred.
14. Remove approximately half of the liquid contents of the well at this point by aspiration and top up well with fresh tissue collection medium.
15. After overnight incubation, repeat this process of removing approximately half of the medium and topping up with fresh medium for 3–5 days. This ensures that the replicating cancer cells are not completely starved of any potentially secreted growth factors and also ensures a relatively consistent supply of fresh medium and antibiotics. After 3–5 days of replacing some medium, one should notice significant in vitro replication and the medium should be relatively clear of any non-adherent debris. One may also observe early colony formation in culture, which can be reassuring (*see* **Note 7**).

3.4 Confirmation of Diagnosis in Patient-Derived Cell Cultures

After successful establishment of cell cultures, some investigators may be interested in diagnostic confirmation. We have utilized several methods including RT-qPCR assays designed to detect select diagnostic markers at the genome level, immunoblotting and/or immunostaining to detect diagnostic markers at the protein level, and finally histopathology and NGS sequencing of originally isolated specimen and established cell cultures to gain confidence in source of material. Any of these procedures can be performed using standardized methodology that is beyond the scope of this chapter and will not be covered herein.

The degree to which cellular replication must occur is to be defined by the end user. Once sufficient material is obtained, cell lines can be treated as any other cell line and can be cryopreserved or used directly for experimental purposes. Cells can be directly infected with selected OVs and experimental readouts can be defined as per the user's interest. In our experience, not all cell lines are tumorigenic in NOD/SCID and/or CD-1 nude mouse models; however, others are and tend to closely represent their original tumor source histologically.

3.5 Use of Oncolytic Viruses in Patient-Derived Primary Cell Cultures

Established short-term cell cultures from patients can be utilized for experimental purposes in a similar fashion to immortalized cell lines. Each of the presented models has a characteristic sensitivity to various OV platforms. In our laboratory, commonly used virus platforms include various herpesviruses, poxviruses, and rhabdoviruses. Each of these families of viruses are able to readily infect and replicate in the primary cultures that we have generated, and various experimental approaches can be performed to suit the user's end needs (as outlined elsewhere in this textbook). Representative photomicrographs of patient-derived primary cell cultures created in our laboratory infected with green fluorescent protein-expressing virus platforms can be observed in Fig. 2.

4 Notes

1. Obtaining institutional ethics approval can be challenging and time-consuming. We have found it best and most practical to obtain an institution-wide ethics approval for the comprehensive use of surgical specimens from patients with various cancers for diverse experimental purposes across many research groups. This is a critical step to take in order to ensure the ethical procurement of such specimens.
2. When obtaining malignant ascites, it is possible to have a source of up to several liters of fluid. If this is to happen, we advise scaling up from multiple 50 mL conical tubes to larger centrifuge-ready containers. Using larger, and therefore fewer, containers limits the necessary handling of cellular material and therefore the chance of contamination and spills. It is not typically possible to predict the amount of material from a given specimen that will be required to generate a viable culture; however, we have had success with very little material (approximately 15 mL of cell dense ascites) and believe that this protocol will be reproducible in such cases.
3. It is important to add antibiotics in original tissue collection medium as specimens derived from surgical procedures are often placed in "clean" or aseptic containers and are not sterile by definition. Furthermore, if tissue undergoes a gross dissection by a pathologist or pathology assistant, their working

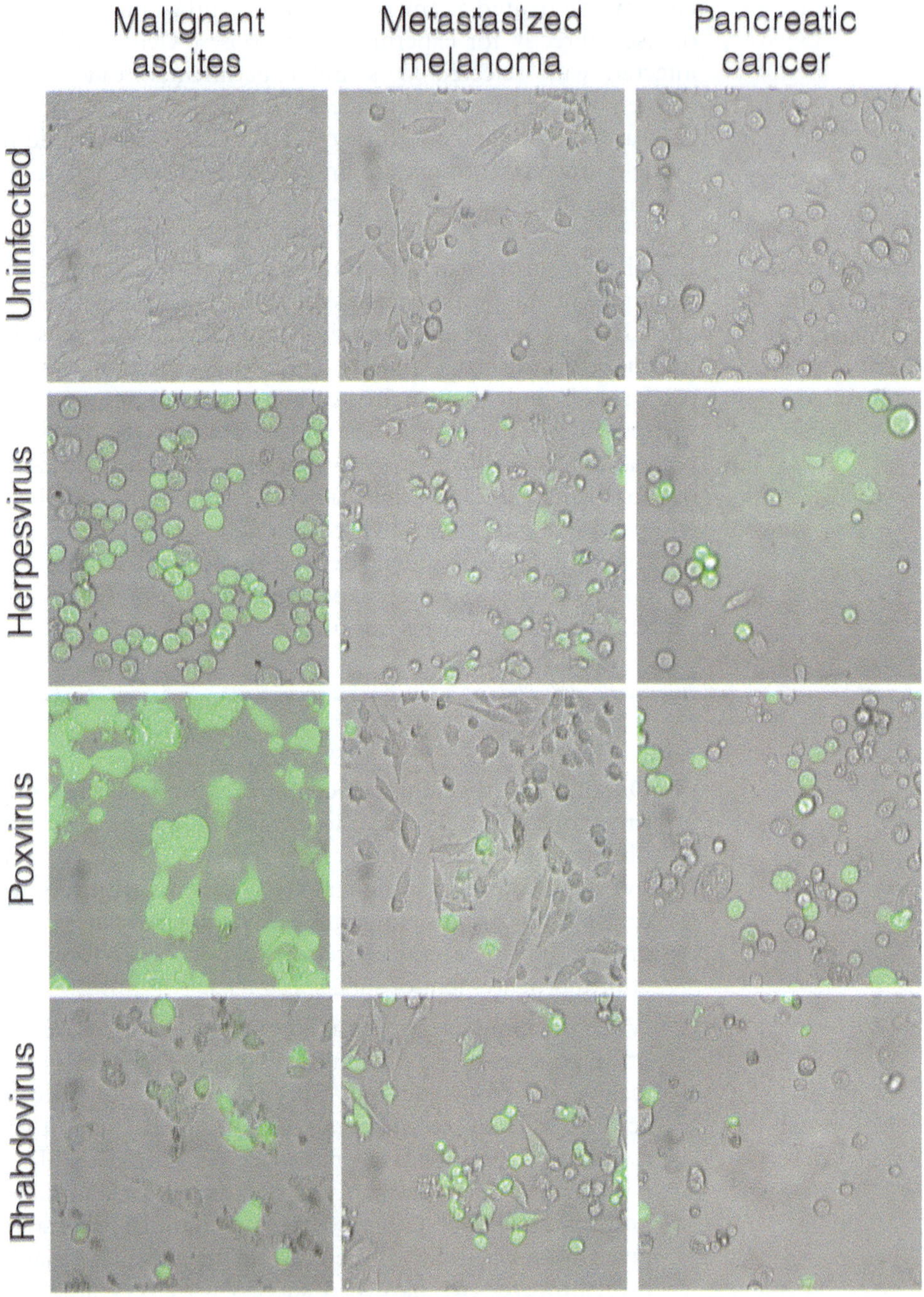

Fig. 2 Representative photomicrographs of green fluorescent protein-labeled oncolytic viruses following infection of primary cultures generated from malignant ascites of ovarian cancer, metastasized melanoma, and fine needle aspiration biopsy of pancreatic cancer. Commonly used oncolytic viruses from the herpesvirus, poxvirus, and rhabdovirus families can readily infect human tissue-derived cultures that have been generated using the methodology described herein

environments are often not equipped to provide a truly sterile environment and we have found that any precautions that can be taken to increase the relative sterility of tissue collection does indeed help in the downstream generation of replicating models.

4. If one of the goals of experiments is to publish data collected using these patient-derived cell cultures, it may be important to accurately document specimen origin. As such, this could include accurate photography, both in the operating theater as well as at the stage of gross dissection. Investigators should prepare to obtain this photographic data, which includes appropriate patient consent and having specific photography equipment available.

   It is not typically possible to predict the amount of material from a given specimen that will be required to generate a viable culture; however, we have had success in generating a viable culture with as little as 50 mm^3 of specimen when an animal amplification step is performed (*see* Fig. 1), and we believe that this protocol will be reproducible in such cases.

5. If attempting to transfer homogenized material via pipette results in plugging of pipette tip with homogenized tissue, a portion of the pipette tip can be cut off using a scalpel blade. This should facilitate easier transfer of homogenized material.

6. After overnight incubation of homogenized tumor specimen, one can expect significant non-adherent cellular debris. Depending on the characteristics of the patient and the original tumor, this can sometimes include a rich source of tumor-infiltrating lymphocytes that can be collected, propagated in culture, and used experimentally as a source of homologous patient-derived tumor-infiltrating lymphocytes.

7. It is prudent to identify the appearance of in vitro replicating fibroblasts since they can be mistaken for the replication of desired cancer cells. Depending on the nature of the original specimen, it is possible that over time and with increasing passages, the replication of fibroblasts outpaces and overwhelms the replication of cancer cells in a culture.

Acknowledgments

This work was partially supported by operating grants held by Dr. John C. Bell. This work was also partially funded by CIHR grant number RN313689-377104 and Canadian Cancer Society Research Institute research grants (2013-702292 and 2018-0665). B.A.K. was supported by a Vanier Canada Graduate Scholarship (2014-2017). A.S. and M.E.W. were independently supported by

Ontario Graduate Scholarships. We acknowledge the technical assistance of our colleagues at the Ottawa Hospital Research Institute, especially Anabel Bergeron, Christiano Tanese de Souza, and Brian Laight. We also acknowledge the many patients who have contributed to our work in an effort to assist in finding better ways to treat and eradicate cancer.

References

1. Achard C, Surendran A, Wedge ME, Ungerechts G, Bell J, Ilkow CS (2018) Lighting a fire in the tumor microenvironment using oncolytic immunotherapy. EBioMedicine 31:17–24. https://doi.org/10.1016/j.ebiom.2018.04.020
2. Keller BA, Bell JC (2016) Oncolytic viruses-immunotherapeutics on the rise. J Mol Med (Berl) 94(9):979–991. https://doi.org/10.1007/s00109-016-1453-9
3. Chambers AF (2009) MDA-MB-435 and M14 cell lines: identical but not M14 melanoma? Cancer Res 69(13):5292–5293. https://doi.org/10.1158/0008-5472.CAN-09-1528
4. Korch C, Hall EM, Dirks WG, Ewing M, Faries M, Varella-Garcia M, Robinson S, Storts D, Turner JA, Wang Y, Burnett EC, Healy L, Kniss D, Neve RM, Nims RW, Reid YA, Robinson WA, Capes-Davis A (2018) Authentication of M14 melanoma cell line proves misidentification of MDA-MB-435 breast cancer cell line. Int J Cancer 142(3):561–572. https://doi.org/10.1002/ijc.31067
5. Stepanenko A, Andreieva S, Korets K, Mykytenko D, Huleyuk N, Vassetzky Y, Kavsan V (2015) Step-wise and punctuated genome evolution drive phenotype changes of tumor cells. Mutat Res 771:56–69. https://doi.org/10.1016/j.mrfmmm.2014.12.006
6. Stepanenko AA, Dmitrenko VV (2015) HEK293 in cell biology and cancer research: phenotype, karyotype, tumorigenicity, and stress-induced genome-phenotype evolution. Gene 569(2):182–190. https://doi.org/10.1016/j.gene.2015.05.065
7. Bhakoo KK, Williams SR, Florian CL, Land H, Noble MD (1996) Immortalization and transformation are associated with specific alterations in choline metabolism. Cancer Res 56(20):4630–4635
8. Yoshioka K, Atsumi Y, Nakagama H, Teraoka H (2015) Development of cancer-initiating cells and immortalized cells with genomic instability. World J Stem Cells 7(2):483–489. https://doi.org/10.4252/wjsc.v7.i2.483
9. Novak P, Jensen TJ, Garbe JC, Stampfer MR, Futscher BW (2009) Stepwise DNA methylation changes are linked to escape from defined proliferation barriers and mammary epithelial cell immortalization. Cancer Res 69(12):5251–5258. https://doi.org/10.1158/0008-5472.CAN-08-4977
10. Degerman S, Landfors M, Siwicki JK, Revie J, Borssen M, Evelonn E, Forestier E, Chrzanowska KH, Ryden P, Keith WN, Roos G (2014) Immortalization of T-cells is accompanied by gradual changes in CpG methylation resulting in a profile resembling a subset of T-cell leukemias. Neoplasia 16(7):606–615. https://doi.org/10.1016/j.neo.2014.07.001
11. Wilding JL, Bodmer WF (2014) Cancer cell lines for drug discovery and development. Cancer Res 74(9):2377–2384. https://doi.org/10.1158/0008-5472.CAN-13-2971
12. Hall M, Liu H, Malafa M, Centeno B, Hodul PJ, Pimiento J, Pilon-Thomas S, Sarnaik AA (2016) Expansion of tumor-infiltrating lymphocytes (TIL) from human pancreatic tumors. J Immunother Cancer 4:61. https://doi.org/10.1186/s40425-016-0164-7
13. Meng Q, Liu Z, Rangelova E, Poiret T, Ambati A, Rane L, Xie S, Verbeke C, Dodoo E, Del Chiaro M, Lohr M, Segersvard R, Maeurer MJ (2016) Expansion of tumor-reactive T cells from patients with pancreatic cancer. J Immunother 39(2):81–89. https://doi.org/10.1097/CJI.0000000000000111
14. Chacon JA, Sarnaik AA, Pilon-Thomas S, Radvanyi L (2015) Triggering co-stimulation directly in melanoma tumor fragments drives CD8(+) tumor-infiltrating lymphocyte expansion with improved effector-memory properties. Oncoimmunology 4(12):e1040219. https://doi.org/10.1080/2162402X.2015.1040219
15. Dudley ME, Wunderlich JR, Shelton TE, Even J, Rosenberg SA (2003) Generation of tumor-infiltrating lymphocyte cultures for use in adoptive transfer therapy for melanoma patients. J Immunother 26(4):332–342

Chapter 19

Considerations for Clinical Translation of MG1 Maraba Virus

Caroline J. Breitbach

Abstract

Oncolytic viral immunotherapy based on the MG1 Maraba platform has undergone extensive preclinical evaluation, resulting in the advancement of two programs into clinical trials. MG1 Maraba encoding tumor antigens (tumor associated antigens or viral antigens) are used to boost antitumor immunity, while MG1 Maraba infects tumors, causes oncolysis and transforms the tumor microenvironment. An overview of MG1 Maraba clinical development is outlined here, along with general considerations relating to the design of clinical trials for complex biologic products such as oncolytic viral immunotherapies. These include choice of patient population, optimized treatment regimen, and endpoints which provide early signals of activity and inform the late-stage development path of these agents with novel mechanisms of action.

Key words Oncolytic virus, Maraba virus, Clinical translation, Clinical trial design, Heterologous prime–boost

1 Introduction to MG1 Maraba Oncolytic Viral Immunotherapy

Oncolytic viral immunotherapies are a promising new class of therapeutics for cancer patients. MG1 Maraba is a novel oncolytic virus that was selected from an intensive screening process and engineered to kill cancer cells; it is a highly selective and potent novel OV capable of infecting and killing human and murine tumor cells both in vitro and in vivo [1]. One of the major mechanisms rendering tumor cells vulnerable to viral oncolysis is the lack of responsiveness to the type I interferon (IFN)-mediated protection within the tumor itself, a defect that is readily exploitable by the IFN-sensitive MG1 Maraba virus [2]. MG1 Maraba targets tumors through multiple complementary mechanisms-of-action, including (1) direct lysis of cancer cells, (2) modulation of the tumor microenvironment (TME), and (3) generation of tumor antigen-specific T cells. The presence of MG1 Maraba in tumors transforms the TME, creating an inflammatory state that turns immunologically "cold" tumors "hot" to make them more susceptible to immune-mediated attack, and induces oncolysis by releasing novel tumor-

Christine E. Engeland (ed.), *Oncolytic Viruses*, Methods in Molecular Biology, vol. 2058,
https://doi.org/10.1007/978-1-4939-9794-7_19,

specific antigens, allowing for antigen spread, or generating T cells that target a broader array of antigens within the tumor. The transition of the TME to a more proinflammatory state better enables the influx and activation of T cells. It was recently demonstrated that MG1 treatment resulted in T cell infiltration in the syngeneic 4T1 triple negative breast cancer model [3]. In addition, TME changes were noted following MG1 Maraba administration in a mouse model of prostate cancer (TRAMP-C2), a classically immunologically 'cold' tumor type. Treatment with MG1 Maraba resulted in tumors with upregulation of T cell markers, T cell activation markers, immune checkpoints, and genes involved in antigen processing, indicative of a genetic signature of T cell attack and enhanced antigen presentation [4].

Though the potential of OVs to induce therapeutic immune responses against cancer antigens has been demonstrated with Imlygic (Amgen), the first oncolytic virus to be approved in the USA and Europe, data from preclinical models and patients suggest that the natural induction of anti-tumor immunity is not robust with a pure oncolytic approach. It appears that only in a minority of cases is the infection of the tumor bed and the concomitant release of endogenous tumor antigens sufficient to trigger a complete tumor response. Therefore, we utilized the ability to genetically engineer MG1 Maraba by encoding selected tumor antigens within the virus backbone, resulting in the generation of a robust immune response directed against a specific antigen or antigens expressed on the patient's cancer cells.

This is accomplished by using heterologous prime–boost technology, a method that has been shown to be effective in the setting of infectious disease vaccines. Essentially two immunologically distinct viruses expressing the same antigen are used in sequence first to prime and then to boost an immune response against the encoded antigen. For instance, in cancer models an adenovirus vaccine has been used to locally prime an immune response against a defined tumor cell antigen. This is followed by systemic therapy with an oncolytic virus encoding the same antigen. This strategy has proven to induce large antitumor immune responses, (e.g., 30% of circulating T cells; [5]). In the setting of vaccination against a foreign antigen (the E6 and E7 of the oncogenic human papilloma virus [HPV], immune responses of even greater magnitude (approximately 60% of circulating T cells; [6]) was demonstrated. Beyond tumor-associated or viral antigens, patient specific neoantigens may also be encoded in the MG1 oncolytic viral immunotherapy platform. Finally, the mechanism by which the OV expressing a tumor antigen induces immune responses of large magnitude has been recently elucidated [7].

By generating tumor antigen-specific T cells, modifying the TME and inducing inflammation, MG1 Maraba may enable therapies such as immune checkpoint inhibitors, which rely on the presence of T cells in a proinflammatory environment, to be successful. This provides strong rationale for the combination of MG1 Maraba and anti-PD-1/L1 antibody therapies. Our preclinical models have demonstrated synergistic activity when MG1 is used in combination with these checkpoint inhibitors (unpublished data).

These preclinical data provided strong rationale for clinical evaluation of MG1 Maraba expressing tumor associated and viral antigens in clinical trials. These trials of MG1 alone or in combination with PD-1/L1 immune checkpoint axis blockers are currently underway (NCT02285816, NCT03618953, NCT03773744, NCT02879760). The following section will outline some key considerations which informed the design of these clinical trials.

2 Clinical Trial Design Considerations

2.1 Patient Population

Firstly, the patient population appropriate for initial clinical study of MG1 Maraba needs to be considered. For first-in-man clinical studies in which the safety of a novel product candidate is to be defined, evaluation of patients who have failed standard of care treatments is customary. If a signal of antitumor activity is observed in this setting, there may be a clear and rapid path to approval in this setting of high unmet medical need.

However, targeting late-stage patients comes at cost of less fit patients who may not be as likely to benefit from the mechanism of action (e.g., immunosuppressed patients may not respond to agent designed to stimulate antitumor immunity), live long enough to benefit from therapy (especially if response is delayed), or may poorly tolerate therapy (in particular with OV therapies that may lead to untoward toxicities in immunosuppressed, heavily pretreated patients). For example, in the pivotal Phase 3 study in melanoma patients leading to the approval of Imlygic (Amgen) [8], the study met its primary endpoint of improvement in durable response rate vs treatment with GM-CSF protein [9]. However, an exploratory subset analysis revealed a clear overall survival advantage found in treatment-naïve but not pretreated melanoma [10]. Arguably the outcome of the study may not have been as compelling had it included only pretreated patients or end-stage patients exclusively.

For the MG1 Maraba oncolytic viral immunotherapy approach, in which MG1 Maraba is engineered to express a tumor antigen, the choice of antigen drives the selection of patients as well. For initial evaluation of MG1 Maraba, the Melanoma-associated antigen 3 (MAGE-A3) was selected. The prior extensive safety database of MAGE-A3 vaccination could be leveraged in the first clinical evaluation of MG1 Maraba [11]. MAGE-A3 is a cancer testes antigen which has been evaluated extensively as a target for cancer vaccination as it is expressed in a broad array of solid tumor types, including melanoma and lung cancer [12].

In addition, an MG1 Maraba expressing both E6 and E7 of HPV serotypes 16 and 18 has advanced into clinical development. This is based on preclinical data demonstrating very robust immune responses against these viral antigens which are known drivers of malignancy [6]. Hence, patients with metastatic HPV associated tumors, including all cervical and anal cancers, as well as a subset of patients with head and neck cancer, vaginal, vulvar and penile cancers are appropriate for clinical evaluation of MG1-E6E7.

With the approval of immune checkpoint inhibitor therapies, particularly those blocking the PD-1/PD-L1 axis, in many advanced cancer indications, the setting in which the safety and efficacy of oncolytic viral immunotherapies is evaluated must be carefully considered. Some tumor types, including microsatellite stable colorectal cancer, prostate cancer and pancreatic cancer, are considered immunologically 'cold' tumors as single agent immune checkpoint therapy has not proven to be beneficial in these patients. An oncolytic viral immunotherapy such as MG1 Maraba expressing tumor antigens may sensitize these tumors to immune checkpoint therapy by transforming the TME and inducing antitumor immunity that can be "enabled" with the addition of the immune checkpoint inhibitor therapy. In contrast, tumor types in which a subset of patients benefit (e.g., NSCLC, melanoma, and renal cell cancer) may be a setting in which MG1 Maraba could either increase the proportion of patients that respond (in the immune checkpoint inhibitor naïve setting) or resensitize patients who have previously failed such therapies.

2.2 Treatment Regimen

Firstly, with development of oncolytic viral immunotherapies, the route of administration must be considered. Currently, the regimen of the only approved agent in the USA, Imlygic (Amgen), consists of multiple intratumoral injections of the therapy in accessible melanoma lesions. This administration concentrates Imlygic within injected tumors and relies on the abscopal effect for clearance of noninjected tumors. Alternatively, intravenous infusion of oncolytic immunotherapies can be considered. Several oncolytic agents, including vaccinia [13] and adenovirus [14] were shown to be delivered to tumors following systemic administration. This regimen has not been evaluated in late-stage clinical trials to date. A unique consideration for MG1 Maraba oncolytic viral immunotherapy pertains to the fact that the immune boost mechanism is dependent on MG1 access B cells in the splenic follicle. Therefore, in mouse models, the immune boost is only achieved following intravenous administration [7]. Hence, an intravenous infusion is incorporated into all regimens being clinically tested. These include multiple intravenous infusions (up to four on a twice-weekly frequency) as well as a combination of intravenous infusion followed by intratumoral injections (thereby achieving the immune boost and concentrating subsequent doses within the tumor bed).

Strategies are being pursued to maximize the amount of input MG1 Maraba dose within tumor beds, even after intravenous dosing. The majority of adverse events post MG1 Maraba dosing are after the first infusion; the second and subsequent doses are better tolerated. Hence, a 'step-up' dosing approach is being tested, in which the second and subsequent doses are escalated while keeping the first dose constant. This approach, termed 'step-up dosing' is being evaluated, with doses of up to 1×10^{12} PFU being shown to be well-tolerated to date.

For the first in-human study of MG1 Maraba, a clinical trial incorporating both evaluation of the immune boost encoding antigen (MAGE-A3) along with the priming vaccination (adenovirus expressing MAGE-A3) was required. Hence, MG1-MAGEA3 administered by intravenous infusion was dose escalated on its own. In parallel, patients were dosed with the adenovirus priming vaccination (Ad-MAGEA3). Though adenovirus vaccines had previously been evaluated in clinical trials, this construct expressing the MAGE-A3 antigen had not been yet tested clinically. Upon completion of these two arms, MG1-MAGEA3 was dose escalated once again, this time in combination with the Ad-MAGEA3 priming vaccination. The Ad-MAGEA3 dose is administered by intramuscular injection, followed 2 weeks later with MG1-MAGEA3 treatment initiation.

Following initial evaluation of the safety of Ad/MG1-MAGEA3, three clinical trials evaluating MG1 in combination with immune checkpoint inhibitor antibodies (blocking the PD-1/L1 axis) have been initiated. In these trials, MG1 expressing the MAGE-A3 antigen or HPV antigens (E6 and E7 of HPV serotypes 16 and 18), is dose escalated in combination with the immune checkpoint inhibitor antibody. The standard dose of pembrolizumab (Merck) or atezolizumab (Roche) is being utilized. Preclinical studies should define the optimal dosing regimen to evaluate clinically. In particular, the impact of concurrent anti-PD-1 antibody dosing on safety, immune response and efficacy should be evaluated.

Other combination therapy approaches may also be considered, with agents including chemotherapy or targeted therapy. Importantly, at minimum, given the impact of these various agents on cell metabolism and cell replication, it must be established that OV replication and gene expression is not inhibited by potential combination agents. Notably, despite being potentially important to establish safety and compatibility, single arm combination studies including OV and a second agent must be interpreted with caution. Although the results of such studies can be compared to historical data or controls, the limitations of such analysis must be clearly acknowledged and follow-on studies employing an appropriate control arm conducted.

Although combination therapy has traditionally been conducted in oncology using investigational plus approved agents (e.g., novel agent plus standard of care), the FDA has published guidelines for codevelopment of new investigational drugs (testing 2 or more novel unapproved agents) that may be particularly applicable to OV as, for instance, part of a vaccine cocktail (e.g., prime-boost strategies) (www.fda.gov/downloads/drugs/guidancecomplianceregulatoryinformation/guidances/ucm236669.pdf+&cd=1&hl=en&ct=clnk&gl=us). Apart from the complexities of obtaining agreement for codevelopment should the agents be controlled by separate commercial entities, codevelopment of unapproved novel agents presents significant challenges compared to that of the traditional single agent pathway. Clearly delineating the relative contribution to safety and efficacy as well as the optimal dose and schedule of the individual components is of obvious import. This, as FDA suggests, may require brief initial evaluation of each agent on separate dose escalation arms, if possible and ethical, even if preclinical evidence strongly suggests the combination as the only relevant approach. Alternatively, a brief serial exposure to each component of the combination followed by combination thereafter may be employed to establish relative safety and activity. Regardless, a solid rationale for codevelopment based on preclinical and early clinical data should be utilized when considering the merits of this approach.

2.3 Endpoints

Despite the primary endpoint of early-phase clinical trials being safety and defining the maximum tolerated/feasible dose, these studies can also be used to understand and validate the mechanism of action of oncolytic viral immunotherapies. In particular, extensive correlate testing can be incorporated during the initial dose escalation and expansion stages of phase 1/2 trials. These include (1) monitoring for changes in the TME, (2) measuring induction of antitumor immune responses, and (3) monitoring for evidence of viral replication.

Firstly, we have designed all early-phase clinical studies of MG1 Maraba to include collection of serial biopsies, that is, collecting tumor material before and after prime–boost treatment. Only with collection of multiple biopsies over time can changes in the TME be measured. Tumor tissue is collected both for gene expression analysis (custom Nanostring probe set as well as RNASeq analysis) as well as immunohistochemical analysis. Gene expression analysis can reveal whether a response to viral infection has been mounted (e.g., by measuring interferon responsive genes such as IFIT1, IFIT2, IFITM1, and IFI35) and also monitor for T cell presence and activation (e.g., CD8, granzyme B, and IFN-γ). This data, coupled with evidence of T cell infiltration on immunohistochemistry, could inform how MG1 Maraba modifies the TME over time.

In addition, extensive immune analyses are performed on peripheral blood. Immune response induction to the encoded antigen (i.e., MAGE-A3 or HPV E6/E7) are monitored both by intracellular cytokine staining (ICS) flow cytometric analysis as well as ELISpot. ICS flow analysis, when performed without any ex vivo expansion of T cells, can inform the frequency of T cells reactive to the antigen of interest. Furthermore, cell phenotyping can be performed, including evaluation of T cell polyfunctionality (by monitoring for production of IFN-γ, TNFα and IL-2). In addition, ELISpot analysis can be performed to identify T cells specific to the encoded antigen, as well as additional tumor-associated antigens expressed by the patient's tumor (as identified by gene expression analysis).

Finally, early-phase clinical trials can be designed to monitor for presence of the oncolytic viral immunotherapy in the tumor, as well as replication. The Nanostring probe set can be customized to include probes for both the MG1 Maraba genome and transcript and tumor can be collected for infectious virus titer analysis. Furthermore, evidence for virus replication in patients can be monitored by measuring MG1 Maraba genomes in the systemic circulation serially after dosing. After acute clearance of MG1 Maraba post infusion, delayed reemergence of genomes may be observed. This is an indication of MG1 Maraba replication in patients, and represents a unique pharmacokinetic property of replicating oncolytic viral immunotherapies. Given that multiple serial blood collections are more feasible than collection of biopsies at multiple timepoints, this indirect measure of replication within days to weeks post dosing is a useful tool to monitor for replication in patients at multiple timepoints.

Early-phase clinical trials are not powered to define efficacy of a novel therapeutic agent. However, early signals of antitumor activity will drive the design of larger studies; therefore endpoints relating to efficacy should be carefully considered. Traditional RECIST imaging endpoints [15], monitoring for growth of existing tumors as well as development of new tumors, should be complemented by imaging modalities that take into account the possibility of transient tumor flare, consistent with the mechanism of action of oncolytic viral immunotherapies. Indeed, delayed response and even initial tumor flare has been documented after treatment with immune checkpoint inhibitor therapy (e.g., ipilimumab and nivolumab [16, 17]). Hence, continued radiographic assessment beyond initial progression and the implementation of alternative radiographic systems including the immune response criteria (e.g., iRECIST) [18] may be of value.

Given the complexity of evaluating antitumor response by radiographic imaging, considering the evaluation of a tumor type in which a tumor marker (e.g., CEA, PSA, CA19-9) can be monitored is warranted in order to reliably gauge antitumor activity early in the development process. The demonstration of a reduction in tumor size or marker level following single agent treatment is much more valuable than single arm data on OS or PFS for a small cohort of patients compared to historical outcomes. The latter endpoints are much more susceptible to selection bias (e.g., a very healthy population) in early phase studies. For example, brain tumor, in particular glioblastoma multiforme, radiographic response assessment is confounded by the significant impact of edema, necrosis, and inflammation on imaging interpretation [19]. Similar challenges are present with tumors affecting bone and hollow viscera [20]. Furthermore, adjuvant therapy approaches (treatment following resection of all measurable disease with primary endpoints of OS or PFS) require randomized populations and large sample sizes.

Summarized in this chapter is the approach to the early-phase clinical development of MG1 Maraba based oncolytic viral immunotherapy, a therapeutic modality designed to induce tumor oncolysis, modulate the TME and generate tumor antigen-specific T cells. MG1 Maraba expressing various tumor antigens (e.g., tumor-associated antigens or viral antigens) have entered early clinical development with a focus on defining safety and mechanism of action. These early trials will form the basis for future late-stage development of the MG1 Maraba platform.

References

1. Andtbacka RH et al (2015) Talimogene laherparepvec improves durable response rate in patients with advanced melanoma. J Clin Oncol 33:2780–2788
2. Lu S (2009) Heterologous prime-boost vaccination. Curr Opin Immunol 21:346–351
3. Bourgeois-Daigneault MC et al (2018) Neoadjuvant oncolytic virotherapy before surgery sensitizes triple-negative breast cancer to immune checkpoint therapy. Sci Transl Med 10(422)
4. Atherton MJ et al (2018) Transforming the prostatic tumor microenvironment with oncolytic virotherapy. Oncoimmunology 7: e1445459
5. Pol JG et al (2014) Maraba virus as a potent oncolytic vaccine vector. Mol Ther 22:420–429
6. Atherton MJ et al (2017) Customized viral immunotherapy for HPV-associated cancer. Cancer Immunol Res 5:847–859
7. Bridle BW et al (2016) Privileged antigen presentation in splenic B cell follicles maximizes T cell responses in prime-boost vaccination. J Immunol 196:4587–4595
8. Kaufman HL, Bines SD (2010) OPTIM trial: a phase III trial of an oncolytic herpes virus encoding GM-CSF for unresectable stage III or IV melanoma. Future Oncol 6:941–949
9. Ingemar Andtbacka RH et al (2013) ASCO. J Clin Oncol 31(Suppl)
10. Kaufman H et al (2014) Primary overall survival (OS) for OPTiM, a randomized phase III trial of talimogene laherparepvec (T-VEC) versus subcutaneous (SC) granulocyte-macrophage colony-stimulating factor (GM-CSF) for the treatment (tx) of unresected stage IIIB/C and IV melanoma. J Clin Oncol 32:abstr 9008a
11. Vansteenkiste JF et al (2016) Efficacy of the MAGE-A3 cancer immunotherapeutic as adjuvant therapy in patients with resected MAGE-A3-positive non-small-cell lung cancer (MAGRIT): a randomised, double-blind, placebo-controlled, phase 3 trial. Lancet Oncol 17:822–835

12. De Plaen E et al (1994) Structure, chromosomal localization, and expression of 12 genes of the MAGE family. Immunogenetics 40:360–369
13. Breitbach CJ et al (2011) Intravenous delivery of a multi-mechanistic cancer-targeted oncolytic poxvirus in humans. Nature 477:99–102
14. Garcia-Carbonero R et al (2017) Phase 1 study of intravenous administration of the chimeric adenovirus enadenotucirev in patients undergoing primary tumor resection. J Immunother Cancer 5:71
15. Eisenhauer EA et al (2009) New response evaluation criteria in solid tumours: revised RECIST guideline (version 1.1). Eur J Cancer 45:228–247
16. Chow LQ (2013) Exploring novel immune-related toxicities and endpoints with immune-checkpoint inhibitors in non-small cell lung cancer. Am Soc Clin Oncol Educ Book
17. Hodi FS et al (2014) Evaluation of immune-related response criteria (irRC) in patients (pts) with advanced melanoma (MEL) treated with the anti-PD-1 monoclonal antibody MK-3475. J Clin Oncol 32:abstr 3006^
18. Wolchok JD et al (2009) Guidelines for the evaluation of immune therapy activity in solid tumors: immune-related response criteria. Clin Cancer Res 15:7412–7420
19. Hygino da Cruz LC Jr, Rodriguez I, Domingues RC, Gasparetto EL, Sorensen AG (2011) Pseudoprogression and pseudoresponse: imaging challenges in the assessment of posttreatment glioma. AJNR Am J Neuroradiol 32:1978–1985
20. Scher HI et al (2008) Design and end points of clinical trials for patients with progressive prostate cancer and castrate levels of testosterone: recommendations of the prostate cancer clinical trials working group. J Clin Oncol 26:1148–1159

Chapter 20

Fluorescence In Situ Hybridization (FISH) Detection of Viral Nucleic Acids in Oncolytic H-1 Parvovirus-Treated Human Brain Tumors

Irina Kiprianova, Alexandra Just, Barbara Leuchs, Jean Rommelaere, and Assia L. Angelova

Abstract

Fluorescence in situ hybridization (FISH) is a specific, sensitive, accurate, and reliable technique widely applied in both research and clinic. Here we describe the detailed protocol of a FISH method established by us to serve the scientific purposes of the first oncolytic parvovirus clinical trial (ParvOryx01). This trial was launched in Germany in 2011. After trial completion in 2015, results were published in Molecular Therapy in 2017. The primary purpose of the trial was to evaluate the safety of an oncolytic parvovirus, H-1PV (ParvOryx), in recurrent glioblastoma patients. In addition, the efficiency of H-1PV tumor targeting after intratumoral or systemic virus administration was assessed by FISH detection of viral nucleic acids (genomic single-stranded DNA, mRNA and parvovirus double-stranded replicative forms) in formalin-fixed paraffin-embedded glioblastoma tissues resected at day 10 after ParvOryx treatment. The FISH method allowed the detection—for the first time in humans—of H-1PV replication markers in brain tumors of parvovirus-treated patients. A protocol combining mRNA FISH with simultaneous immunofluorescent staining for tumor and tumor microenvironment markers was also developed and is described here, in order to better characterize H-1PV cellular targets and H-1PV treatment-associated tumor microenvironment changes.

Key words Oncolytic H-1 parvovirus (H-1PV), Recurrent glioblastoma, First oncolytic parvovirus clinical trial, Paraffin sections, Oncolytic parvovirus replication markers, Tumor microenvironment, Fluorescence in situ hybridization (FISH), Combined FISH/immunofluorescence (IF)

1 Introduction

Fluorescence in situ hybridization (FISH) was originally developed in the 70s of the last century, with the purpose of visualizing Drosophila polytene chromosomes and the so-called "lampbrush" chromosomes in oocytes of frogs and salamanders [1]. The first hybridization probes consisted of ^{3}H-labeled rRNA, and the resulting hybrids were detected using autoradiography [2]. The introduction of fluorescent probes about 10 years later represented a major methodological breakthrough [3–5]. Since then, the FISH

Christine E. Engeland (ed.), *Oncolytic Viruses*, Methods in Molecular Biology, vol. 2058,
https://doi.org/10.1007/978-1-4939-9794-7_20,

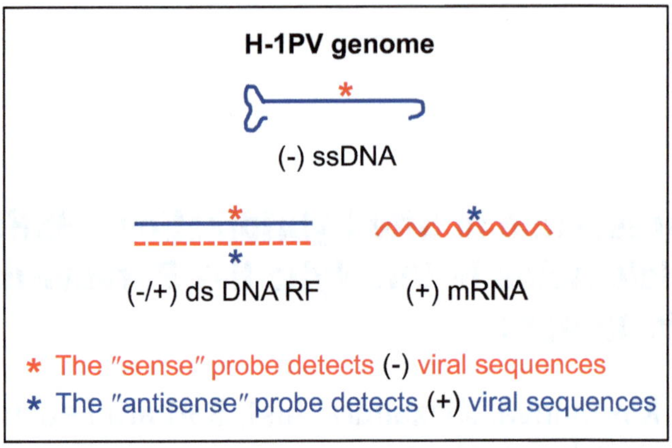

Fig. 1 FISH detection of parvovirus nucleic acids: molecular targets of hybridization and the corresponding hybridization probes. *ss* single-stranded, *ds* double-stranded, *RF* replicative form

technique rapidly developed and became a method of choice for single molecule in situ visualization.

Here we describe the in situ detection of specific viral nucleic acid sequences in formalin-fixed, paraffin-embedded (FFPE) tumor tissues resected from oncolytic parvovirus-treated glioblastoma patients within the frame of the first-in-man oncolytic parvovirus clinical trial, ParvOryx01 [6]. The investigational drug—H-1PV (ParvOryx), an oncolytic rat parvovirus with preclinically proven oncosuppressive activities [7, 8]—is a small, nonenveloped single-stranded (ss) DNA virus which does not naturally infect humans. Its ~5 kb genome encodes at least three nonstructural (NS1, NS2, SAT) and two capsid (VP1, VP2) proteins. Transcription from NS- and VP/SAT-coding sequences is driven by two promoters, P4 and P38, respectively. Viral genomic DNA replicates via a "rolling-circle" mechanism, involving the covalent crosslinking of complementary DNA strands through terminal hairpins [9]. These parvovirus double-stranded replicative forms (ds RFs) as well as the P4/P38-driven transcripts accumulating in H-1PV-infected tumor cells were taken as biomarkers of intratumoral parvovirus replication and were in particular the molecular targets for FISH detection (Fig. 1), as described in detail below.

2 Materials

Dry-heat sterilize all glassware in a hot air oven (200 °C) for 3.5 h. Prepare all solutions using ultrapure, nuclease-free, diethyl pyrocarbonate (DEPC)-treated H_2O. On the first procedure day, use RNase-free reagents and equipment and ensure an RNase-free work environment (wear sterile gloves, apply RNaseZAP™/

RNase AWAY® decontamination solution and RNase decontamination wipes to laboratory work surfaces). When disposing of waste materials, strictly follow waste disposal regulations.

Tumor tissues for analysis were obtained from ParvOryx01 patients at the time of glioblastoma resection, at day 10 after parvovirus intratumoral or intravenous administration. The samples were fixed, embedded in paraffin, sectioned into 5 μm-thick sections, and mounted onto SuperFrost Plus™ adhesion slides at the Department of Pathology (University Hospital, Heidelberg, Germany).

2.1 Deparaffinization and Rehydration (See Note 1)

1. Xylene.
2. Ethanol gradient: Prepare 100%, 90%, 70%, and 50% ethanol in individual staining dishes.
3. DEPC-H_2O: Add 1 ml DEPC to 1000 ml double-distilled H_2O and incubate overnight at room temperature. Autoclave the solution twice at 121 °C for 80 min to ensure DEPC inactivation and sterility.

2.2 Washing (Applicable to Multiple Procedure Steps)

1. Phosphate-buffered saline (PBS): Optimally, use Dulbecco's PBS (DPBS) without calcium and magnesium.
2. Tris-buffered saline (TBS): Dissolve 87.66 g NaCl and 12.11 g Tris base in DEPC-H_2O. Fill up to 1000 ml, adjust pH to 7.4 with HCl and autoclave before use.
3. PBS-Tween® 20 (PBST): 0.1% Tween® 20 in PBS.

2.3 Antigen Retrieval (See Note 2) and Fixation

1. 10 mM citrate buffer, pH 6.0: Use commercially available 10× stock solutions.
2. Proteinase K: *see* **Note 3**.
3. 1 mM glycine in saline–sodium citrate (SSC) buffer: Add 150 μl 1 M glycine to 150 ml 2× SSC.
4. SSC buffer, 20× stock solution: 3 M NaCl, 0.3 M sodium citrate. Adjust pH to 7.0 and store at room temperature. Alternatively, you may use any commercially available SSC 20× stock solutions.
5. Washing buffer (*see* **items 1–3** in Subheading 2.2).
6. 4% paraformaldehyde (PFA) in DPBS (*see* **Note** 4): Dissolve 20 g PFA in 500 ml DPBS and clear the solution for 1 h in a water bath (70 °C).
7. Washing buffer (*see* **items 1–3** in Subheading 2.2).

2.4 Acetylation and Prehybridization

1. Acetylation solution: Mix 2 ml triethanolamine, 750 μl acetic anhydride, and 150 ml DEPC-H_2O. Use only acetylation solution that was freshly prepared on the day of use.
2. Washing buffer (*see* **items 1–3** in Subheading 2.2).

Table 1
LNA™ oligonucleotide hybridization probe sequences

Probe name	Probe sequence[a]	Probe target
NS1_"sense"	5′DIGAATTCGCTAGGTTCAATGCGGCT3′DIG	(−) DNA
VP_"sense"	5′DIG TGACCTACCAACATCAGATACA3′DIG	
NS1_"antisense"	5′DIG TCAGCACACAACAGATGGCAT3′DIG	(+) mRNA
VP_"antisense"	5′DIG TACTATCCAGAGCAACCATCAT3′DIG	(+) DNA (RF)

[a]Probe sequences derive from H-1PV NS1 and VP protein-coding sequences. The sense probe sequence is reverse-complement to the (−) strand viral genomic DNA. The antisense probe sequence is reverse-complement to H-1PV (+) mRNA and the (+) strand complementary DNA (*see* also Fig. 1). The probes were custom-designed and synthesized by Exiqon/QIAGEN, and double digoxigenin (DIG)-labeled at their 5′ and 3′ ends

3. Prehybridization solution: 50% deionized formamide, 5× SSC, 5× Denhardt's solution, yeast tRNA, salmon sperm DNA, and DEPC-H_2O. For 10 ml, mix 5 ml 100% deionized formamide, 2.5 ml 20× SSC, 200 μl Denhardt's 50× stock, 100 μl yeast tRNA, 100 μl salmon sperm DNA, and 2.1 ml DEPC-H_2O.

2.5 Hybridization

1. Hybridization probes: Locked nucleic acid (LNA™) probes custom-designed by Exiqon (now a QIAGEN company) were used. These modified RNA nucleotides (https://www.qiagen.com/us/shop/genes-and-pathways/exiqon-lna-products/lna-oligos-reagents/) contain an additional bridge between the 2′ oxygen and the 4′ carbon which locks the ribose moiety in a 3′-endo conformation. The LNA™ nucleotides are mixed with DNA or RNA residues to generate the desired LNA™ oligonucleotide hybridization probe with increased target specificity and high-potency binding. LNA™ oligonucleotide probe sequences used by us for the in situ detection of parvovirus nucleic acids are given in Table 1 (*see* also Fig. 1). All probes were double digoxigenin (DIG)-labeled at their 5′ and 3′ ends and fluorescence-detected through a DIG-specific antibody (*see* below).
2. Hybridization solution: Dilute 30 μM LNA™ probe stocks 1:200 in prehybridization solution (*see* **item 3** in Subheading 2.4).
3. Parafilm® laboratory film.

2.6 Post-hybridization and Probe Visualization

1. Stringent washing solution: For 500 ml, mix 250 ml deionized formamide, 30 ml 20× SSC and 220 ml H_2O (*see* **Note 5**).
2. 0.2× SSC: Dilute 20× SSC stocks (*see* **item 4** in Subheading 2.3).
3. 1% hydrogen peroxide (H_2O_2) in PBS.

4. Blocking solution: Use Roche Blocking Reagent for nucleic acid hybridization and detection. Dissolve the Blocking Reagent to a 10% final concentration in maleic acid buffer.
5. Maleic acid buffer: 100 mM maleic acid, 150 mM NaCl, pH 7.5. For 100 ml, dissolve 1.16 g maleic acid and 0.87 g NaCl. Adjust pH with NaOH and autoclave the solution before use.
6. Sheep anti-DIG horseradish peroxidase (HRP)-conjugated antibody: Dilute 1:100 in blocking solution.
7. 4′,6-diamidino-2-phenylindole (DAPI) for nuclear DNA labeling.
8. PerkinElmer TSA® Plus Cyanine 3 detection kit for signal amplification.
9. DAKO fluorescence mounting medium.

2.7 Equipment

1. RNase-free (dry heat-sterilized) glassware.
2. Sterile disposable plasticware.
3. RNase-free filter tips.
4. Fume hood.
5. 600 W microwave oven.
6. Hybridization chamber.
7. Vortex mixer.
8. Benchtop centrifuge.
9. Multiblock heater.
10. Water bath.
11. Fluorescent microscope and image analysis software.

3 Methods

FISH protocol establishment and validation, as well as the subsequent analysis of patients' samples were performed in the Division of Tumor Virology at the German Cancer Research Center (DKFZ, Heidelberg, Germany). Method modification was based on Silahtaroglu et al. [10], and was first tested in (a) fixed H-1PV-infected newborn kidney (NBK) cells, and (b) FFPE tumor sections from human glioma xenografts treated with H-1PV after implantation in immunodeficient rats. Method validation included H-1PV-positive, H-1PV-negative, "no hybridization probe", and "no fluorescent dye" controls, and was carried out in repeated experiments by two independent operators (inter assay). In both fixed cultured cells and human glioma xenograft-derived FFPE tissue sections, FISH proved to be a specific and sensitive technique which achieves reliable in situ detection of parvovirus DNA replication and expression biomarkers.

Each patients' sample batch analysis always included human glioma xenograft tissue from rats intratumorally treated with H-1PV (positive control), and human glioma xenograft tissue from rats mock-treated with PBS (negative control). Mismatch probes (custom designed by Exiqon/QIAGEN; with 3–5 swapped nucleotides) were also used in parallel, to systemically control method specificity (Fig. 2). Furthermore, when an additional tissue pretreatment step with RNase A (100 μg/ml in 2× SSC, 30 min at 37 °C) was included (before the enzyme digestion step, *see* **step 4** in Subheading 3.2), substantial reduction of the signals generated by the "antisense" probe was observed (Fig. 3), demonstrating that the probe predominantly reveals viral mRNA transcript synthesis (*see also* Fig. 1).

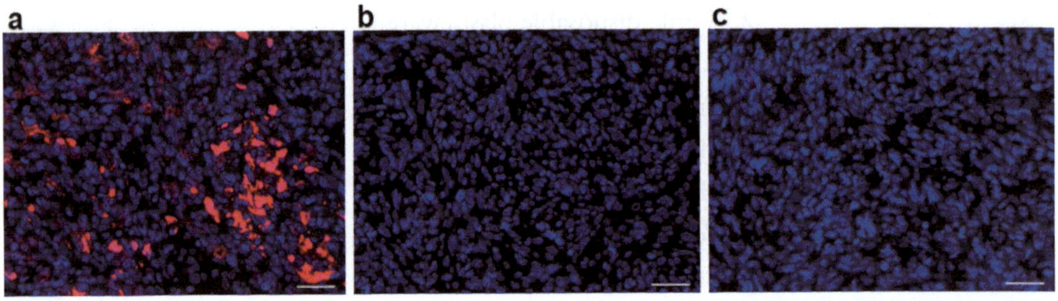

Fig. 2 FISH detection of parvovirus mRNA in human glioma xenograft treated with H-1PV after implantation in rats. (**a**) The LNA™ "antisense" probe binds to H-1PV transcripts generating strong cyanine 3-visualized positive signal (red) which is not observed when no probe (**b**) or a mismatch probe (**c**) is used. Scale bars, 50 μm

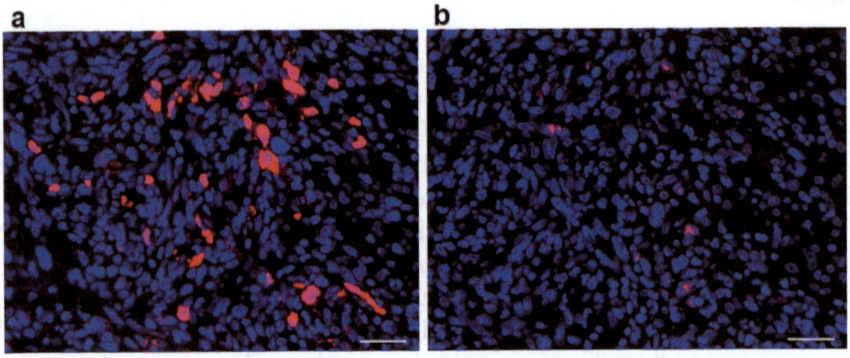

Fig. 3 FISH detection of parvovirus mRNA in human glioma xenograft treated with H-1PV after implantation in rats. (**a**) The LNA™ "antisense" probe binds to H-1PV transcripts generating strong cyanine 3-visualized positive signal (red). (**b**) Substantial reduction of the positive signal is observed when a prehybridization tissue pretreatment with RNase A is applied. Scale bars, 50 μm

The FISH procedure is carried out on three sequential days, the first of which encompasses all critical steps and requires an RNase-free environment. The second and third days include probe visualization and coverslip mounting onto the ready-for-observation slides.

First Day (*see* **Note 6**). ***Work RNase-free!***

3.1 Deparaffinization and Rehydration

1. Deparaffinize the test tissue sections by immersing the slides in xylene (2 × 10 min).
2. Rehydrate by sequential immersion of slides in 100%, 90%, 70% and 50% ethanol (2 × 5 min for each rehydration step).
3. Immerse the slides for 5 min in DEPC-H_2O.

3.2 Antigen Retrieval and Fixation

1. Immerse the slides in citrate buffer and boil them 2 × 7.5 min in a 600 W microwave oven (*see* **Note 7**).
2. Let the slides cool down to room temperature for 20 min.
3. Wash slides for 5 min in PBS.
4. Immerse the slides in proteinase K solution for 8 min exactly.
5. Immerse the slides in 1 mM glycine solution to stop proteolysis and avoid overdigestion.
6. Fix the slides for 5 min in 4% PFA.
7. Extensively wash 3 × 5 min in TBS.

3.3 Acetylation and Prehybridization

1. Acetylate the slides for 30 min in acetylation solution (*see* **item 1** in Subheading 2.4) to reduce nonspecific background.
2. Wash slides (2 × 5 min) in TBS.
3. Prepare a humidified chamber by coating the bottom of a 150 mm petri dish with wet paper (soak paper with DEPC-H_2O) and placing a grid on top. Carefully distribute approximately 100–150 μl (depending on the section size) of prehybridization solution over the surface of each tissue section.
4. Incubate the prehybridized slides in a humidified chamber for 1.5 h at room temperature.

3.4 Hybridization

1. Dilute the hybridization probes in hybridization solution to the final concentration required (*see* **item 2** in Subheading 2.5 and **Note 8**).
2. Shortly vortex and spin in a benchtop centrifuge.
3. Warm the probe for 5 min at 80 °C and place on ice immediately.
4. Carefully distribute approximately 100–150 μl of the hybridization probe over the section surface, cover with Parafilm® and place into a humidified chamber.

5. Leave the humidified chamber in the hybridization oven overnight at 54 °C or 63 °C (*see* **Note 9**), for "sense" or "antisense" probe hybridization, respectively.

Second Day

3.5 Post-hybridization and Probe Visualization

1. Carefully remove the Parafilm® from the slides.
2. Perform stringent (high temperature, low salt) washing. Wash 3 × 10 min in stringent washing solution in a water bath (*see* **Note 10**).
3. Wash the slides for 15 min in 0.2× SSC.
4. Wash the slides for 5 min in PBST.
5. Block endogenous peroxidase activity through slide immersion in 1% H_2O_2 for 30 min at room temperature (*see* **Note 11**).
6. Wash slides for 5 min in PBS, then 2 × 5 min in PBST.
7. Encircle the sections with a PAP pen.
8. Apply blocking solution over the section surface and incubate in a humidified chamber for 30 min at room temperature.
9. Remove blocking solution by carefully tapping the slides and immediately apply the anti-DIG antibody diluted 1:100 in blocking solution. Incubate overnight at 4 °C in a humidified chamber.

Third Day

3.6 Post-hybridization and Probe Visualization (Continued)

10. Wash the slides 3 × 7 min in PBST, followed by 5 min in PBS.
11. Add DAPI and TSA® Plus Cyanine 3 Amplification Reagent (according to manufacturer's instructions) to TSA® Plus Cyanine 3 Amplification Diluent. Then apply approximately 100–150 μl per slide and incubate the slides for 30 min at room temperature in a dark humidified chamber.
12. Wash slides 3 × 10 min with PBS.
13. Mount coverslips onto the slides using DAKO fluorescence mounting medium.
14. The above described FISH method allowed the detection—for the first time in a human patient—of oncolytic H-1 parvovirus replication markers in H-1PV-treated brain tumors (Fig. 4).

3.7 FISH Combination with Protein Immunostaining

Furthermore, an extension to the above described protocol was made, in order to combine detection of parvovirus nucleic acids with tumor and tumor microenvironment marker visualization. The development of this approach was based on the careful analysis by Nehmé et al. of the technical hurdles encountered when simultaneously applying in situ mRNA hybridization and

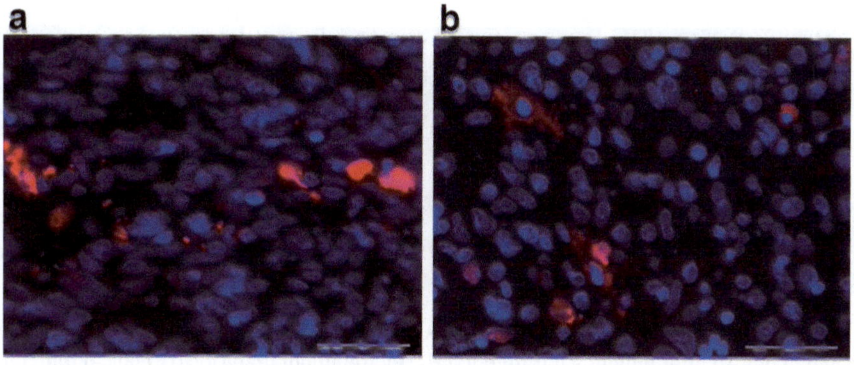

Fig. 4 FISH detection of parvovirus RNA transcripts in glioblastoma tissue resected at day 10 after patient's intratumoral H-1PV treatment (ParvOryx01 clinical trial, ClinicalTrials.gov Identifier: NCT01301430). (**a**) LNA™ "antisense" probe reveals the synthesis of viral (+) strand mRNA (red). This leads to the expression of the oncotoxic H-1PV NS1 protein (**b**), detected by immunofluorescence (NS1-specific monoclonal antibody, red). Scale bars, 50 μm

immunostaining [11]. It has to be noted that these authors have used frozen coronal sections from healthy rats with a thickness of 20 μm. In contrast, within the frame of the ParvOryx01 clinical trial-accompanying research, we succeeded in adapting the combined FISH/IF procedure to 5 μm-thick FFPE human brain tumor (glioblastoma) sections. Moreover, we notably applied this combined technique to visualize not only nuclear and cytoplasmic but also receptor-linked transmembrane cellular proteins (e.g., the pan-leukocytic marker CD45). The antibody directed against the specific cellular marker of interest was added at the protocol step described in Subheading 3.5, **step 9**, that is, simultaneously with the anti-DIG antibody. Visualization was achieved by adding an appropriate fluorophore-conjugated secondary antibody diluted in PerkinElmer Amplification Diluent at the protocol step described in Subheading 3.6, **step 11**. Importantly, when combining FISH with IF, the biggest concern is to preserve the antigenic site of the targeted protein, considering the necessity of proteolytic tissue pretreatment and the high temperatures required for hybridization. The two methods' compatibility should therefore be tested for each individual FISH and IF target and the optimal conditions should be determined in each individual case. A useful protocol of centromeric X/Y chromosome region FISH with concomitant IF in human pancreas FFPE has been recently shared by Ye and Gillespie [12]. However, it can only be applied to DNA, and not to RNA targets, since the proteinase digestion step is fully omitted (*see* **Note 3**). Our own experience showed that at least three glioblastoma microenvironment markers, the microglia/macrophage marker CD68, the glial fibrillary acidic protein (GFAP), and the leukocytic marker CD45 remain unaffected by proteinase K pretreatment and can be therefore successfully visualized simultaneously with H-1PV mRNA FISH detection (Fig. 5) [6, 13].

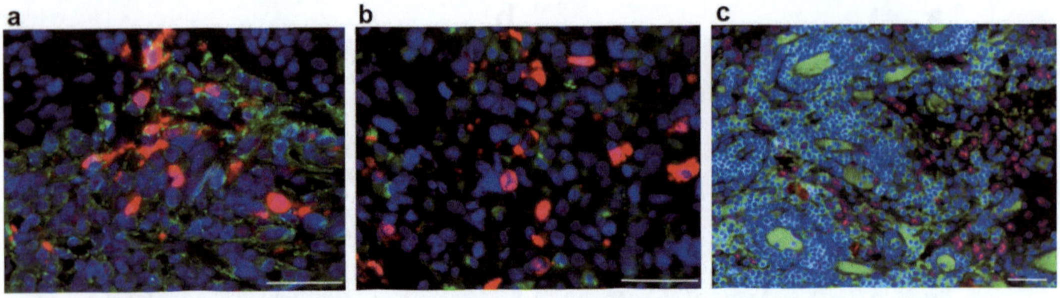

Fig. 5 Combined FISH/IF detection of parvovirus mRNA and tumor microenvironment markers in glioblastoma tissue resected at day 10 after patient's intratumoral H-1PV treatment (ParvOryx01 clinical trial, ClinicalTrials.gov Identifier: NCT01301430). (**a**) H-1PV RNA transcripts (FISH, red) + GFAP (IF, green). (**b**) H-1PV RNA transcripts (FISH, red) + CD68 (IF, green). (**c**) $CD45^+$ leukocytic glioblastoma infiltration (IF, green) in H-1PV RNA transcript-positive (FISH, red) tumor regions. Scale bars, 50 μm

3.8 Conclusion

In conclusion, the first-in-human oncolytic parvovirus clinical trial-accompanying research provided evidence that FISH is a reliable method for parvovirus nucleic acid detection in resected tumors. Furthermore, through a double FISH/IF staining, we demonstrated that within the highly heterogeneous glioblastoma cellular environment, H-1PV is selectively expressed in tumor—but not in tumor-associated microglia/macrophage—cell populations [6].

The protocols established and described here are applicable to both paraffin-embedded and cryopreserved tumor tissue sections and are of great value when patient-derived post-oncolytic parvovirus treatment operational material needs to be analyzed. Combining FISH (to detect H-1PV expression markers) with IF (to target specific host cell markers), although technically challenging, offers the intriguing possibility of visualizing parvovirus-infected tumor cells in the context of their complex in vivo tumor microenvironment.

4 Notes

1. Perform each deparaffinization and rehydration step in a separate staining dish using a fresh solution. All steps involving xylene or ethanol must be conducted under a fume hood in order to prevent any hazardous events. Strictly follow waste disposal regulations when disposing of used chemicals.
2. RNA targeting was optimized using the most commonly applied antigen unmasking approach, that is, heat-induced antigen/epitope retrieval (HIAR). For our particular research purposes, citrate buffer-HIAR, coupled with enzyme digestion (*see* **Note 3**), proved to be the most efficient prehybridization tissue treatment. However, if different tissue types and/or

molecular RNA targets are to be analyzed, other (EDTA- or Tris/EDTA-based) HIAR buffers may be worth testing.

3. The optimal proteinase K digestion conditions which ensure the generation of high hybridization signals while avoiding undesirable tissue morphology disruption are tissue type-specific and should be determined through proteinase K titration. From our experience, in human glioblastoma-derived FFPE tissues, the strongest hybridization signal is obtained through sample digestion with proteinase K (1 μg/ml) for 8 min.

   Nota bene—if a combined FISH/IF approach is intended, consider revising the proteinase K digestion step since it is not compatible with all proteins' antigenic site preservation! However, as proteinase K pretreatment is required for the generation of a strong mRNA FISH labeling, the optimal strategy is to substitute proteinase K with 0.6% Triton X-100 in PBS (30 min at room temperature). In contrast, glioblastoma tissue pretreatment under the same conditions with 0.05% Tween® 20 in PBS fails to reveal any positive H-1PV mRNA FISH signal (Angelova and Just, unpublished observations).

4. The fixative must be freshly prepared on the day of use. PFA is hazardous upon inhalation or exposure of eyes and skin. Gloves and safety glasses should be worn for protection. Protocol steps involving PFA must be carried out in a fume hood.

5. The stringent washing solution does not need to be RNase-free but requires preheating to reach the respective probe-specific optimal hybridization temperature (seal the bottleneck with Parafilm® and let the solution warm up by keeping it overnight—while hybridization is running—in the hybridization chamber).

6. Do not use a PAP pen on the first day of the procedure since it is not compatible with high temperatures. Alternatively, you may use the improved Super PAP Pen from ThermoFisher which is especially formulated to withstand the high temperatures required during the hybridization step.

7. Use plastic cuvettes without a lid. After the first boiling round, fill up with citrate buffer, if necessary.

8. For optimal signal strength achievement, use a mix of equal amounts of H-1PV NS1- and VP-detecting LNA™ probes.

9. During the prehybridization step, think about preheating the hybridization oven by adjusting the temperature required.

10. Both the stringent washing solution and the water bath must be preheated to the hybridization temperature (here, 54 °C or 63 °C for the "sense" or "antisense" probe, respectively).

11. Quenching endogenous peroxidase is a critical step in the TSA® Cyanine 3 Signal Amplification kit-based probe

visualization. Since the activation and the covalent binding of the fluorophore in this kit are catalyzed by HRP, blocking endogenous peroxidase activity is essential for background noise reduction.

Acknowledgments

Andreas Unterberg (principal investigator) and Karsten Geletneky conducted the first oncolytic parvovirus clinical trial (ParvOryx01) in the Department of Neurosurgery of Heidelberg University Hospital. ParvOryx01and the trial-accompanying laboratory research were sponsored by ORYX GmbH & Co.KG (Baldham/Munich, Germany). Assia Angelova and Alexandra Just were financially supported by ORYX funding. We thank Milena Barf for assistance with protein immunostaining. We acknowledge image analysis assistance by Felix Bestvater, Damir Crunic and Manuela Brom (Light Microscopy unit, DKFZ, Heidelberg). We are grateful to Jörg Schlehofer (Division of Tumor Virology, DKFZ, Heidelberg) and Andrii Domanskyi (Institute of Biotechnology, University of Helsinki) for helpful discussions.

References

1. Gall JG (2016) The origin of in situ hybridization—a personal history. Methods 98:4–9
2. Gall JG, Pardue ML (1969) Formation and detection of RNA-DNA hybrid molecules in cytological preparations. Proc Natl Acad Sci U S A 63:379–383
3. Rudkin GT, Stollar BD (1977) High resolution detection of DNA-RNA hybrids in situ by indirect immunofluorescence. Nature 265:472–473
4. Bauman JGJ, Wiegant J, Borst P et al (1980) A new method for fluorescence microscopical localization of specific DNA sequences by in situ hybridization of fluorochromelabelled DNA. Exp Cell Res 128:485–490
5. Langer-Safer PR, Levine M, Ward DC (1982) Immunological method for mapping genes on Drosophila polytene chromosomes. Proc Natl Acad Sci U S A 79:4381–4385
6. Geletneky K, Hajda J, Angelova AL et al (2017) Oncolytic H-1 parvovirus shows safety and signs of immunogenic activity in a first phase I/IIa glioblastoma trial. Mol Ther 25:2620–2634
7. Rommelaere J, Geletneky K, Angelova AL (2010) Oncolytic parvoviruses as cancer therapeutics. Cytokine Growth Factor Rev 21:185–195
8. Angelova AL, Geletneky K, Nüesch JPF et al (2015) Tumor selectivity of oncolytic parvoviruses: from in vitro and animal models to cancer patients. Front Bioeng Biotechnol 3:55
9. Cotmore SF, Tattersall P (1996) Parvovirus DNA replication. In: Depamphilis ML (ed) DNA replication in eukaryotic cells, vol 31. Cold Spring Harbor Laboratory Press, Cold Spring Harbor, NY
10. Silahtaroglu NM, Tommerup N, Vissing H (2003) FISHing with locked nucleic acids (LNA): evaluation of different LNA/DNA mixmers. Mol Cell Probes 17:165–169
11. Nehmé B, Henry M, Mouginot D (2011) Combined fluorescent in situ hybridization and immunofluorescence: limiting factors and a substitution strategy for slide-mounted tissue sections. J Neurosci Methods 196:281–288
12. Ye J, Gillespie KM (2016) Fluorescence in situ hybridization with concomitant immunofluorescence in human pancreas. Methods Mol Biol 1433:153–158
13. Geletneky K, Nüesch JPF, Angelova A, Kiprianova I, Rommelaere J (2015) Doublefaceted mechanism of parvoviral oncosuppression. Curr Opin Virol 13:17–24

Chapter 21

Mathematical Modeling of Oncolytic Virotherapy

Johannes P. W. Heidbuechel, Daniel Abate-Daga, Christine E. Engeland, and Heiko Enderling

Abstract

Mathematical modeling in biology has a long history as it allows the analysis and simulation of complex dynamic biological systems at little cost. A mathematical model trained on experimental or clinical data can be used to generate and evaluate hypotheses, to ask *"what if"* questions, and to perform in silico experiments to guide future experimentation and validation. Such models may help identify and provide insights into the mechanisms that drive changes in dynamic systems. While a mathematical model may never replace actual experiments, it can synergize with experiments to save time and resources by identifying experimental conditions that are unlikely to yield favorable outcomes, and by using optimization principles to identify experiments that are most likely to be successful. Over the past decade, numerous models have also been developed for oncolytic virotherapy, ranging from merely theoretic frameworks to fully integrated studies that utilize experimental data to generate actionable hypotheses. Here we describe how to develop such models for specific oncolytic virotherapy experimental setups, and which questions can and cannot be answered using integrated mathematical oncology.

Key words Mathematical modeling, Oncology, Virus, Oncolytic virotherapy, Combination immunotherapy

1 Introduction

Oncolytic virotherapy has generated a wealth of promising data but has also suffered from unfortunate setbacks of unsuccessful or unsafe treatment protocols (reviewed in [1]). It has come a long way from anecdotal observations of viral infections associated with tumor remissions [2] over the application of infectious body fluids to cancer patients [3] to modern clinical application based on rationally designed and scientifically sound trials [4, 5]. With the approval of T-VEC (Imlygic™) by the US Food and Drug Administration and the European Medicines Agency in 2015 [6], oncolytic virotherapy has entered clinical practice in the Western world. Still, considerable efforts are necessary to enhance efficacy of oncolytic viruses while retaining specificity and safety [7].

Christine E. Engeland (ed.), *Oncolytic Viruses*, Methods in Molecular Biology, vol. 2058,
https://doi.org/10.1007/978-1-4939-9794-7_21,

In contrast to the beginnings of research in the oncolytic virotherapy field, the efficacy of oncolytic viruses (OVs) is not anymore regarded to only depend on direct tumor cell killing in the course of viral replication. Rather, it relies on multifaceted modes of action [8–11], including, but not limited to, induction of antitumor immune responses via stimulation of both innate and adaptive immunity, bystander killing by cytokine secretion, and targeting of tumor vasculature [12]. This multitude of mechanisms, together with the wide variety of potentially oncolytic viruses available today [13, 14], the possibilities offered by genetic engineering [1, 15–17], and the range of conceivable regimens for combination therapy [18] (e.g., immune checkpoint inhibitors [19–21], CAR T cells [22–24], chemotherapeutics [25–27], radiation [28–30]) requires careful choice of individual treatment options. Exhaustive testing in preclinical experiments (not to speak of clinical trials) would not be feasible given the plethora of possible combinations and schedules in terms of dosing, timing, and route of administration. Mathematical modeling, rationally based on (pre)clinical data, can accelerate the progress of oncolytic virotherapy by identifying crucial parameters, by generating new hypotheses to be tested experimentally, and by predicting therapeutic outcomes in silico. Such an integrated, iterative workflow wherein experimental data continuously refines quantitative models and vice versa was recently described [31].

Numerous mathematical models have been developed to simulate tumor growth and treatment responses [32, 33], including the dynamics of oncolytic virotherapy with various levels of detail and complexity [34–57]. Here we discuss a template protocol for developing mathematical models of oncolytic virotherapy with increasing complexity based on data generated from different experiments. The following seven steps provide the best practice of the art of mathematical modeling of biological systems:

1. Formulate the biological question the model should help address.
2. Determine the model components.
3. Qualitatively describe the biological system.
4. Quantitatively describe the biological system, develop a mathematical model.
5. Simulate, solve, and analyze the model.
6. Validate model results.
7. Relate results back to the question.

Here we focus on the protocol for **steps 2–4**. For a detailed description of the remaining modeling steps we refer to mathematical biology textbooks [58–62].

2 Materials

2.1 Experimental Data

Mathematical modeling can reduce experimental costs by simulating the outcome of a given setup, but strongly depends on the quality of available information. To establish a meaningful model, training data needs to be sufficiently detailed, derived from quantitative measurements, and relevant for the respective research question. Sometimes, preexisting data can be useful if fulfilling the aforementioned criteria. In general, collaborations between computational mathematicians on one and "wet lab" experimentalists on the other side are highly encouraged already prior to data acquisition to design informative experiments. Exemplary experimental data are described in Subheading 3 of this chapter.

2.2 Model Variables

To mathematically describe the changes in a complex adaptive dynamic system, the system variables whose quantities change need to be identified. Whilst it is mathematically tempting to include as many of the variables currently known to contribute to the dynamics of oncolytic virotherapy, best practice should limit model variables to biological quantities that can be sequentially measured. We demonstrate below the construction of three different classes of models based on the spatiotemporal resolution of different biological experimental setups. If the derived simple mathematical model cannot reproduce the experimentally observed system dynamics, additional mechanisms and variables may be included. Better model fit to data due to increased degrees of freedom may, however, come at the cost of reduced predictability. Various statistical approaches including information criteria analysis and identifiability studies help determine the necessary and warranted complexity [63, 64].

2.3 Model Interaction Terms and Parameters

Unlike statistical modeling that aims to correlate random variables (such as biomarkers) with another set of random variables (such as clinical outcomes), differential equations describe the mechanisms by which variables change over time, and model analysis or numerical simulations allow predicting outcomes. The mechanisms that determine changes in variables need to be identified. As with model variables, emphasis should be paid to mechanisms that can be identified. For example, while it is known that the number of cancer cells changes due to cell proliferation and cell death, measurements of total cell number at discrete times may only inform a rate of overall population growth (or decline) that is the combination of the underlying mechanisms (net growth rate = cell proliferation events − cell death events).

2.4 Flow Diagram of the System

The effects of a system variable on its own dynamics and the interaction with other variables are routinely depicted in flow diagrams as graphical representation of the complex system. In

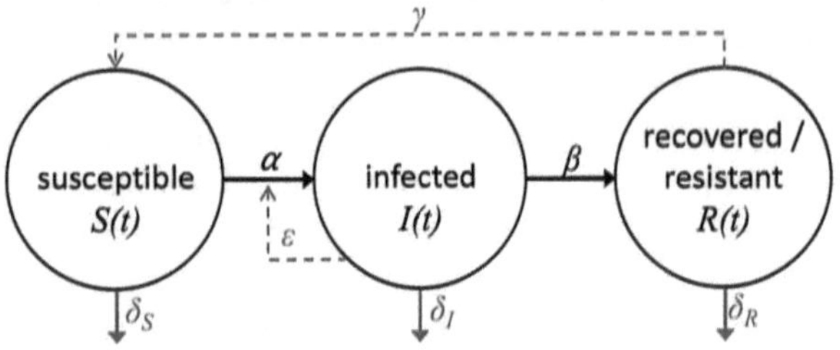

Fig. 1 Flow diagram of the simple SIR model. The three system variables are the population of susceptible individuals at time t, $S(t)$, infected individuals, $I(t)$, and recovered or resistant individuals, $R(t)$. Susceptible individuals become infected with rate α per unit time, and infected individuals can recover or become resistant with rate β per unit time. Additional model mechanisms (shown in light gray) may include the rate of recovery γ per unit time at which recovered cells can become susceptible again, the rate δ at which any individual dies, and the rate at which infected individuals can infect susceptible individuals, ε per unit time. Numerous other mechanisms can be added to the simple SIR model to answer specific questions

example, the frequently studied class of susceptible—infected—resistant (SIR) models to simulate infectious diseases [65] is summarized in the flow diagram in Fig. 1. Such models can be expanded by introduction of additional variables (e.g., representing chemical compounds or parts of the immune system) (*see* Subheading 3.4).

2.5 Numerical Simulation

To simulate mathematical models, numerous numerical schemes have been developed that can be implemented with most if not all available computer programming languages (C++, Java, Python, Julia, R, etc.). Different open-source or commercial software packages exist with a variety of solvers preimplemented and ready to use (including Matlab, Maple, and Mathematica). The chosen simulation framework often depends on personal preference and convenience. Simulation results and biological conclusions should be robust, reproducible and independent of the chosen numerical implementation.

3 Methods

A myriad of experimental approaches is routinely deployed to investigate tumor-virus interactions and the potential of virotherapy against cancers, dependent on the question the experiment is aimed to answer. Different experimental setups yield different data, and mathematical models to help answer biological questions should be built to utilize these data (Fig. 2).

Common experimental data include total cell count in vitro or tumor volume measurements in vivo over time, without resolution

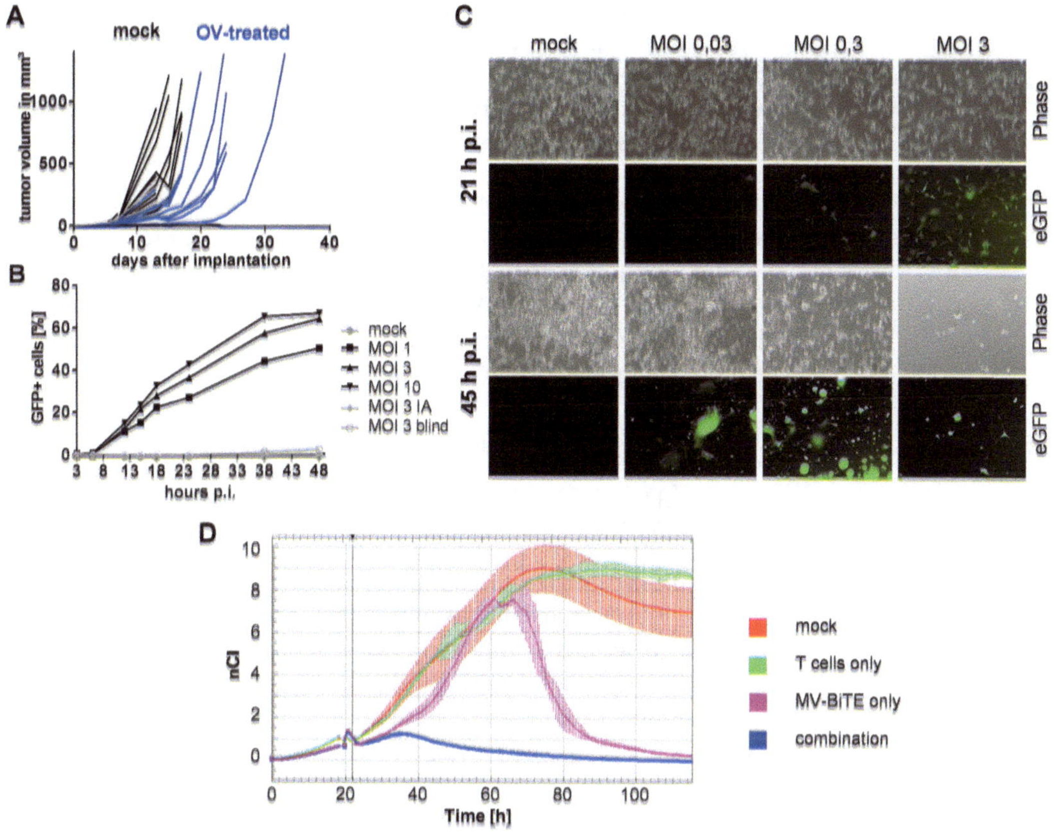

Fig. 2 Exemplary experimental data for mathematical modeling of oncolytic virotherapy. (**a**) Longitudinal tumor volumes in individual mice treated with intratumoral injections of carrier fluid (mock, in black) or oncolytic measles virus (MV) (OV-treated, in blue). (**b**) Flow cytometry analysis of reporter gene expression by primary chronic lymphocytic leukemia cells over time post infection (p.i.) with MV encoding eGFP (MV-eGFP) at indicated multiplicities of infection (MOI). IA—virus inactivated by UV-irradiation, blind—virus with surface protein mutations abrogating receptor binding. (**c**) Fluorescence microscopy of Mel-888 (human melanoma cell line) infected with MV-eGFP. (**d**) Cell viability assay of human pancreatic adenocarcinoma (HPAC) cells inoculated with an MV encoding a bispecific T cell engager (MV-BiTE) and/or ex vivo stimulated, polyclonal human T cells. Cell attachment to the plate bottom as surrogate for the relative amount of viable tumor cells (normalized cell index, nCI) was determined via impedance measurement with xCELLigence real-time cell analyzer technology

3.1 Mathematical Modeling of Aggregate Cell Numbers or Tumor Volumes Over Time

of cell proliferation or death rates, phenotype heterogeneity (susceptible, infected, resistant) or viral load (Fig. 2a). Such data can be modeled using an ordinary differential equation to model the change in tumor volume, $V(t)$, over time, t, with (Eq. 1):

$$\overbrace{\frac{\mathrm{d}V}{\mathrm{d}t}}^{\text{Change in tumor volume over time}} = \overbrace{\lambda V}^{\text{Growth rate proportional to current volume}}, \qquad (1)$$

where λ is the growth rate of the tumor volume per unit time. From Fig. 2a it is clear that the growth rate of untreated tumor volumes is larger than that of OV-treated tumor volumes. Fitting an exponential trendline to the original data (or performing a linear regression) yields the respective growth rates, with the untreated growth rate being larger than the growth rate of the treated tumor volume (Fig. 3). This model is the simplest model that could fit the data, but is a gross oversimplification of reality. Treatment with oncolytic viruses does not reduce the growth rate; rather it induces cell death and volume reduction. Although not explicitly measured, we can derive the death rate due to the oncolytic virus, γ per unit time, to extend the model as follows (Eq. 2):

$$\overbrace{\frac{\mathrm{d}V}{\mathrm{d}t}}^{\text{Change in tumor volume over time}} = \overbrace{\lambda V}^{\text{Growth rate proportional to current volume}} - \overbrace{\gamma V}^{\text{Death rate proportional to current volume}}. \qquad (2)$$

Here, the death rate due to the oncolytic virus, γ per unit time, is the difference in the growth rates of untreated and treated volume growth, that is, $\gamma = \lambda_{\text{mock}} - \lambda_{\text{treated}}$ (in example, $\gamma = 0.2$ in Fig. 3). Although useful for demonstrative purposes, the developed models cannot make predictions beyond the nature of the experiments they have been calibrated with. To use mathematical models to answer more specific questions, more data is required. Each model is only as good as the data it is based upon. It is

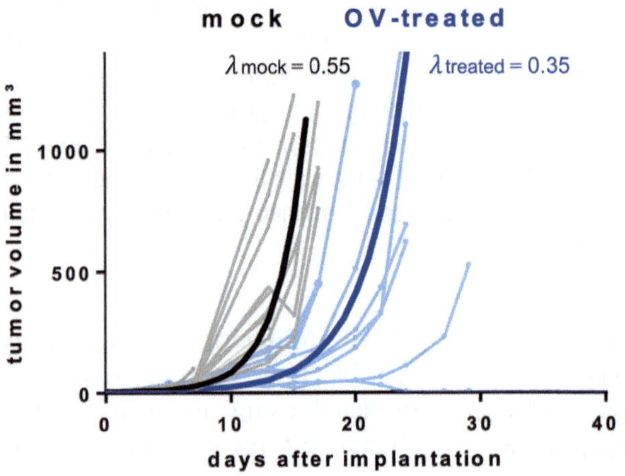

Fig. 3 Model fit to the experimental data. Shaded thin curves are data from Fig. 2a. Thick saturated curves show calculated solutions to (Eq. 1) with corresponding growth rates. Estimated median exponential growth rate for untreated (mock) tumor volume is $\lambda = 0.55$ per day, and estimated median exponential growth rate for treated tumor volume is $\lambda = 0.35$ per day

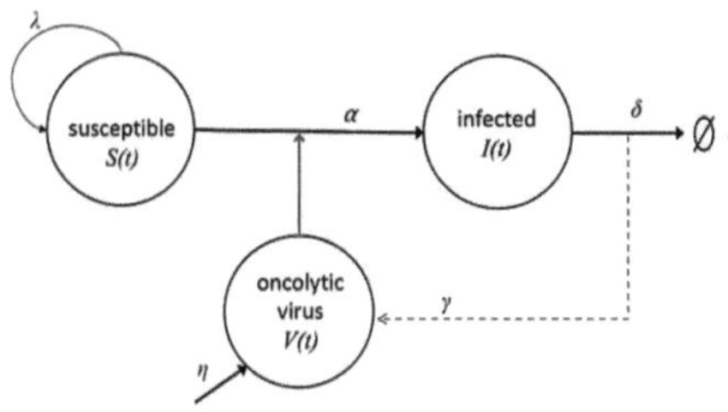

$$\overbrace{\frac{dS}{dt}}^{\text{change in susceptible cells over time}} = \overbrace{\lambda S}^{\text{growth rate}} - \overbrace{\alpha SV}^{\text{infection rate proportional to virus load}}$$

$$\overbrace{\frac{dI}{dt}}^{\text{change in infected cells over time}} = \overbrace{\alpha SV}^{\text{infection of susceptible cells}} - \overbrace{\delta I}^{\text{virus-induced death}}$$

$$\overbrace{\frac{dV}{dt}}^{\text{change in virus over time}} = \overbrace{\eta}^{\text{injection}} + \overbrace{\gamma\delta I}^{\text{release from dying infected cells}}$$

Fig. 4 Flow diagram and equations of the uninfected cell–virus–infected cell model

conceivable that the virus-related death rate (parameter γ in the above example) is dependent on the dose of injected virus particles. Multiple experiments with different doses could inform the dose-dependency of this parameter, and help predict responses to doses not evaluated experimentally yet.

3.2 Mathematical Modeling of Cell Infection Over Time

Predicting the total number of cells (Fig. 2b), or proportion of tumor volume, that are infected by the virus over time requires additional complexity in the mathematical model. A variation of the SIR model (Fig. 1) can simulate the number of uninfected and infected cells in a population. For simplicity, let us consider the reduced SI model (only susceptible cells and infected cells), but with the extension of a variable for the amount of free virus. Then, a fraction of the population of susceptible cells becomes infected proportionally to their interaction with the free virus population. When infected cells die, virus is released to further infect susceptible cells (Fig. 4). This leads to the following system of nonlinear coupled differential equations:

3.3 Mathematical Modeling of the Spatiotemporal Evolution of Tumor-Virus Interactions

The above models assume the tumor and virus populations to exist in a so-called well-mixed system, that is each cell and virus can potentially interact with one another regardless of their spatial location. Cell infection and virus release dynamics are clearly spatial phenomena (Fig. 2c). To mathematically formulate spatiotemporal dynamics, partial differential equations are commonly used. Here, the number of cells or tumor volume, *V*, depends not only on time, *t*, but also the spatial location, *x*, (one-dimensional for simplicity), that is, *V*(*x*,*t*), and spatial displacement of cells and viruses. The above system of equations then becomes Eq. 3.

$$\frac{\overbrace{\partial S(x,t)}^{\text{Change in susceptible cells over time and space}}}{\mathrm{d}t} = \overbrace{f(S)}^{\text{Spatial displacement}} + \overbrace{\lambda S}^{\text{Growth rate}} - \overbrace{\alpha SV}^{\text{Infection rate proportional to virus load}}$$

$$\frac{\overbrace{\partial I(x,t)}^{\text{Change in infected cells over time and space}}}{\mathrm{d}t} = \overbrace{f(I)}^{\text{Spatial displacement}} + \overbrace{\alpha SV}^{\text{Infection of susceptible cells}} - \overbrace{\delta I}^{\text{Virus-induced death}}$$

$$\frac{\overbrace{\partial V(x,t)}^{\text{Change in virus over time and space}}}{\mathrm{d}t} = \overbrace{f(V)}^{\text{Spatial displacement}} + \overbrace{\eta}^{\text{Injection}} + \overbrace{\gamma\delta I}^{\text{Release from dying infected cells}} \tag{3}$$

The function $f\{S, I, V\}$ often takes the form of diffusion (e.g., $f(V) = d_V \nabla^2 V$ with d_V being the rate of virus dispersal) or directed migration (e.g., $f(V) = -\xi\, \nabla\cdot(V \nabla S)$ where ξ is the rate at which viruses adhere to susceptible cells per unit space and unit time). Such mathematical models have more interesting dynamics, and simulation results depend on the location of the initial virus infection, viral cytotoxic efficacy, and the density of cancer cells. Figure 5 shows an example of a slightly more complex model of Choindroitinase ABC I (Chase ABC)-mediated enhancement of oncolytic virus spread and antitumor efficacy in glioma [66].

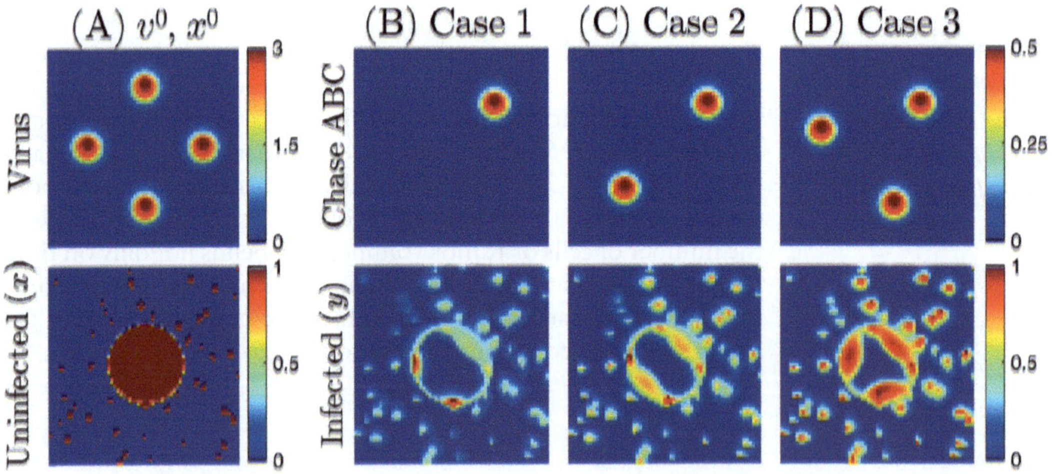

Fig. 5 Example of spatiotemporal partial differential equation model of the bacterial enzyme Chase ABC that promotes the spread of an oncolytic virus to infect and eradicate glioma cells. (**a**) Initial locations and densities of virus and uninfected cells. (**b–d**) Top: Initial locations and densities of Chase ABC following injection at one (**a**), two (**c**), and three (**d**) sites, respectively. Bottom: The locations and densities of infected cells after $t = 60$ h. Reproduced from Kim et al. (2014) [66] under the Creative Commons Attribution License

3.4 Model Extensions

In the era of immunotherapy and discovery of the synergistic potential of immunotherapy with oncological therapies such as radiation or chemotherapy, oncolytic virotherapy is poised to be thoroughly analyzed through the lenses of antitumor immunity. Mathematical models of tumor-virus-immune dynamics could include, for example, a T cell compartment capable of recognizing and killing susceptible and infected tumor cells at specific rates, β_S and β_I, respectively (Fig. 6a). Upon immunogenic tumor cell death, tumor-specific T cells are recruited, for example, by antigen-presenting cells at rates μ (Fig. 6b). As a next layer of complexity,

A

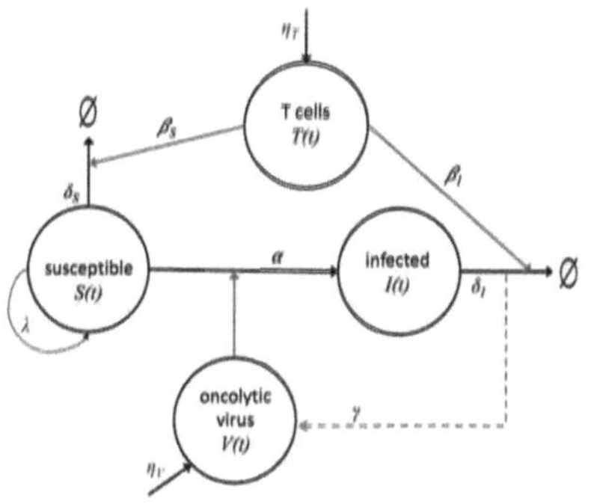

$$\frac{dS}{dt} = \lambda S - \alpha S V - \overbrace{\beta_S T S}^{\text{immunogenic cell death}}$$

$$\frac{dI}{dt} = \alpha S V - \delta I - \overbrace{\beta_I T I}^{\text{immunogenic cell death}}$$

$$\frac{dV}{dt} = \eta_V + \gamma \delta I$$

$$\overbrace{\frac{dT}{dt}}^{\text{change in T cell over time}} = \overbrace{\eta_T}^{\text{arrival from circulation}}$$

B

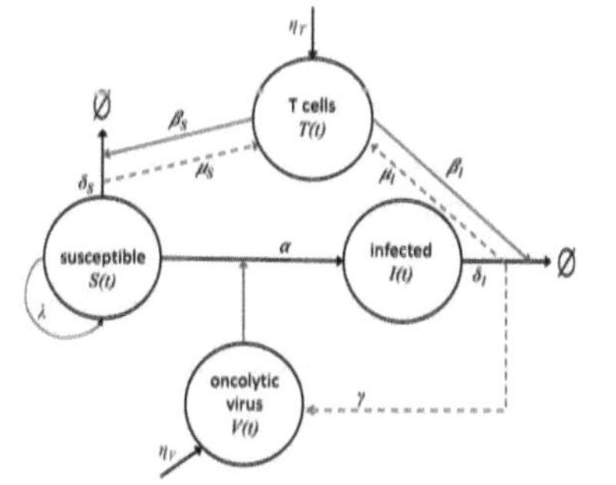

$$\frac{dS}{dt} = \lambda S - \alpha S V - \beta_S T S$$

$$\frac{dI}{dt} = \alpha S V - \delta I - \beta_I T I$$

$$\frac{dV}{dt} = \eta_V + \gamma \delta I$$

$$\frac{dT}{dt} = \eta_T + \overbrace{\mu_S \beta_S T S}^{\text{recruitment after susceptible cell death}} + \overbrace{\mu_I \beta_I T I}^{\text{recruitment after infected cell death}}$$

C

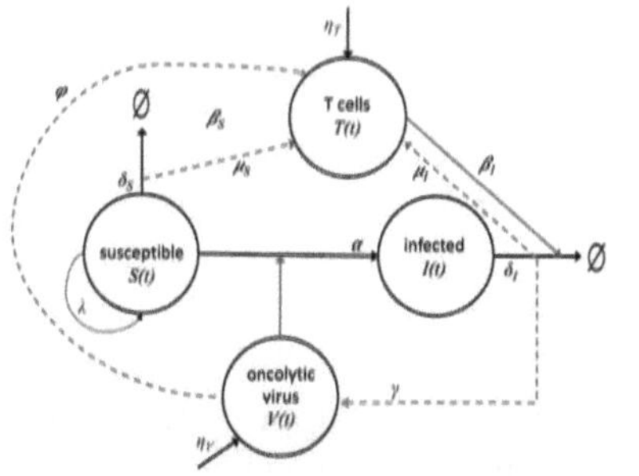

$$\frac{dS}{dt} = \lambda S - \alpha S V - \beta_S T S$$

$$\frac{dI}{dt} = \alpha S V - \delta I - \beta_I T I$$

$$\frac{dV}{dt} = \eta_V + \gamma \delta I$$

$$\frac{dT}{dt} = \eta_T + \mu_S \beta_S T S + \mu_I \beta_I T I + \overbrace{\varphi V T}^{\text{recruitment proportional to viral load (OV-BiTE)}}$$

Fig. 6 Flow diagrams and corresponding equations of mathematical models with increasing complexity. (**a**) Introduction of T cells that arrive from circulation and kill susceptible and infected cells. (**b**) Additional recruitment of T cells due to immunogenic cell death. (**c**) Stimulation of T cells proportional to oncolytic viral load (visualization of OV-encoded bispecific T cell engagers (BiTEs)). Overbraces in equation highlight additional terms compared to models introduced above

the recognition of target cells and thus the rate of tumor cell killing can be enhanced by adding OV-encoded bispecific T cell engagers (BiTEs) [23, 67–71], artificial antibodies designed to cross-link T cells to tumor cells, to the system (Fig. 6c, *see* Fig. 2d for exemplary data).

4 Conclusion

Mathematical modeling becomes increasingly integrated into biomedical research, and oncology in particular [72–74]. In a truly integrated approach, experimental studies yield data and hypotheses that can be tested with a mathematical model, and model predictions in turn can guide future experimentation for validation [75]. When developing scientific models, including in vitro, in vivo, and mathematical models, one must always remember that every model is an abstraction of reality and wrong by definition [76]. Making models increasingly complex may not increase their usefulness; in fact, it may decrease model usability due to heightened uncertainty. For models to make useful predictions, model development and parameterization need to rely on data that are obtainable, quantifiable, and sufficiently detailed to inform quantitative model mechanisms. Only a calibrated model should then be used to make predictions that, of course, need to be validated with test data that were not included in model training. Iterative approaches of in vitro as well as in vivo experimentation and in silico testing can inform rational design of both preclinical and clinical studies, leading to more refined research plans. Especially with a plethora of promising immunotherapeutic combinations conceivable, multidisciplinary collaborative efforts to mathematically model oncolytic virotherapy regimens will expedite the development and impact of the field.

Acknowledgments

J.P.W.H. receives a PhD stipend by the Helmholtz International Graduate School for Cancer Research at the German Cancer Research Center (DKFZ). C.E.E. receives funding from the Wilhelm Sander-Stiftung (Grant 2018.058.1). H.E. and D.A.D. thank Dr. Alexander R. A. Anderson and the Moffitt Physical Sciences in Oncology Center for the organization and support of the fourth Moffitt IMO workshop on Viruses and Cancer, where parts of the idea of this project were conceived.

References

1. Kelly E, Russell SJ (2007) History of oncolytic viruses: genesis to genetic engineering. Mol Ther 15(4):651–659. https://doi.org/10.1038/sj.mt.6300108
2. Dock G (1904) The influence of complicating diseases upon leukaemia. Am J Med Sci 127:563–592
3. Hoster HA, Zanes RP Jr, Von Haam E (1949) Studies in Hodgkin's syndrome; the association of viral hepatitis and Hodgkin's disease; a preliminary report. Cancer Res 9(8):473–480
4. Pol J, Buque A, Aranda F, Bloy N, Cremer I, Eggermont A, Erbs P, Fucikova J, Galon J, Limacher JM, Preville X, Sautes-Fridman C, Spisek R, Zitvogel L, Kroemer G, Galluzzi L (2016) Trial watch-oncolytic viruses and cancer therapy. Oncoimmunology 5(2): e1117740. https://doi.org/10.1080/2162402x.2015.1117740
5. Msaouel P, Opyrchal M, Dispenzieri A, Peng KW, Federspiel MJ, Russell SJ, Galanis E (2018) Clinical trials with oncolytic measles virus: current status and future prospects. Curr Cancer Drug Targets 18(2):177–187. https://doi.org/10.2174/1568009617666170222125035
6. Pol J, Kroemer G, Galluzzi L (2016) First oncolytic virus approved for melanoma immunotherapy. Oncoimmunology 5(1):e1115641. https://doi.org/10.1080/2162402X.2015.1115641
7. Russell SJ, Peng KW (2017) Oncolytic virotherapy: a contest between apples and oranges. Mol Ther 25(5):1107–1116. https://doi.org/10.1016/j.ymthe.2017.03.026
8. Lichty BD, Breitbach CJ, Stojdl DF, Bell JC (2014) Going viral with cancer immunotherapy. Nat Rev Cancer 14(8):559–567. https://doi.org/10.1038/nrc3770
9. Workenhe ST, Mossman KL (2014) Oncolytic virotherapy and immunogenic cancer cell death: sharpening the sword for improved cancer treatment strategies. Mol Ther 22(2):251–256. https://doi.org/10.1038/mt.2013.220
10. Cassady KA, Haworth KB, Jackson J, Markert JM, Cripe TP (2016) To infection and beyond: the multi-pronged anti-cancer mechanisms of oncolytic viruses. Viruses 8(2). https://doi.org/10.3390/v8020043
11. Achard C, Surendran A, Wedge ME, Ungerechts G, Bell J, Ilkow CS (2018) Lighting a fire in the tumor microenvironment using oncolytic immunotherapy. EBioMedicine 31:17–24. https://doi.org/10.1016/j.ebiom.2018.04.020
12. Arulanandam R, Batenchuk C, Angarita FA, Ottolino-Perry K, Cousineau S, Mottashed A, Burgess E, Falls TJ, De Silva N, Tsang J, Howe GA, Bourgeois-Daigneault MC, Conrad DP, Daneshmand M, Breitbach CJ, Kirn DH, Raptis L, Sad S, Atkins H, Huh MS, Diallo JS, Lichty BD, Ilkow CS, Le Boeuf F, Addison CL, McCart JA, Bell JC (2015) VEGF-mediated induction of PRD1-BF1/Blimp1 expression sensitizes tumor vasculature to oncolytic virus infection. Cancer Cell 28(2):210–224. https://doi.org/10.1016/j.ccell.2015.06.009
13. Miest TS, Cattaneo R (2014) New viruses for cancer therapy: meeting clinical needs. Nat Rev Microbiol 12(1):23–34. https://doi.org/10.1038/nrmicro3140
14. Ungerechts G, Bossow S, Leuchs B, Holm PS, Rommelaere J, Coffey M, Coffin R, Bell J, Nettelbeck DM (2016) Moving oncolytic viruses into the clinic: clinical-grade production, purification, and characterization of diverse oncolytic viruses. Mol Ther 3:16018. https://doi.org/10.1038/mtm.2016.18
15. Twumasi-Boateng K, Pettigrew JL, Kwok YYE, Bell JC, Nelson BH (2018) Oncolytic viruses as engineering platforms for combination immunotherapy. Nat Rev Cancer 18(7):419–432. https://doi.org/10.1038/s41568-018-0009-4
16. Russell SJ, Peng KW, Bell JC (2012) Oncolytic virotherapy. Nat Biotechnol 30(7):658–670. https://doi.org/10.1038/nbt.2287
17. Heidbuechel JPW, Engeland CE (2019) Paramyxoviruses for tumor-targeted immunomodulation: design and evaluation ex vivo. J Vis Exp (143):e58651. https://doi.org/10.3791/58651
18. Martin NT, Bell JC (2018) Oncolytic virus combination therapy: killing one bird with two stones. Mol Ther 26(6):1414–1422. https://doi.org/10.1016/j.ymthe.2018.04.001
19. Zamarin D, Holmgaard RB, Subudhi SK, Park JS, Mansour M, Palese P, Merghoub T, Wolchok JD, Allison JP (2014) Localized oncolytic virotherapy overcomes systemic tumor resistance to immune checkpoint blockade immunotherapy. Sci Transl Med 6(226):226ra232. https://doi.org/10.1126/scitranslmed.3008095
20. Engeland CE, Grossardt C, Veinaide R, Bossow S, Lutz D, Kaufmann JK,

Shevchenko I, Umansky V, Nettelbeck DM, Weichert W, Jager D, von Katie C, Ungerechts G (2014) CTLA-4 and PD-L1 checkpoint blockade enhances oncolytic measles virus therapy. Mol Ther 22(11):1949–1959. https://doi.org/10.1038/mt.2014.160

21. Ribas A, Dummer R, Puzanov I, VanderWalde A, Andtbacka RHI, Michielin O, Olszanski AJ, Malvehy J, Cebon J, Fernandez E, Kirkwood JM, Gajewski TF, Chen L, Gorski KS, Anderson AA, Diede SJ, Lassman ME, Gansert J, Hodi FS, Long GV (2017) Oncolytic virotherapy promotes intratumoral T cell infiltration and improves anti-PD-1 immunotherapy. Cell 170 (6):1109–1119.e1110. https://doi.org/10.1016/j.cell.2017.08.027
22. Ajina A, Maher J (2017) Prospects for combined use of oncolytic viruses and CAR T-cells. J Immunother Cancer 5(1):90. https://doi.org/10.1186/s40425-017-0294-6
23. Wing A, Fajardo CA, Posey AD Jr, Shaw C, Da T, Young RM, Alemany R, June CH, Guedan S (2018) Improving CART-cell therapy of solid tumors with oncolytic virus-driven production of a bispecific T-cell engager. Cancer Immunol Res 6(5):605–616. https://doi.org/10.1158/2326-6066.cir-17-0314
24. Nishio N, Diaconu I, Liu H, Cerullo V, Caruana I, Hoyos V, Bouchier-Hayes L, Savoldo B, Dotti G (2014) Armed oncolytic virus enhances immune functions of chimeric antigen receptor-modified T cells in solid tumors. Cancer Res 74(18):5195–5205. https://doi.org/10.1158/0008-5472.CAN-14-0697
25. Bressy C, Benihoud K (2014) Association of oncolytic adenoviruses with chemotherapies: an overview and future directions. Biochem Pharmacol 90(2):97–106. https://doi.org/10.1016/j.bcp.2014.05.003
26. Wennier ST, Liu J, McFadden G (2012) Bugs and drugs: oncolytic virotherapy in combination with chemotherapy. Curr Pharm Biotechnol 13(9):1817–1833
27. Fillat C, Maliandi MV, Mato-Berciano A, Alemany R (2014) Combining oncolytic virotherapy and cytotoxic therapies to fight cancer. Curr Pharm Des 20(42):6513–6521
28. Li H, Peng KW, Russell SJ (2012) Oncolytic measles virus encoding thyroidal sodium iodide symporter for squamous cell cancer of the head and neck radiovirotherapy. Hum Gene Ther 23 (3):295–301. https://doi.org/10.1089/hum.2011.128
29. Opyrchal M, Allen C, Iankov I, Aderca I, Schroeder M, Sarkaria J, Galanis E (2012) Effective radiovirotherapy for malignant gliomas by using oncolytic measles virus strains encoding the sodium iodide symporter (MV-NIS). Hum Gene Ther 23(4):419–427. https://doi.org/10.1089/hum.2011.158
30. Mansfield D, Pencavel T, Kyula JN, Zaidi S, Roulstone V, Thway K, Karapanagiotou L, Khan AA, McLaughlin M, Touchefeu Y, Seth R, Melcher AA, Vile RG, Pandha HS, Harrington KJ (2013) Oncolytic vaccinia virus and radiotherapy in head and neck cancer. Oral Oncol 49(2):108–118. https://doi.org/10.1016/j.oraloncology.2012.07.019
31. Santiago DN, Heidbuechel JPW, Kandell WM, Walker R, Djeu J, Engeland CE, Abate-Daga D, Enderling H (2017) Fighting cancer with mathematics and viruses. Viruses 9 (9):239. https://doi.org/10.3390/v9090239
32. Michor F, Beal K (2015) Improving cancer treatment via mathematical modeling: surmounting the challenges is worth the effort. Cell 163(5):1059–1063. https://doi.org/10.1016/j.cell.2015.11.002
33. Araujo RP, McElwain DL (2004) A history of the study of solid tumour growth: the contribution of mathematical modelling. Bull Math Biol 66(5):1039–1091. https://doi.org/10.1016/j.bulm.2003.11.002
34. Dingli D, Cascino MD, Josic K, Russell SJ, Bajzer Z (2006) Mathematical modeling of cancer radiovirotherapy. Math Biosci 199 (1):55–78. https://doi.org/10.1016/j.mbs.2005.11.001
35. Friedman A, Tian JP, Fulci G, Chiocca EA, Wang J (2006) Glioma virotherapy: effects of innate immune suppression and increased viral replication capacity. Cancer Res 66 (4):2314–2319. https://doi.org/10.1158/0008-5472.CAN-05-2661
36. Tian JP (2011) The replicability of oncolytic virus: defining conditions in tumor virotherapy. Math Biosci Eng 8(3):841–860. https://doi.org/10.3934/mbe.2011.8.841
37. Rommelfanger DM, Offord CP, Dev J, Bajzer Z, Vile RG, Dingli D (2012) Dynamics of melanoma tumor therapy with vesicular stomatitis virus: explaining the variability in outcomes using mathematical modeling. Gene Ther 19(5):543–549. https://doi.org/10.1038/gt.2011.132
38. Crivelli JJ, Foldes J, Kim PS, Wares JR (2012) A mathematical model for cell cycle-specific cancer virotherapy. J Biol Dyn 6 Suppl 1:104–120. https://doi.org/10.1080/17513758.2011.613486
39. Jacobsen K, Pilyugin SS (2015) Analysis of a mathematical model for tumor therapy with a fusogenic oncolytic virus. Math Biosci 270

(Pt B):169–182. https://doi.org/10.1016/j.mbs.2015.02.009

40. Paiva LR, Binny C, Ferreira SC Jr, Martins ML (2009) A multiscale mathematical model for oncolytic virotherapy. Cancer Res 69 (3):1205–1211. https://doi.org/10.1158/0008-5472.CAN-08-2173
41. Paiva LR, Silva HS, Ferreira SC, Martins ML (2013) Multiscale model for the effects of adaptive immunity suppression on the viral therapy of cancer. Phys Biol 10(2):025005. https://doi.org/10.1088/1478-3975/10/2/025005
42. Camara BI, Mokrani H, Afenya EK (2013) Mathematical modeling of glioma therapy using oncolytic viruses. Math Biosci Eng 10 (3):565–578. https://doi.org/10.3934/mbe.2013.10.565
43. Kim PS, Crivelli JJ, Choi IK, Yun CO, Wares JR (2015) Quantitative impact of immunomodulation versus oncolysis with cytokine-expressing virus therapeutics. Math Biosci Eng 12(4):841–858. https://doi.org/10.3934/mbe.2015.12.841
44. Macnamara C, Eftimie R (2015) Memory versus effector immune responses in oncolytic virotherapies. J Theor Biol 377:1–9. https://doi.org/10.1016/j.jtbi.2015.04.004
45. Wang Z, Guo Z, Peng H (2016) A mathematical model verifying potent oncolytic efficacy of M1 virus. Math Biosci 276:19–27. https://doi.org/10.1016/j.mbs.2016.03.001
46. Barish S, Ochs MF, Sontag ED, Gevertz JL (2017) Evaluating optimal therapy robustness by virtual expansion of a sample population, with a case study in cancer immunotherapy. Proc Natl Acad Sci U S A 114(31):E6277–E6286. https://doi.org/10.1073/pnas.1703355114
47. Malinzi J, Ouifki R, Eladdadi A, Torres DFM, White JKA (2018) Enhancement of chemotherapy using oncolytic virotherapy: mathematical and optimal control analysis. Math Biosci Eng 15(6):1435–1463. https://doi.org/10.3934/mbe.2018066
48. Jenner AL, Yun CO, Kim PS, Coster ACF (2018) Mathematical modelling of the interaction between cancer cells and an oncolytic virus: insights into the effects of treatment protocols. Bull Math Biol 80(6):1615–1629. https://doi.org/10.1007/s11538-018-0424-4
49. Boemo MA, Byrne HM (2019) Mathematical modelling of a hypoxia-regulated oncolytic virus delivered by tumour-associated macrophages. J Theor Biol 461:102–116. https://doi.org/10.1016/j.jtbi.2018.10.044
50. Ratajczyk E, Ledzewicz U, Leszczynski M, Friedman A (2017) The role of TNF-alpha inhibitor in glioma virotherapy: a mathematical model. Math Biosci Eng 14(1):305–319. https://doi.org/10.3934/mbe.2017020
51. Timalsina A, Tian JP, Wang J (2017) Mathematical and computational modeling for tumor virotherapy with mediated immunity. Bull Math Biol 79(8):1736–1758. https://doi.org/10.1007/s11538-017-0304-3
52. Ashyani A, RabieiMotlagh O, Mohammadinejad HM (2018) A mathematical approach to effects of CTLs on cancer virotherapy in the second injection of virus. J Theor Biol 453:78–87. https://doi.org/10.1016/j.jtbi.2018.05.018
53. Malinzi J, Eladdadi A, Sibanda P (2017) Modelling the spatiotemporal dynamics of chemovirotherapy cancer treatment. J Biol Dyn 11 (1):244–274. https://doi.org/10.1080/17513758.2017.1328079
54. Friedman A, Lai X (2018) Combination therapy for cancer with oncolytic virus and checkpoint inhibitor: a mathematical model. PLoS One 13(2):e0192449. https://doi.org/10.1371/journal.pone.0192449
55. Mahasa KJ, Eladdadi A, de Pillis L, Ouifki R (2017) Oncolytic potency and reduced virus tumor-specificity in oncolytic virotherapy. A mathematical modelling approach. PLoS One 12(9):e0184347. https://doi.org/10.1371/journal.pone.0184347
56. Rodriguez-Brenes IA, Hofacre A, Fan H, Wodarz D (2017) Complex dynamics of virus spread from low infection multiplicities: implications for the spread of oncolytic viruses. PLoS Comput Biol 13(1):e1005241. https://doi.org/10.1371/journal.pcbi.1005241
57. Wodarz D, Hofacre A, Lau JW, Sun Z, Fan H, Komarova NL (2012) Complex spatial dynamics of oncolytic viruses in vitro: mathematical and experimental approaches. PLoS Comput Biol 8(6):e1002547. https://doi.org/10.1371/journal.pcbi.1002547
58. Murray JD (2013) Mathematical biology II. Springer, New York, NY
59. Edelstein-Keshet L (1988) Mathematical models in biology. SIAM, Philadelphia, PA
60. Ciarletta PH, Hillen T, Othmer H, Trucu D (2016) Mathematical models and methods for living systems. Springer, Cham
61. Otto SPD, Day T (2011) A biologist's guide to mathematical modeling in ecology and evolution. Princeton University Press, Princeton
62. Kuang YN, Nagy JD, Eikenberry SE (2016) Introduction to mathematical oncology. CRC Press, Boca Raton, FL

63. Aho K, Derryberry D, Peterson T (2014) Model selection for ecologists: the worldviews of AIC and BIC. Ecology 95(3):631–636
64. Eisenberg MC, Jain HV (2017) A confidence building exercise in data and identifiability: modeling cancer chemotherapy as a case study. J Theor Biol 431:63–78. https://doi.org/10.1016/j.jtbi.2017.07.018
65. Anderson RM, May RM (1992) Infectious diseases of humans: dynamics and control. Oxford university press, Oxford
66. Kim Y, Lee HG, Dmitrieva N, Kim J, Kaur B, Friedman A (2014) Choindroitinase ABC I-mediated enhancement of oncolytic virus spread and anti tumor efficacy: a mathematical model. PLoS One 9(7):e102499
67. Yu F, Wang X, Guo ZS, Bartlett DL, Gottschalk SM, Song XT (2014) T-cell engager-armed oncolytic vaccinia virus significantly enhances antitumor therapy. Mol Ther 22 (1):102–111. https://doi.org/10.1038/mt.2013.240
68. Fajardo CA, Guedan S, Rojas LA, Moreno R, Arias-Badia M, de Sostoa J, June CH, Alemany R (2017) Oncolytic adenoviral delivery of an EGFR-targeting T-cell engager improves antitumor efficacy. Cancer Res 77(8):2052–2063. https://doi.org/10.1158/0008-5472.can-16-1708
69. Freedman JD, Hagel J, Scott EM, Psallidas I, Gupta A, Spiers L, Miller P, Kanellakis N (2017) Oncolytic adenovirus expressing bispecific antibody targets T-cell cytotoxicity in cancer biopsies. EMBO Mol Med 9 (8):1067–1087. https://doi.org/10.15252/emmm.201707567
70. Speck T, Heidbuechel JPW, Veinalde R, Jaeger D, von Kalle C, Ball CR, Ungerechts G, Engeland CE (2018) Targeted BiTE expression by an oncolytic vector augments therapeutic efficacy against solid tumors. Clin Cancer Res 24(9):2128–2137. https://doi.org/10.1158/1078-0432.CCR-17-2651
71. Freedman JD, Duffy MR, Lei-Rossmann J, Muntzer A, Scott EM, Hagel J, Campo L, Bryant RJ, Verrill C, Lambert A, Miller P, Champion BR, Seymour LW, Fisher KD (2018) An oncolytic virus expressing a T-cell engager simultaneously targets cancer and immunosuppressive stromal cells. Cancer Res 78:6852. https://doi.org/10.1158/0008-5472.CAN-18-1750
72. Gatenby RA, Maini PK (2003) Mathematical oncology: cancer summed up. Nature 421 (6921):321. https://doi.org/10.1038/421321a
73. Anderson AR, Quaranta V (2008) Integrative mathematical oncology. Nat Rev Cancer 8 (3):227–234. https://doi.org/10.1038/nrc2329
74. Altrock PM, Liu LL, Michor F (2015) The mathematics of cancer: integrating quantitative models. Nat Rev Cancer 15(12):730–745. https://doi.org/10.1038/nrc4029
75. McGuire MF, Enderling H, Wallace DI, Batra J, Jordan M, Kumar S, Panetta JC, Pasquier E (2013) Formalizing an integrative, multidisciplinary cancer therapy discovery workflow. Cancer Res 73(20):6111–6117. https://doi.org/10.1158/0008-5472.CAN-13-0310
76. Box GEP (1976) Science and statistics. J Am Stat Assoc 71(356):791–799. https://doi.org/10.1080/01621459.1976.10480949

Index

Christine E. Engeland (ed.), *Oncolytic Viruses*, Methods in Molecular Biology, vol. 2058,
https://doi.org/10.1007/978-1-4939-9794-7,

G

H

I

P

MIX
Papier aus verantwortungsvollen Quellen
Paper from responsible sources
FSC® C105338

If you have any concerns about our products,
you can contact us on
ProductSafety@springernature.com

In case Publisher is established outside the EU,
the EU authorized representative is:
Springer Nature Customer Service Center GmbH
Europaplatz 3, 69115 Heidelberg, Germany

Printed by Libri Plureos GmbH
in Hamburg, Germany